CRTD-Vol. 27

PROCEEDINGS OF THE 1993 INTERNATIONAL FORUM ON

DIMENSIONAL TOLERANCING AND METROLOGY

presented at
The 1993 International Forum on Dimensional Tolerancing and Metrology
Dearborn, Michigan
June 17–19, 1993

sponsored by
ASME—
Center for Research and Technology Development
Council on Codes and Standards
National Science Foundation

edited by
Vijay Srinivasan
IBM T. J. Watson Research Center / Columbia University

Herbert B. Voelcker
Cornell University

THE AMERICAN SOCIETY OF MECHANICAL ENGINEERS
UNITED ENGINEERING CENTER / 345 EAST 47TH STREET / NEW YORK, NEW YORK 10017

ISBN No. 0-7918-0697-9

PREFACE
to
THE PROCEEDINGS
of
THE 1993 INTERNATIONAL FORUM
on
DIMENSIONAL TOLERANCING & METROLOGY

Herbert Voelcker
Cornell University
Chairman, Forum Steering Committee

Vijay Srinivasan
IBM T. J. Watson Research Center
Chairman, Forum Editorial Committee

This *Proceedings* contains papers covering most of the presentations and group discussions at a recent conference on dimensional tolerancing and metrology. Tolerancing and metrology are in flux. The methods used currently in industry mainly evolved from shop practice, and require a level of human interpretation that is not compatible with modern CMM-based inspection procedures and CAD-based design. There has been a surge of activity over the past five years aimed at improving tolerancing and metrology through mathematization of the national tolerancing standard, standardization of the measurement methods used for determining conformance to the standard, and more generally, providing firmer foundations for tolerancing and metrology at large. This *Proceedings* provides a 'window' into the current state of affairs and some of the changes that are afoot.

*

To provide some perspective for the papers that follow, let's return to basics: *why* are tolerancing and metrology important? Tolerancing is important because manufacturing and assembly processes are inherently imprecise, and produce parts and products that vary. Designers use tolerances to specify how much variability can be accomodated without degrading a product's 'function' or losing the ability to assemble parts interchangeably. Dimensional metrology provides means to check conformance to tolerance specifications, and also provides data to control manufacturing and assembly processes.

Current tolerancing techniques evolved relatively recently from machine-shop and drafting-room practice; their roots lie in gaging techniques from the early 1800s that enabled parts to be made precisely enough to be assembled interchangeably. In America, ANSI Standard Y14.5M-1982 specifies the language for expressing tolerances; very similar languages covered by ISO standards are used in Europe and Asia. There is no accepted mathematical theory of dimensional tolerancing, and the current Y14.5 standard is best viewed as a collection of sensible principles, defined informally with prose and graphics through special-case examples. There is no metrology standard that specifies procedures for determining conformance to Y14.5 specifications.

The informal definitions in Y14.5 and the lack of standardized measurement procedures caused few problems in the pre-1980 era, when design, manufacturing planning, and a good deal of physical manufacturing were done manually by skilled artisans; when ambiguities arose, the artisans conferred and used commonsense knowledge to resolve them. However, the widespread introduction of coordinate measuring machines (CMMs) and CAD systems that began in the 1970s caused some serious underlying problems to surface in the 1980s. As the decade unfolded, growing cries were heard about a 'Metrology Crisis'. A nadir of sorts was reached in 1988, when Richard Walker issued his now-famous GIDEP (Government-Industry Data Exchange Program) Alert; this reported major inconsistencies in CMM part assessement data, and effectively stopped the use of CMMs for a short period on mainly military jobs.

As Richard Hook explains in the first paper below, the 'Metrology Crisis' reflects two underlying phenomena. One is 'methods divergence' – the fact that measurements made by different methods (e.g. manual micrometer, functional gage, CMM), or processed by different algorithms (as in CMMs), often yield different results. The second might be called 'specification ambiguity', i.e. ambiguity in the definition of what is to be measured. The latter is often attributable to the tolerancing standard itself, whereas the former reflects the lack of well specified procedures for assessing conformance to the standard. As Hook also explains, representatives from the Y14.5 (dimensioning and tolerancing) and B89 (dimensional metrology) standards committees began to meet informally in the mid-'80s to discuss remedial action.

These matters 'went public' in a 1988 NSF-sponsored workshop (the 'Orlando Workshop') organized by Vijay Tipnis. About 80 representatives from three 'communities' – standards, industrial users, and research/academe – held open discussions on many aspects of tolerancing and metrology, and produced a report with many recommendations. Amongst the most important are the following (in paraphrase).

- Y14.5 should be 'mathematized' and generalized, so that
- provably correct measurement procedures can be designed to assess conformance.
- More general approaches to tolerancing should be sought, so that parts and products not now covered by tolerancing standards can be brought under a single umbrella.
- Education and research in tolerancing and metrology should be expanded markedly.

The Orlando Workshop set what was, in effect, a work agenda. The standards community responded vigorously by establishing the Y14.5.1 committee to mathematize Y14.5, and the B89.3.2 committee to address measurement methods. Both committees have worked diligently, and the fruits of their efforts – emerging new standards – are discussed in papers below. Research has also gained momentum, but there has been little progress on the educational front.

Five years have passed since the 'Orlando Agenda' was set, and the 1993 Forum was conceived in part as a five-years-later stock-taking: what have we accomplished to date, how do we view the issues today, and what should we do next? The Forum was not designed to focus explicitly on these questions, however; its organization reflects the Steering Committee's conception of the logical structure of the field. Thus the sessions cover

Standards
Principles of Tolerancing
Dimensional Metrology
Tolerancing & Metrology in the Product Life Cycle
Computational Methods in Tolerancing & Metrology
CAD/CAM Systems
Codification, Education, and Certification.

These topical sessions were interspersed with discussion sessions, and capped with a 'Current State of Affairs' panel and some *ad hoc* panels.

In retrospect, we believe that the Forum succeeded only partially in answering the stock-taking questions, and in posing sharply the major outstanding issues. (Little was said, for example, about how to *use* tolerances, e.g. assign them in design.) Also, the tenor of the meeting was subdued relative to that at Orlando. The Orlando Workshop ended on a high note: the divergence problems plaguing shop-floor practitioners had been brought into public view, generic solutions had been prescribed (mathematize Y14.5, standardize measurement methods, promote education, ...), and communities with no experience of working together – notably the research and standards communities – had discovered they could at least talk to one another. There has been solid accomplishment in the intervening five years, with the new Y14.5.1 draft standard and the emerging B89.3.2 standard as 'tangibles', but we have also found that the issues are genuinely *difficult,* that it takes time for communities with different value systems to learn to work together, and that an uncertain economy restricts the travel funds and released time needed for joint standards development and research. In summary, 'ultimate' solutions are not in sight, and we seem to be in the early stages of a major, long-term effort aimed at more effective management and control of manufacturing variability through systematic, science-based methods.

One promising consequence of the Forum is the establishment of a standing ASME Research Committee on Tolerancing and Metrology. This Committee will, amongst other activities, organize a Forum follow-on meeting a few years hence.

In conclusion, we wish to thank sincerely for their effort and support

- the Sponsors of the Forum: the National Science Foundation, the ASME Council on Codes and Standards, the Brown & Sharpe Manufacturing Company, and Applied Computer Solutions, Inc.;
- the able staff of the ASME Center for Research and Technology Development (CRTD): Howard Clark, Raymond Art, Norma Johnston, and Lisa Strachan;
- all members of the hardworking and congenial Forum Steering Committee (listed below), and our industrious Reporters, Steffen Parratt and Dean Robinson.

Herb Voelcker and Vijay Srinivasan

Steering Committee for the Forum

CONTENTS

CRTD-Vol. 27, International Forum
on Dimensional Tolerancing and Metrology
ASME 1993

Keynote Address
INTERACTION OF DIMENSIONING, TOLERANCING, AND METROLOGY

Richard Hook
METCON, Incorporated
Coventry, Rhode Island

ABSTRACT

This Forum is concerned with two activities that need to be better understood and integrated. They are *specifying* dimensions and tolerances for workpiece designs to capture and preserve design intent, and *measuring* actual workpieces to determine conformance with the specifications and design intent. The need for standards to govern dimensioning and tolerancing became apparent more than fifty years ago, but companion standards for measurement were not thought necessary until the 1980s, when coordinate measuring machines brought 'measurement divergence' to the fore as a central problem. Work aimed at resolving measurement divergence and related matters accelerated after a 1988 Workshop brought the problems into public view, and new tolerancing and metrology standards are now emerging. This Forum will review the accomplishments to date, and consider what additional steps may be needed.

INTRODUCTION

This Forum is concerned with the study of dimensions and tolerances as tools to express design intent, with measurement techniques for assessing manufactured parts, and with the integration of specification and measurement. One might think that at least the measurement part should be easy, because we have many good measuring devices. We shall see, however, that different devices have characteristically different methods for making measurements, and these methods may affect the results to such an extent that the measured results may not be the results the designer had in mind at all.

STANDARDS

Attempts to standardize design notation for dimensions began in the 1920s and '30s. Later, more refined approaches included tolerances, but little has been done until very recently to standardize methods for measurement.

In the mid-1930s, work was getting under way in England to identify preferable means of designating dimensions; this culminated eventually in British Standard 308. At about the same time, the Handbook for Draftsmen appeared in America, and shortly thereafter the American Society of Mechanical Engineers (ASME) published a first drafting standard, Z14.1, under the American Standards Association. Ten years later, in 1945, Ordnance Standard 30-1-7 appeared. By 1957, the American National Standards Institute (ANSI) had replaced the earlier ASA, and information from all of the above standards, plus some new knowledge, was incorporated in the ASME/ANSI Y14.5 Standard.

The aforementioned standards provided a "language" to achieve uniformity in the stating of dimensions and tolerances with respect to workpiece features and references, but none purported to specify how *measurements* were to be made. The designers of the time were well aware that fixed gages, calipers, and micrometers were used for measurement, and probably the original standards writers, in defining how dimensions and tolerances were to be specified, also had in mind how they would be measured in terms of the equipment available at the time. Coordinate measuring machines were not yet on the scene. When they did appear – a few in the 1960s, more in the 1970s, and a flood in the 1980s – they complicated matters considerably.

MEASURING METHODS AND THEIR DIVERGENCE

We said earlier that the measuring method used affects the measured results. Therefore, while the designer may know what is wanted physically, functionally, and dimensionally, he or she must also be aware of the measurement methods used to assess the part because these will affect the results.

Figure 1 illustrates some of the issues with a very simple example: measuring a cylinder diameter by three methods – a fixed gage, a micrometer, and a coordinate measuring machine (CMM).

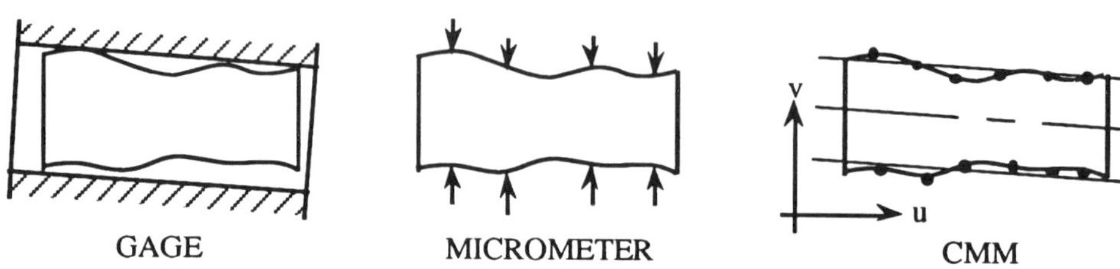

FIGURE 1: TECHNOLOGIES FOR CONFORMANCE ASSESSMENT.

- With the fixed gage, one is assured that the maximum diameter of the cylinder is within the prescribed limits, but information about other diameters or surface condition is not available.
- With the micrometer, one knows something about the diameter being measured but nothing about the rest of the cylinder. However, the more diametrical measurements that are made the more one knows about the work piece.
- With the coordinate measuring machine, the surface is characterized by a true cylindrical form fitted to the data by a least-squares or other 'best-fit' criterion. The calculated surface may contain some of the measured points, while others will lie above or below it.

It should be obvious that a designer specifying dimensions and tolerances per Y14.5 and thinking they will be measured by gages, when in fact they will be measured by CMMs, will have to deal with numbers representing something quite different than what he or she has in mind.

The differences in the measuring methods illustrated above, and the different results they yield, are what we have come to call *measurement divergence (or methods divergence)* ... and these examples are just "the tip of the iceberg". With technological advances, we can predict that more measurement processes will emerge and impose their own characteristics on measured results. This will be particularly true as we specify tighter tolerances and increased precision in measurement.

METROLOGISTS' CREDOS

In 1883, with 59 years of experience behind him, the British physicist Lord Kelvin (1824-1907) made the following observation.

"I often say that when you can measure
what you are speaking about and express it in numbers,
you know something about it;
but when you cannot measure it;
when you cannot express it in numbers,
your knowledge is of a meagre and unsatisfactory kind,
it may be the beginning of knowledge,
but you have scarcely, in your thoughts,
advanced to the stage of science, whatever the matter may be."

Metrologists have taken Kelvin's statement as the cornerstone on which they build their contributions. But 110 years have passed since Kelvin philosophized, our world is more complicated, and we know now that the results one obtains vary with the measurement process used. Let me suggest, therefore, that Kelvin's statement be modified as follows to reflect the impact of methodology.

I often say that when you can measure
what you are speaking about, and express it in numbers,
and understand the significance and meaning of the numbers,
you know something about it;
but when you cannot measure it;
when you cannot express it in *meaningful* numbers,
your knowledge is of a meager and unsatisfactory kind,
it may be the beginning of knowledge,
but you have scarcely, in your thoughts,
advanced to the stage of science, whatever the matter may be.

RESOLVING MEASUREMENT DIVERGENCE

In the early 1980s, when CMMs were proliferating rapidly and becoming the measurement system of choice, members of ASME Working Group B89.1.12[1] became aware that results based on fitting curves and surfaces to CMM data are quite different from results obtained through the methods implied by Y14.5 (gages, manual instruments). It appeared that both B89.1.12 and Y14.5 were faced with a measurement divergence problem that could best be solved by working together. The National Bureau of Standards (now NIST) provided neutral ground for an historic first meeting of the two groups, convened eleven years ago by Tom

[1] ASME Working Group B89.1.12 (currently Sub-Committee B89.4) developed the Performance Evaluation Standard ASME/ANSI B89.1.12M – Methods for Performance Evaluation of Coordinate Measuring Machines.

Charlton, to discuss divergence and possible courses of constructive action. Y14.5 and B89.1.12 have continued to work cooperatively toward a mutually satisfactory solution since that time.

Five years ago a Workshop was convened in Orlando, Florida, by the ASME, with full funding from the National Science Foundation, to explore various problems in tolerancing (limited geometric coverage, some apparently ambiguous definitions, links to process precision and process control, and so forth), as well as the measurement divergence problem. Of the numerous suggestions for future action that emerged from the workshop, two have made substantial progress.

The first was the formation of Subcommittee Y14.5.1, Mathematical Definition of Y14.5 Dimensioning and Tolerancing Principles. This group includes members from both the B89.1.12 and Y14.5 organizations. Its assignment was to 'mathematize' Y14.5 to reduce ambiguity and improve compatibility between Y14.5 specifications and CMM type measurements, as well as to consider the impact of computer processed information on the factory floor. The Y14.5.1 committee recently issued a draft of Y14.5.1M, Mathematical Definition of Dimensioning and Tolerancing Principles, for consideration by both Y14.5 and B89.

The second activity was focused on organizing under B89 a group to work in parallel with Y14.5.1. This has emerged as WG B89.3.2, which is concerned with standardization of measurement methods. Substantial progress has been made. While it would be impossible to develop a single method, or even a reasonable number of measurement procedures, given the great variety of methods available, it is possible to establish a "measurement plan" which defines all things to be considered before measuring to ensure that the end result will be what the designer had in mind.

THE 1993 FORUM

Five years have passed since the Orlando meeting and it is now appropriate to explore the positive results of that meeting, and to consider further improvements in both dimension and tolerance specification and measurement processes.

In addition, the world is shrinking, the marketplace for business is globally expanding, and international (ISO) standards are of growing importance and warrant our attention.

Toward these ends, a stimulating agenda has been generated for this Forum and global experts have convened to participate in what we hope will be an important exploration for improvement of both today's and tomorrow's specification and measurement-process needs.

CRTD-Vol. 27, International Forum
on Dimensional Tolerancing and Metrology
ASME 1993

STRATEGY FOR MANAGING STANDARDS

Paul E. McKim
Technical Service Division
Caterpillar Incorporated
Peoria, Illinois

ABSTRACT

The voluntary standards system in the United States faces many issues that must be resolved. Many companies are looking to strategic use of standards as part of their business strategy to enhance competitiveness. This requires easy access to existing standards, and improved standards setting environment, and a commitment to develop and use technically correct standards.

The United States voluntary standards development process has the potential to bring the right people together in a forum to develop strategies that will produce standards that meet specific needs.

ASME committees such as Y14, and B89 are developing standards that meet strategic needs. The committees have made progress in developing better, more technically correct standards in the field of dimensional tolerancing and metrology. The committees need to continue this spirit of cooperation as we move toward more internationalization of standards setting.

REFERENCES

International Standards and Trade, *Hearing before the Subcommittee on Technology and Competitiveness of the Committee on Science, Space and Technology*, U.S. House of Representatives, March 4, 1992, U.S. Government Printing Office.

ANSI 1993 Annual Public Conference, March 18-19, 1993, Washington D.C.

"Global Connections: The Quest for Worldwide Standardization", American National Standards Institute, Special advertising section, Fortune Magazine, November 1991.

CRTD-Vol. 27, International Forum
on Dimensional Tolerancing and Metrology
ASME 1993

THE NEW ASME Y14.5M STANDARD ON DIMENSIONING AND TOLERANCING

Alvin G. Neumann
Technical Consultants Inc.
Longboat Key, Florida

Abstract

The current American national standard for dimensioning and tolerancing is the ANSI Y14.5M, 1982. This document has been under review for the past 11 years. During this period, twenty-three full subcommittee meetings and many working group meetings have been held in various cities around the country. As a result of these meetings a new draft standard has been prepared and will soon be available for public review.

Many changes have taken place since the publication of the standard in 1982. There have been major advances in the computerization of product engineering, manufacturing and quality control. There has been an explosive growth in the quest for knowledge in the area of product definition, total quality management and world-class engineering. Around the globe, private organizations, government agencies and universities are studying dimensioning and tolerancing to better understand how to define, verify and manufacture product.

This presentation will explore some of the proposed changes in the new draft document. We will examine the impact these proposed revisions will have on dimensioning and tolerancing in engineering, manufacturing and quality.

The rationale and logic for some of the revisions will be explained. This will include the coordination and connections with the other related draft standards including the new ASME Y14.5.1 on "Mathematical Definition of Dimensioning and Tolerancing Principles, the ASME Y14.5.2 on "Certification of GD&T Professionals." and the ASME B89 committee on "Dimensional Metrology" In our quest for global competitiveness, international papers were presented and decisions were made to more closely align in the international arena. The connection and inter-relationship with the international standards will be discussed.

Introduction

The current American national standard in the United States for dimensioning and tolerancing is the ANSI Y14.5M, 1982. This document has been around now for almost eleven years. The document is even older when you consider that there is a review and approval process that takes place before a draft can be published. The review and approval process took almost two years for the 1982 standard. In effect then, the current ANSI Y14.5M, 1982 document is over thirteen years old.

Can you remember what it was like thirteen years ago? Designers were still drawing on drafting boards with straight edges and triangles. There were some CAD systems around but the drawing packages we had were very primitive by today's standards.

Can you remember what it was like in quality control and inspection? Statistical Process Control was just getting popular here in the U.S. Almost everyone verified parts on a surface plate or with hard gaging. Coordinate measuring machines were around but they were not common in the workplace.

Can you remember what it was like in manufacturing? NC machines were popular but the interest in CNC equipment was just getting started.

Think for a moment what has happened since 1982. Statistical process control, design of experiments, team engineering, concurrent engineering, simultaneous engineering, world class quality these are just a sampling of some of the buzz-words and concepts that have evolved since the ANSI Y14.5M, 1982 standard has been published. We certainly have come a long way.

When you look back at the current dimensioning and tolerancing standard we can see that it has really stood the test of time. The principles and framework that are documented in the Y14.5M, 1982 and earlier standards provide a solid framework from which we can build for the future. There are a few places in the 1982 standard that need a little tweaking, clarification and adjustment but the basic principles and foundation are on solid ground.

The New ASME Y14.5M Standard

The Y14.5M subcommittee has been working on a new document since the 1982 document has been published. The chairman is Mr. Lowell Foster and the two vice chairman are Mr. Carl Lance and Mr. Don Buchman. This document has been under review and revision for the past 11 years. During this period, twenty-four full subcommittee meetings and many working group meetings have been held in various cities around the country.

As a result of these meetings, a new draft standard has been prepared and will soon be available for public review. There are many improvements, enhancements and clarifications of the basic principles.

The Y14.5M subcommittee met during the week of June 14, 1993 in Dearborn, Michigan to review the comments to our new draft. We believe that all of the comments have been resolved. The next step is to go to public review. The new ASME Y14.5M draft is expected to be available for public review at the end of July. The public will have 60 days to review the comments and send in their comments. The committee has scheduled another meeting in October in Tucson, AZ to review the comments. If all goes well, we expect to respond to the public comments soon after and publish the new Y14.5M document either late this year or early next year.

The current ANSI Y14.5M, 1982 standard is based on a storehouse of knowledge and wisdom that has been accumulated through trial and error over the years since Eli Whitney and John Hall made the first interchangeable parts. These concepts, practices and "tricks of the trade" have been handed down from father to son. The Y14.5M, 1982 standard has collected, refined and documented these concepts. It relies on a lot of common sense and the practical aspects of inspection equipment, manufacturing equipment and functional gages. As new concepts were developed they had to pass the "litmus test" which requires functional gages to check concept viability before being adopted into the standard.

Many changes have taken place since the publication of the standard in 1982. There have been major advances in the computerization of product engineering, manufacturing and quality control. There has been an explosive growth in the quest for knowledge in the area of product definition, total quality management and world -class engineering. Around the globe, private organizations, government agencies and universities are studying dimensioning and tolerancing to better understand how to define, verify and manufacture product.

Orlando International Tolerancing Workshop and ASME Y14.5.1 Mathematical Definitions Standard

Computers have changed many things in the world. They have also changed the way we look at and verify tolerances. The International Tolerancing Workshop that was held in Orlando in the fall of 1988 was a turning point in the conventional wisdom on tolerancing and it was to change forever how we looked at tolerancing parts.

At the Orlando workshop, there were many questions on how to mathematically interpret tolerances per the ANSI Y14.5M, 1982 standard. In particular there were many questions on how to establish a datum reference plane on a part. The ANSI Y14.5, 1982 standard explains that a part surface is placed on a datum simulator such as a surface plate and contacts on the high points of that surface. Measurements are then taken from the simulated plane and not the part.

Questions were raised that if the part surface were irregular or convex and were to "rock" then which high points were to make contact. The Y14.5M, 1982 standard explains that "if a part wobbles, it may be adjusted to an optimum condition". This is exactly what a part inspector might do if he or she were working with measurement equipment on a surface plate.

As we all know, this answer did not go over very well with people who were trying to put some mathematical rigor to many of the concepts that were outlined in the Y14.5M document. The Y14.5M concepts worked very well when we worked on surface plates and hard gages but as we investigated the principles using a mathematical microscope some of the concepts were not very clear. It was not only the fact that the concepts were not clear but additional clarification and definition was needed for work and study to progress in the continuing improvement of computerization in engineering, manufacturing and quality systems.

It seems that some people and organizations, needing more in-depth explanation of the concepts, were making up there own definitions and principles. The worry was that without a clear mathematical definition of Y14.5M, the hard work completed over the years would fragment, lose direction and become a total mess.

A mathematical definition to the Y14.5M principles was required immediately. At the time, the Y14.5M committee realized that adding the requirements of a mathematical base to the existing Y14.5M was beyond its scope. The users of Y14.5M are designers, draftsman, quality engineers, manufacturing engineers, tool makers and inspectors. They need the practical common sense (surface plate, machine table, and functional gage) explanations of tolerancing. In addition, the Y14.5M was nearing completion. It was time a new document was published. To review and mathematize everything that had been completed over the previous 8 years would have been a monumental task. There was not enough time.

It was at the Mechanical Tolerancing Workshop that the sister document to Y14.5M was born. This brand new document is the Y14.5.1 Mathematical Definitions of Dimensioning and Tolerancing Principles. Mr. Richard Walker is the Chairman and Dr. Ted Hopp is the Vice Chairman.

Harmonization of Y14.5M and Y14.5.1

In the beginning, as you might imagine, there was a lot of friction between the two groups. After all, the Y14.5M was the premier document that had established and documented dimensioning and tolerancing principles and practices for the past 40 years. There was a lot at stake. The Y14.5M committee needed assurances that the new Y14.5.1 group were going to take their task with vigor and sincerity. A large influence for making the two groups work together was Mr. Paul McKim the past Y14 chairman and Mr. Frank Bakos the past vice chairman. (now present chairman of ASME Y14).

Since 1988 both the Y14.5 and the Y14.5.1 committee has been hard at work sharing information, sharing members, attending joint meetings and working together to produce the "mother" of all documents relating to geometric dimensioning and tolerancing.

The new ASME Y14.5.1 standard is designed to put some mathematical rigor to the ASME Y14.5M standard consistent with its current practices and principles. The new ASME Y14.5.1 standard has created very explicit mathematical definitions for use in many area such as Computer Aided Design (CAD), Computer Aided Manufacturing (CAM) and Coordinate Measuring Machines (CMMs). The Y14.5M and Y14.5.1 committee members have worked closely to revise and adjust their documents as necessary to ensure compatibility and continuity.

This new Draft ASME Y14.5.1 Mathematical Definitions document is completed and is also presently going through the public review process with the Y14.5M. The two documents are expected to be published simultaneously.

The new ASME Y14.5M and its sister document the ASME Y14.5.1 dimensioning and tolerancing standards are state of the art. The committee members of both documents have followed and incorporated many changes as a direct result of the major advances that computers have had in product engineering, manufacturing and quality control. The new documents are designed to take us forward to the next century and make United States more competitive in the world marketplace. The two new standards will allow us to define and document our product so it may be interpreted universally. It will allow our products to be interchangeable with other parts all over the globe.

Enhancements and Revisions to Y14.5M

There are many enhancements and revisions to the Y14.5M standard. A detailed list of the changes can be found in Appendix A in the new standard. Following is a listing of some of the major revisions and enhancements.

Symbology

There has been some changes to the symbology in the Y14.5M document. These changes have been necessary to insure our compatibility with ISO.

Symmetry and Concentricity

The old symmetry symbol has been resurrected from the ANSI Y14.5, 1973 standard. See figure 1. Symmetry has a new definition that controls the features median points to a datum axis or median plane. The concentricity definition has also been clarified to also control a features median points to an axis.

Datum Feature Symbol

The datum feature symbol has been changed from a rectangular box to a square box that is attached to the datum with a triangular foot. This was done to more closely align ourselves in the international arena. The new datum feature symbol is in accordance with ISO. See figure 2.

Modifying Symbols

There have been a few new modifying symbols added such as free state, statistical process control, tangent plane and controlled radius. See figure 3. The first three symbols can be added behind the feature tolerance in a feature control frame. The RFS symbol has been removed from the standard because it is no longer needed. The old rule 2 from the Y14.5M, 1982 version has been revised to coincide with ISO and now all position callouts are implied RFS unless otherwise specified.

Size

Size is an area that the Y14.5M and the Y14.5.1 worked very close. The maximum material limit of size which required perfect form at MMC was no problem. It was the definition of the LMC limit of size that gave us trouble. The Y14.5M traditional definition of LMC was a two point check, similar to a micrometer or caliper check. The Y14.5.1 group had a hard time placing mathematical rigor to this concept. In our last meeting in Chicago, it was decided to go with the proposed Y14.5.1 definition of size which has a spherical LMC limit.

Definitions

There are many new definitions. The definitions have been worked over intensely with the Y14.5M and the Y14.5.1 group to ensure compatibility between the documents. The definitions are too numerous to mention but there are two that come to mind, inner boundary and outer boundary. These two new definitions provide the means to complete tolerance analysis and stack-ups on complex parts.

Composite Tolerancing-Position and Profile

Composite tolerancing has been expanded to allow profile tolerancing as well as position tolerancing. Composite tolerancing is also enhanced by providing a means to not only control the location of features but to orient the features as well. If datum(s) are entered in the feature relating control of a feature control frame, they control not only the location of features to each other but also control the orientation of the features to the specified datum(s). See figure 4.

Boundary Concept

The boundary concept was shown in the Y14.5M, 1982 standard on an elongated hole. In the new standard the boundary concept is expanded to an irregular feature. A profile specification is used to control size and a position (boundary) is used to control the location of the feature. See figure 5.

Harmonization with National and International Standards.

In addition to the harmonization between Y14.5 and Y14.5. there was a close alliance with other national and international standards committees to ensure compatibility between concepts, symbology and documents.

B89 Dimensional Metrology

The ASME B-89 Committee on Dimensional Metrology is especially important in the scheme of things. The chairman on this committee is Mr. Richard Hook and the vice chairman is Mr. Fred Parsons. This series of documents define standards for measurement principles, calibration and the measurement methods and procedures necessary to insure conformance to the product definition specifications.

In working in close contact with the B89 group, we found some problems with how roundness was defined in ASME Y14.5M and how it was verified in B89.3.1. After many conversations and consultations and it was finally resolved.

International Standards Organization

The International Standards Organization (ISO) 9000 series of standards also play a significant role in also a major player in interchangeability and compatibility. The United States has taken a leadership roles in International standards. In fact the United States has, through its member body ANSI, assumed the ISO/TC10/SC5 Secretariat. A ISO standards meeting was held in Beijing, China during the week of May 16th, 1993. In our quest for global competitiveness, international papers were presented and decisions were made to more closely align the U.S. in the international arena. The United States presented 7 position papers of which 5 were accepted to be balloted as new work items for enhancement to the present ISO standard. At the request of the U.S. a working group was established in ISO/TC10/SC5 to develop a mathematical definitions standard similar to the ASME Y14.5.1 document.

ASME Y14.5.2- Certification of Geometric Dimensioning and Tolerancing Professionals

The ASME Y14.5.2- Certifications of Geometric Dimensioning and Tolerancing Professionals is another important part of the geometric tolerancing standardization process. The chairman of this committee is Mr. Frank Bakos. This standard is still in a draft stage. After it is published it will establish guidelines for the testing and certification of personnel in the use and interpretation of the ASME Y14.5M dimensioning and tolerancing standard.

Future

The future looks good. The next revision of the Y14.5 standard is planned to have a mathematical base along with its long tradition of functionality and "common sense engineering". The standards groups are working together. A joint meeting of Y14.5, Y14.5.1, B89 and ISO TC10/SC5 is scheduled for Longboat Key in January of 1994. If everything goes as planned, it will be an historic meeting. It will be the first time that all of these experts from the various standards groups will be in the same place at the same time. I believe that the mathematical base in all of the standards is key for developing quality standards for the United States and the rest of the world. The U.S. currently holds a leadership role in the development of an international standard. The U.S. also has "state of the art" standard documents ready to meet the quality demands of the future.

References

Dimensioning and Tolerancing, ANSI Standard Y14.5M, 1982, American Society of Mechanical Engineers: New York, NY, 1982.

Draft- Dimensioning and Tolerancing, ASME Standard Y14.5M, 199X, American Society of Mechanical Engineers: New York, NY, 199X.

Draft- Mathematical Definition of Dimensioning and Tolerancing Principles, ASME Standard Y14.5.1, 199X, American Society of Mechanical Engineers: New York, NY, 199X.

	TYPE OF TOLERANCE	CHARACTERISTIC	SYMBOL	SEE:
FOR INDIVIDUAL FEATURES	FORM	STRAIGHTNESS	⏤	6.4.1
		FLATNESS	⏥	6.4.2
		CIRCULARITY (ROUNDNESS)	○	6.4.3
		CYLINDRICITY	⌭	6.4.4
FOR INDIVIDUAL OR RELATED FEATURES	PROFILE	PROFILE OF A LINE	⌒	6.5.2 (b)
		PROFILE OF A SURFACE	⌓	6.5.2 (a)
FOR RELATED FEATURES	ORIENTATION	ANGULARITY	∠	6.6.2
		PERPENDICULARITY	⟂	6.6.4
		PARALLELISM	//	6.6.3
	LOCATION	POSITION	⌖	5.2
		CONCENTRICITY	◎	5.11.3
		SYMMETRY	⌯	5.13
	RUNOUT	CIRCULAR RUNOUT	↗ *	6.7.1.2.1
		TOTAL RUNOUT	⌰ *	6.7.1.2.2

* ARROWHEADS MAY BE FILLED OR NOT FILLED 3.3.1

Fig. 1 GEOMETRIC CHARACTERISTIC SYMBOLS

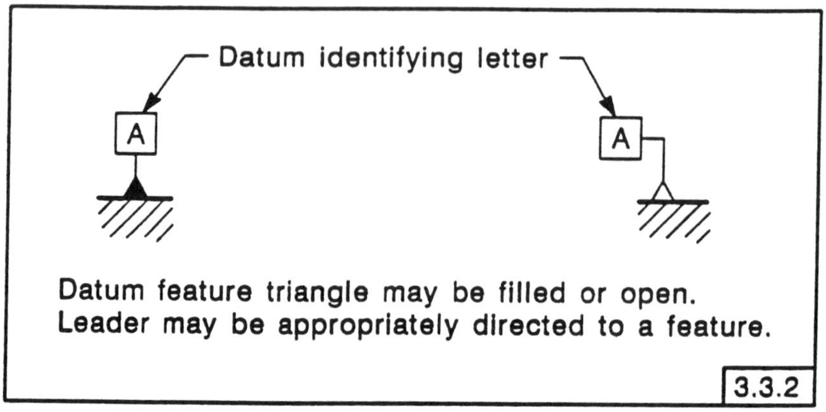

Fig. 2 DATUM FEATURE SYMBOL

TERM	SYMBOL	SEE:
AT MAXIMUM MATERIAL CONDITION	Ⓜ	3.3.5
AT LEAST MATERIAL CONDITION	Ⓛ	3.3.5
PROJECTED TOLERANCE ZONE	Ⓟ	3.3.6
FREE STATE	Ⓕ	3.3.19
TANGENT PLANE	Ⓣ	3.3.20
DIAMETER	Ø	3.3.7
SPHERICAL DIAMETER	SØ	3.3.7
RADIUS	R	3.3.7
SPHERICAL RADIUS	SR	3.3.7
REFERENCE	( )	3.3.8
ARC LENGTH	⌒	3.3.9
CONTROLLED RADIUS	CR	3.3.7
STATISTICAL TOLERANCE	⟨ST⟩	3.3.10
BETWEEN	◄─►	3.3.11

Fig. 3 MODIFYING SYMBOLS

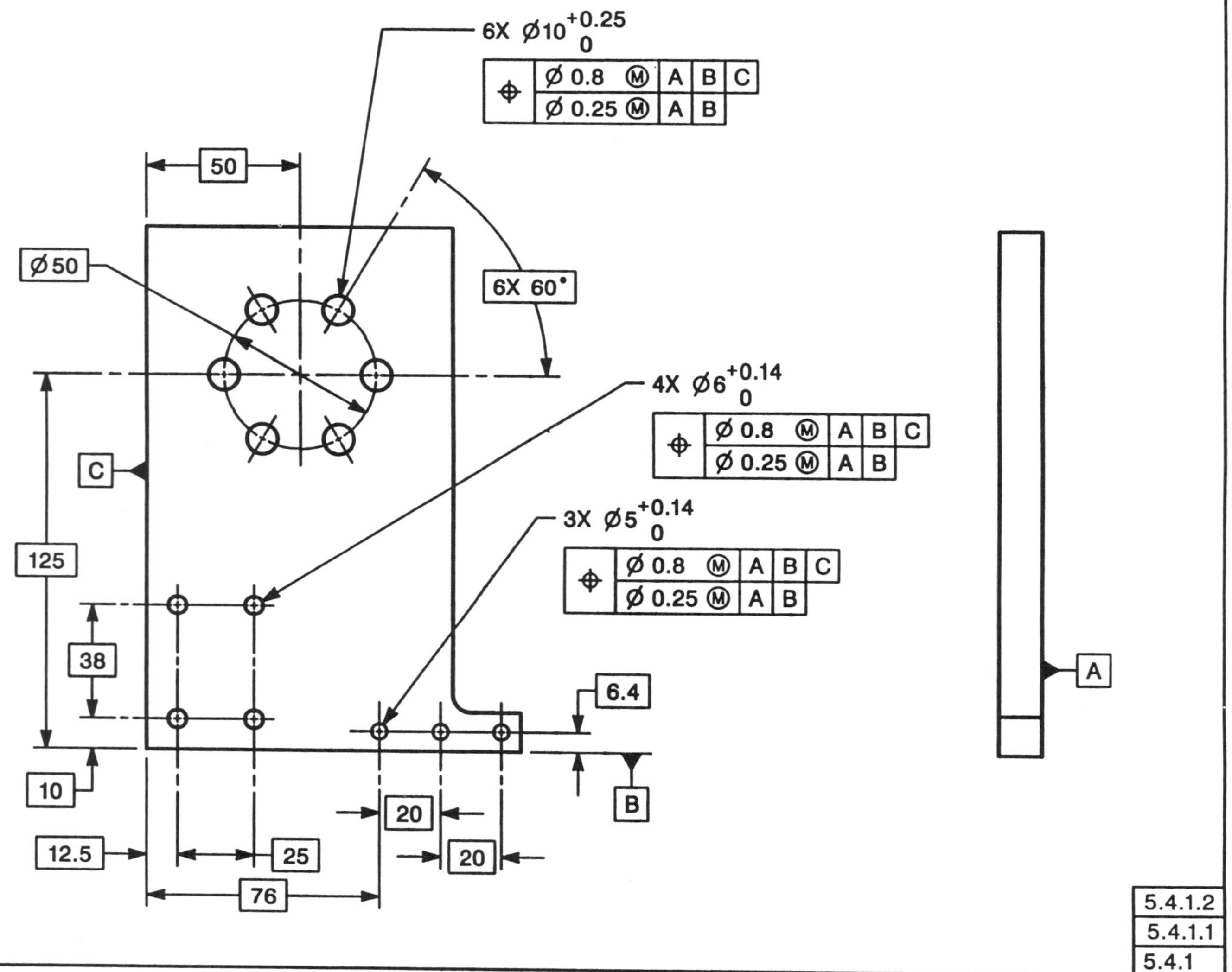

Fig. 4 HOLE PATTERNS OF FIGURE 5-19 WITH SECONDARY DATUMS IN FEATURE-RELATING SEGMENTS OF COMPOSITE FEATURE CONTROL FRAMES

THIS ON THE DRAWING

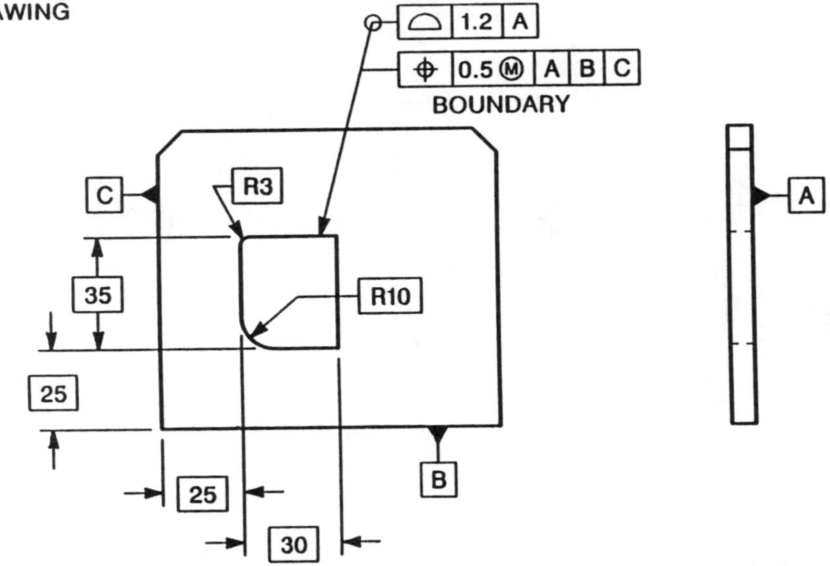

6.5.5.1

MEANS THIS

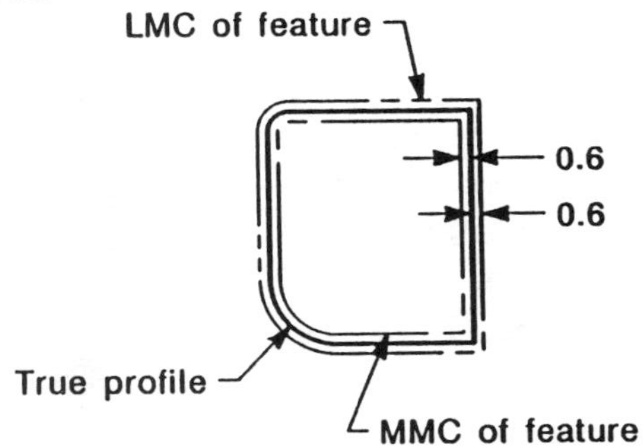

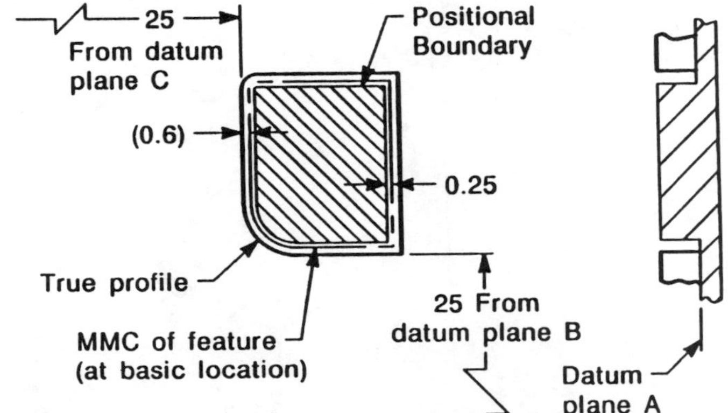

Profile Control

The surface, all around, must lie between two profile boundaries 1.2 apart equally disposed about the true profile.

Postion Control

No portion of the surface may be permitted to lie within the boundary of MMC contour minus the positional tolerance when positioned with respect to datum planes A, B and C.

Fig. 5 BOUNDARY PRINCIPLE USED WITH PROFILE CONTROLS

CRTD-Vol. 27, International Forum
on Dimensional Tolerancing and Metrology
ASME 1993

CREATION AND EVOLUTION OF THE ASME Y14.5.1M STANDARD

Richard K. Walker
Marine Division
Westinghouse Electric Corporation
Sunnyvale, California

Vijay Srinivasan
Department of Manufacturing Research
IBM T. J. Watson Research Center
Yorktown Heights, New York
and
Department of Mechanical Engineering
Columbia University
New York, New York

ABSTRACT

The forthcoming ASME Y14.5M-1993 standard will soon have a mathematical companion in the form of ASME Y14.5.1M-1993 standard. Effort on this mathematical document began in early 1989. This paper describes the origin, workings, and accomplishments of the Y14.5.1 subcommittee that was responsible for this new standard.

INTRODUCTION

Engineers must deal with geometric variations in the products they design, manufacture, sell and service. The ASME Council on Codes and Standards, in coordination with the American National Standards Institute, set up the Y14.5 committee to provide a common drafting and documentation language to help engineers in these tasks. In recent times, this has resulted in successful issuance of dimensioning and tolerancing standards in 1966, 1973 and 1982. The next release of this standard, tentatively titled ASME Y14.5M-1993, is imminent (Neumann, 1993). This time, however, it will be accompanied by a mathematical standard, tentatively titled ASME Y14.5.1M-1993.

The new mathematical companion is the result of nearly five years of work by the ASME Y14.5.1 subcommittee. In this paper we describe the origin, scope, and workings of the subcommittee. We also touch on various challenging issues addressed in the last five years. The paper concludes with some thoughts on what the future might hold for mathematization efforts.

In what follows the term Y14.5.1 is used to refer to the subcommittee and also to the document which will be issued as the ASME Y14.5.1M-1993 standard. The context will indicate which of these two the term refers.

ORIGIN OF Y14.5.1

The original ASME Dimensioning and Tolerancing Committee had its beginning in the 1960's with the formation of the ASME Y14.5 committee. The task of that committee was to develop a dimensioning and tolerancing standard independent of manufacturing and measurement processes. The Y14.5M-1982 standard further states that "this document is not intended as a gaging standard". It is an attempt to create a language strong enough to express unambiguously the design intent to manufacturing and quality organizations.

Along with the development of Y14.5, new measurement methods were being invented and perfected. The advent of numerically controlled measurement instrumentation revolutionized the science of metrology. A need for a performance evaluation standard for coordinate measuring machines (CMMs) arose from their rapid proliferation. In response, ASME Codes and Standards formed the B89.1.12 committee for the performance evaluation of CMMs. B89.1.12 formed a software subcommittee, after issuance of the first ASME B89.1.12 "Performance Evaluation of Coordinate Measuring Machines" standard (1985). One of the tasks of this software subcommittee was to examine the software interpretation by CMM vendors of the Y14.5 standard.

Interpretation of geometric dimensioning and tolerancing (GD&T) by CMMs required a thought process quite different from the traditional world of functional gaging. Moreover, initially software developers of coordinate measurement systems had very limited understanding of GD&T. The software subcommittee wanted CMMs mentioned in the Y14.5 standard: this was viewed as legitimizing the use of CMMs to check for GD&T specifications.

The fact that CMMs did not always interpret GD&T specifications correctly was made public by the issuance of a GIDEP Alert (Walker, 1988). Further discussions with the B89.1.12 and Y14.5 committee members revealed several ambiguities within the Y14.5 standard, which made it impossible to interpret some specifications. Knowledge of these problems made the leaders in B89 and Y14 committees willing and concerned participants in the International Workshop on Mechanical Tolerancing sponsored by ASME and the National Science Foundation in 1988 at Orlando, Florida. A key recommendation of this workshop was that (Tipnis, 1990)

> "An attempt should be made to determine if mathematical definitions could be formulated to put the current standards such as ANSI Y14.5M on a firmer basis to reduce the risk of misinterpretations and ambiguities. These definitions should be representation-free and non-computational to minimize dependencies on specific representations, coordinate systems and choice of origin".

Based on this and other recommendations, leaders of both the Y14.5 and B89 committees formed an ad-hoc committee under Y14 to provide "mathematical definitions of the dimensioning and tolerancing principles" found in Y14.5 standard. This ad-hoc committee soon became the Y14.5.1 subcommittee of Y14 with the above stated mission.

SCOPE OF Y14.5.1

The first meeting of the Y14.5.1 subcommittee was held in January of 1989 at Longboat Key, Florida. Since then, it has met three times per year. Initial discussions covered the overall scope of the final document to be written by the subcommittee, the document format, and other technical matters.

The subcommittee identified four major categories of tolerance definition:

1. the mathematical definition of the tolerance zone,
2. the mathematical definition of conformance to the tolerance,
3. the mathematical definition of the actual value, and
4. the mathematical definition of the measured value.

The subcommittee later decided to drop the measured value definition from its scope to preserve the original intent of creating a GD&T language independent of manufacturing and measuring methods. While defining part conformance to the tolerance, it was assumed that the actual part consists of infinite number of points that make up all the surfaces of the part — in other words, the subcommittee was concerned with the imperfect-form semantics.

The following example on flatness illustrates how Y14.5.1 defines tolerances:

6.4.2 Flatness. *Flatness is the condition of a surface having all elements in one plane. A flatness tolerance specifies a tolerance zone defined by two parallel planes within which the surface must lie.*

Definition: A flatness tolerance specifies that all points of the surface must lie in some zone bounded by two parallel planes which are separated by the specified tolerance.

A flatness zone is a volume consisting of all points $\vec{\mathbf{P}}$ satisfying the condition

$$|\hat{\mathbf{T}} \cdot (\vec{\mathbf{P}} - \vec{\mathbf{A}})| \leq \frac{t}{2}$$

where:

$\hat{\mathbf{T}}$ is the direction vector of the parallel planes defining the flatness zone;

$\vec{\mathbf{A}}$ is a position vector locating the mid-plane of the flatness zone;

t is the size of the flatness zone (the separation of the parallel planes).

Conformance: A feature conforms to a flatness tolerance t_0 if all points of the feature lie within some flatness zone as defined above, with $t = t_0$. That is, there exist $\hat{\mathbf{T}}$ and $\vec{\mathbf{A}}$ such that with $t = t_0$, all points of the feature are within the flatness zone.

Actual value: The actual value of flatness for a surface is the smallest flatness tolerance to which the surface will conform.

The sentences seen above in italic are direct quotations from the Y14.5M standard document, which the Y14.5.1M standard document tracks very closely.

WORKING MODE OF Y14.5.1

The topics addressed by the subcommittee were not always amenable to such quick and clean solutions as the flatness tolerance seen above. Several difficult topics such as datums, boundary determination (i.e. where does a hole stop and the intersecting plane start, in an actual part) and size were identified at fairly early stages of the subcommittee deliberations. In addition, several inconsistencies within the Y14.5 standard were identified.

In order to address these issues the subcommittee split into several working groups, debated the technical problems, and presented possible solutions. The process was formalized by the use of a standardized "red list item" document. For each issue under consideration, the red list item document articulated the area of misunderstanding, presented a possible solution, and demanded a documented resolution. These red list items were formally presented to the Y14.5 committee for comments. The Y14.5 committee would either concur or offer an alternate solution. This process worked reasonably well.

As mentioned earlier, the Y14.5.1 subcommittee met three times a year, each meeting lasting for two or three days. These meetings were held in coordination with the Y14.5 and B89 committee meetings to promote maximum communication among different but related committees. The Y14.5.1 subcommittee meeting would typically open with a plenary session lasting for a few hours during which matters of general concern were discussed. Then the working groups went into break-out sessions to work on their assignments. Usually, the meeting ended with another plenary session to summarize the findings of the working groups for the benefit of all members, and to plan for the next meeting. The working group members continued their work even after returning home from the meeting, often by telephone and computer communication.

The official membership of Y14.5.1 was carefully chosen to represent a broad spectrum of talent and interest. New members were added, and old ones deleted, as the work progressed. The subcommittee meetings, however, were open to anyone who was interested in the mission of Y14.5.1. In fact, important contributions were made by several people who were not official members of the Y14.5.1 subcommittee. Such openness was an important part of the process by which Y14.5.1 evolved into its current state.

CHALLENGES ADDRESSED BY Y14.5.1

As each working group tackled a specific technical issue it became clear that "the most difficult problem facing the group is not so much developing formal mathematics for tolerances, but developing formal mathematics that match engineering common sense" (Hopp, 1990). In this section we will describe some of challenges addressed by the subcommittee.

Datum

The group that worked on datums found its task extremely challenging and continued the work for almost four years. The Y14.5 standard, recognizing the imperfect form of datum features, states that "if irregularities on the surface of a primary or secondary datum feature are such that the part is unstable (that is, it wobbles) when brought into contact with the corresponding surface of a fixture, the part may be adjusted to an optimum position, if necessary, to simulate the datum". Actions such as "wobbling" and "adjusting to an optimum position" were not easily translatable into mathematical procedures. The initial alternatives considered by the working group, and their drawbacks have been described by Hopp (1990).

The proposed resolution to the datum problem is to allow for a non-unique solution. The idea is to define a candidate datum set using a mathematical procedure. Any one member in this candidate set should be acceptable from a functional point of view. However, the set is prevented from becoming too large by imposing some geometric constraints. Any member in the candidate set can be used to assess conformance with the associated part features.

The following excerpt from Y14.5.1 on planar datum features illustrates the way datums are defined:

> **4.3.2 Planar Datum Features.** The candidate datum set for a nominally flat datum feature is defined in a procedural manner. This empirical definition specifies a set of datums which are reasonable from a functional standpoint. If the datum feature is perfectly flat, the candidate datums set consists of only one datum; otherwise it may consist of more than one datum. This is equivalent to "rocking" the datum feature on a perfect surface plate. The definition below limits the amount that the datum feature can "rock" in a manner that is roughly proportional to the form variation of the datum feature.
>
> **Primary Planar Datum Features:** The candidate datum set for a nominally flat primary datum feature is defined by the following procedure:
>
> > Consider a plane P which is an external set of support for the datum feature. Let C be the set of contact points of the datum feature and P.
> >
> > Consider an arbitrary line L in P. Orthogonally project each point on the boundary of the datum feature onto L, giving the line segment L'. Consider regions of L' that are within some fraction x of the endpoints of L'. That is, if the length of L' is n, consider regions of L' within a distance xn of the endpoints of L'. Unless otherwise specified on the drawing, the value of x shall be $\frac{1}{3}$. If all of the orthogonal projections of the points in C are within either single regions, then plane P is rejected as a valid datum plane.
> >
> > Do this for all lines in P. (Note that parallel lines will yield identical results.) If no line rejects P, then P is a candidate datum for the datum feature.
>
> The procedure is illustrated in Figure 4.1, which shows one line direction L. The line segment L' is bounded by the projection of the datum feature onto L. The particular line direction illustrated in the figure does not reject P as a valid datum plane since the projections of the contact points are not all in region 1 or all in region 2 of L'. Note that only the direction of L in the plane P is important. P is a candidate datum for the datum feature if it is not rejected by any line direction in P.

Size

Defining actual local size proved to be another interesting challenge. The Y14.5 committee was asked what function size, as evaluated by a two-point cross sectional measurement, was trying to assure. The overwhelming response was that of strength.

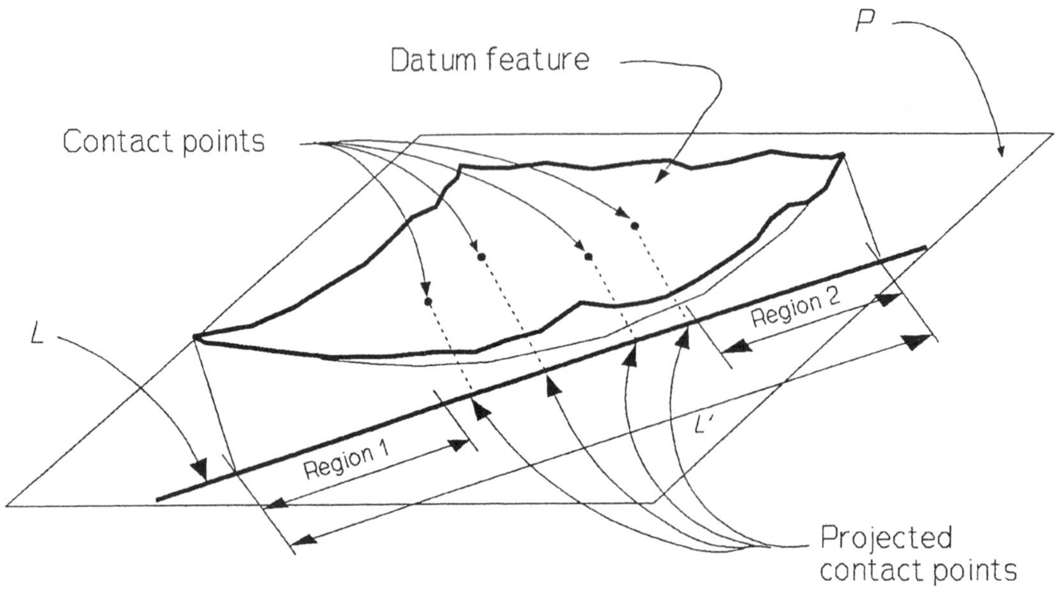

Figure 1: Illustration of Figure 4.1 mentioned in the excerpts from Y14.5.1 on datum definition.

The next question was obvious at this point. Does a two-point measurement assure strength? Two-point measurement of cylinders has the familiar fallacy of missing lobes. So again the subcommittee set out to create a mathematical definition that expressed design intent without precluding the use of cross sectional measurements. Details of the proposed solution and its rationale can be found in (Srinivasan, 1993a & 1993b).

Section Six

Section 6 in the Y14.5M standard deals with tolerances of form, orientation, profile and runout. This is a rich collection of important concepts. Since parallelism and perpendicularity are special cases of orientation, some effort was spent in consolidating these, which formed Appendix A of the Y14.5.1 standard. Mathematizing the principles behind Section 6 moved swiftly after this.

In the Spring of 1993, an important difference between the interpretations of circularity in Y14.5M-1982 and B89.3.1 was brought to light by Mark Nasson (1993). This resulted in a work item for both the Y14.5 and Y14.5.1 committees. The story of how this issue was handled and resolved is a good case study on the internal workings of standardization committees such as Y14.5.1, which we now proceed to describe.

There are two major issues in defining circularity. The first is how to choose the cross-sectional planes, whose intersection with the actual feature will be checked for circularity. The second is whether to constrain the center of the annular tolerance zone within which the actual feature should be contained, or to let the center lie anywhere in the plane. The choices can be made at the following three levels.

1. The cross-sectional planes are generated normal to a straight line, or they are generated normal to a curve.

2. If we choose the straight line to generate the cross sectional planes, then the straight line can be any straight line, or it can be a particular straight line that closely approximates the "center-line" of the actual feature.

3. Within the cross-sectional plane, the center of the annular tolerance zone is constrained to be the point of intersection of the straight line or the curve with the plane, or the center can be left totally unconstrained.

These lead us to the following six choices in the definition of circularity:

1. Find a straight line perpendicular to which all the cross-sectional planes are constructed. Within each plane, the center of the annular tolerance zone must be located at the point of intersection of the straight line with the plane.

2. Find a straight line perpendicular to which all the cross-sectional planes are constructed. Within each plane, the center of the annular tolerance zone can lie anywhere.

3. Find a straight line which somehow approximates the "center-line" of the actual feature. All the cross-sectional planes are constructed perpendicular to this straight line. Within each plane, the center of the annular tolerance zone must be located at the point of intersection of the straight line with the plane.

4. Find a straight line which somehow approximates the "center-line" of the actual feature. All the cross-sectional planes are constructed perpendicular to this straight line. Within each plane, the center of the annular tolerance zone can lie anywhere.

5. Find a curve (which is tangent continuous) through each point of which cross-sectional planes are constructed normal to the curve. Within each plane, the center of the annular tolerance zone must be located at the point of intersection of the curve with the plane.

6. Find a curve (which is tangent continuous) through each point of which cross-sectional planes are constructed normal to the curve. Within each plane, the center of the annular tolerance zone can lie anywhere.

The Y14.5.1 subcommittee created the following six test cases to check the above six choices. A part has been designed to be nominally a cylinder. However, while manufacturing it we produced

A: a perfect right circular cylinder,

B: a perfect right circular cone,

C: a perfect hour-glass,

D: a piece of a perfect torus (garden hose),

E: a perfect right elliptic cylinder,

F: a pack of "poorly stacked coins".

Test Cases	Choices					
	1	2	3	4	5	6
A: cylinder	$\surd$	$\surd$	$\surd$	$\surd$	$\surd$	$\surd$
B: cone	$\surd$	$\surd$	$\surd$	$\surd$	$\surd$	$\surd$
C: hour-glass	$\surd$	$\surd$	$\surd$	$\surd$	$\surd$	$\surd$
D: garden hose	$\times$	$\times$	$\times$	$\times$	$\surd$	$\surd$
E: elliptic cylinder	$\times$	$\surd$	$\times$	$\times$?	$\times$	$\surd$
F: poorly stacked coins	$\times$	$\surd$	$\times$	$\times$?	$\times$	$\surd$

Table 1: Test results. $\surd$- has perfect circularity, $\times$- does not have perfect circularity.

It should be noted that Test Case E (elliptic cylinder) is a special case of Test Case F. Both can be obtained by translational sweeps of a circular disk. The garden hose (Test Case D) should not be confused with Test Case F. A torus cannot be generated by a purely translational sweep of a circular disk. The test question is: Which of the above cases have perfect circularity (i.e., zero circularity tolerance)?

The test results are presented in Table 1. The interpretation of circularity according to Y14.5M-1982 seems to be along Choice #1, whereas the interpretation according to B89.3.1 seems to be along Choice #4. These choices and test results were presented to the Y14.5 committee, which voted for Choice #5. After this, the Y14.5.1 subcommittee produced the following mathematical definition of circularity tolerance:

6.4.3 Circularity (Roundness). *Circularity is a condition of a surface where: (a) for a cylinder or cone, all points of the surface intersected by any plane perpendicular to an axis are equidistance from that axis; (b) for a sphere, all points of the surface intersected by any plane passing through a common center are equidistant from that center. A circularity tolerance specifies a tolerance zone bounded by two concentric circles within which each circular element of the surface must lie, and applies independently at any plane described in (a) and (b) above.*

Definition: A circularity tolerance specifies that all points of each circular element of the surface must lie in some zone bounded by two concentric circles whose radii differ by the specified tolerance. Circular elements are obtained by taking cross-sections perpendicular to some spine. For a sphere, the spine is 0-dimensional (a point), and for a cylinder or cone the spine is 1-dimensional (a simple, non self-intersecting, tangent-continuous curve). The concentric circles defining the circularity zone are centered on, and in a plane perpendicular to, the spine.

A circularity zone at a given cross-section is an annular area consisting of all points $\vec{\mathbf{P}}$ satisfying the conditions

$$\hat{\mathbf{T}} \cdot (\vec{\mathbf{P}} - \vec{\mathbf{A}}) = 0$$

and

$$||\vec{\mathbf{P}} - \vec{\mathbf{A}}| - r| \leq \frac{t}{2}$$

where:

$\hat{\mathbf{T}}$ for a cylinder or cone, is a unit vector that is tangent to the spine at $\vec{\mathbf{A}}$. For a sphere, $\hat{\mathbf{T}}$ is a unit vector that points radially in all directions from $\vec{\mathbf{A}}$;

$\vec{\mathbf{A}}$ is a position vector locating a point on the spine;

r is a radial distance (which may vary between circular elements) from the spine to the center of the circularity zone ($r > 0$ for all circular elements);

t is the size of the circularity zone.

Figure 6.3 illustrates a circularity zone for a circular element of a cylindrical or conical feature.

Conformance: A cylindrical or conical feature conforms to a circularity tolerance t_0 if there exists a 1-dimensional spine such that at each point $\vec{\mathbf{A}}$ of the spine the circular element perpendicular to the tangent vector $\hat{\mathbf{T}}$ at $\vec{\mathbf{A}}$ conforms to the circularity tolerance t_0. That is, for each circular element there exist $\vec{\mathbf{A}}$ and r such that with $t = t_0$ all points of the circular element are within the circularity zone.

A spherical feature conforms to a circularity tolerance t_0 if there exists a point (a 0-dimensional spine) such that each circular element in each cutting plane containing the point conforms to the circularity tolerance t_0. That is, for each circular element there exist $\hat{\mathbf{T}}$, r and a common $\vec{\mathbf{A}}$ such that with $t = t_0$ all points of the circular element are within the circularity zone.

Actual value: The actual value of circularity for a feature is the smallest circularity tolerance to which the feature will conform.

Again, the sentences seen above in italic are direct quotations from the Y14.5M standard document. A short description of the Y14.5.1 definition using the notion of sweeps can be found in (Srinivasan, 1993b).

Boundary Problem

Mathematical definition of the boundary between two intersecting geometries in an actual part remains unsolved to this day. When two or more surface features come together, the question often asked of an actual part is "where does one feature stop and the other begin"?

A real-life interaction between a government inspector and a major jet engine manufacturer illustrates this problem. The company employee was measuring the flatness of a surface with an indicator, and the government inspector told him to traverse closer to the edge. The company employee felt that he was being asked to measure the radius at the edge of the part, and the government inspector felt that it was still part of the surface that should be measured for flatness. There is no clear mathematical definition of how to distinguish between the two features in actual parts.

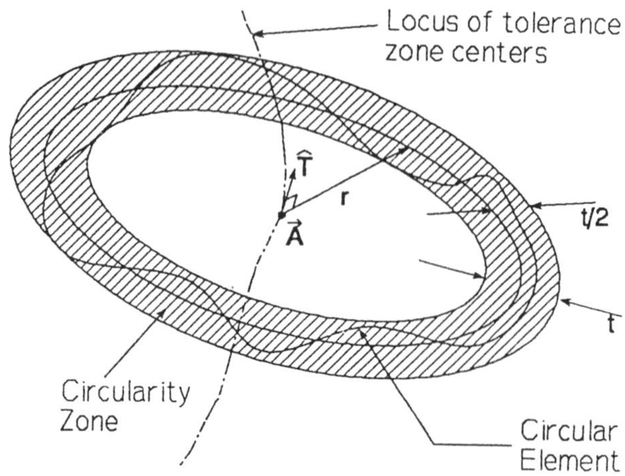

Figure 2: Illustration of Figure 6.3 mentioned in the excerpts from Y14.5.1 on circularity definition.

VALUE OF Y14.5.1

As the preparation of the Y14.5.1 standard is nearing completion, it is time to reflect on its potential value. Two questions can be asked about the Y14.5.1 standard. (1) Does it bridge any technological gap, and if so how? (2) What does it do for the designer, manufacturer, inspector, CAD vendor, and the public at large?

The answer to the first question can be found in the opening sentences of the ASME Y14.5M-1993 standard:

> "This standard establishes uniform practices for stating and interpreting dimensioning, tolerancing, and related requirements for use on engineering drawings and in related documents. For a mathematical explanation of many of the principles in this standard, see ASME Y14.5.1M 'Mathematical Definition of Dimensioning and Tolerancing Principles'. "

It is our hope that the mathematical definitions found in the Y14.5.1 standard will enable a sharper understanding of the meaning behind the Y14.5 standard, and will eliminate any ambiguity in its interpretation.

The answer to the second question depends a lot on how the engineering community responds to the Y14.5.1 document. If this improves the designer's understanding of the GD&T language, then we will see a more widespread and consistent use of GD&T in our engineering drawings and related documents. The same benefit can be expected by the manufacturer. The CAD vendors seem eager to have a mathematical definition of the tolerancing standard: the STEP community has committed to providing "generic resources" for tolerancing products on the basis of the Y14.5.1 standard (Laurence, 1993). The CMM software community will now have a strong basis to design and test their inspection algorithms (Hopp, 1993). The use of traditional measurement tools such as micrometers, dial gages, and functional gages can be continued without any additional burden.

FUTURE OF Y14.5.1

The ASME Y14.5.1M-1993 standard opens with the following sentence:

> "This standard presents a mathematical definition of geometrical dimensioning and tolerancing consistent with the principles and practices of ASME Y14.5M-1993, enabling determination of actual values."

Assuming success in the objective stated above, there will be continued need to sustain the Y14.5.1 efforts in harmony with the Y14.5 efforts.

There still remain several unresolved issues. As mentioned earlier, the boundary problem has so far defied our attempts to provide a solution. The Y14.5.1 document can benefit from revisions based on user reaction: it is only the first attempt in providing a mathematical basis for current GD&T. More research needs to be done to get a better understanding of virtual and resultant conditions. Assembly and other functional tolerancing have not received enough attention. To fully capture such functional intent, it may become necessary to incorporate some syntactic enhancements into the GD&T language. The importance of statistical tolerancing has been barely acknowledged in the standards. The structure, and mode of working, of the Y14.5.1 subcommittee may require some changes to meet these new challenges.

The interaction between Y14.5.1 and measurement committees such as B89.3.2 should be strengthened. It may be argued that the mission of Y14.5.1 is independent of the way in which measurements are made on actual parts. However, the success of the Y14.5.1 standard depends to a large extent on how it is used by the metrological community in developing its measurement practices.

Standardization communities around the world have lately been talking about "international harmonization" of the various national and ISO standards. Recently ISO has recognized that design specifications as they appear on the drawing using GD&T are not traceable to a measurement standard in many cases. Using a general geometric product specification matrix, Bennich (1993) argues that only very few ISO drawing specifications are unambiguous. The U.S. based Y14.5.1 subcommittee was the first organization in the world of its kind to begin work towards an unambiguous specification of product geometry. Much can be gained by sharing its past experience and future efforts with other interested parties around the world. An ISO Joint Harmonization Group has recently been set up by the ISO/TC 3 (CMMs), TC 10/SC 5 (GD&T), and TC 57 (Surface Texture) committees to eliminate ambiguities and contradictions among their standards. This is a welcome move, and may provide a channel to "internationalize" the mission and efforts of Y14.5.1.

ACKNOWLEDGMENT AND A DISCLAIMER

We would like express our deep gratitude to all participants in the ASME Y14.5.1 subcommittee meetings. In addition, we want to thank our colleagues in the Y14, B89, and Y14.5 committees for their patience, support, and understanding. The views expressed in this paper, however, are our own and not that of the official Y14.5.1 subcommittee.

REFERENCES

ANSI B89.1.12M, 1985, "Methods of Performance Evaluation of Coordinate Measuring Machines," The American Society of Mechanical Engineers, New York, NY.

Bennich,P., "Chain of Standards — A New Concept in Tolerancing and Engineering Drawing," this proceedings.

Hopp,T.H., 1990, "The Mathematics of Datums," *ASPE Newsletter*, September Issue.

Hopp,T.H., 1993, "Computational Metrology," this proceedings.

Laurence,N., 1993, "A High Level View of STEP — A Formal Specification of the Information Content of a Product Design," this proceedings.

Nasson,M.A, 1993, "Revelations from the Y14.5.1 Definition of Circularity and Cylindricity," *Presented at Quality Expo International*, Chicago, IL, April 20-22, also available as *Technical Report CSDL-P-3255*, The Charles Stark Draper Laboratory, Cambridge, MA.

Neumann,A.G., 1993, "The New ASME Y14.5M Standard on Dimensioning and Tolerancing," this proceedings.

Srinivasan,V., 1993a, "Recent Efforts in Mathematization of ASME/ANSI Y14.5M Standard," *Proceedings of 3rd CIRP Seminars on Computer Aided Tolerancing*, Éditions Eyrolles, Paris, pp.223-232.

Srinivasan,V., 1993b, "Role of Sweeps in Tolerancing Semantics," this proceedings.

Tipnis,V., 1990, "Research Needs and Technological Opportunities in Mechanical Tolerancing," *ASME CRTD-15*, The American Society of Mechanical Engineers, New York, NY.

Walker,R.K., 1988, "CMM Form Tolerance Algorithm Testing," *GIDEP Alert X1-A1-88-01 and X1-A1-88-01a*, Published by the Government-Industry Data Exchange Program, Pomona, CA.

CRTD-Vol. 27, International Forum
on Dimensional Tolerancing and Metrology
ASME 1993

MEASUREMENT METHODS AND THE NEW STANDARD B89.3.2

Walter Beckwith
Measuring Systems Division
Brown & Sharpe Manufacturing Company
North Kingstown, Rhode Island

Frederick G. Parsons
Department of Engineering
Federal Products Company
Providence, Rhode Island

ABSTRACT

During the ASME/NSF workshop on mechanical tolerancing held in Orlando, Florida September 28 through October 2, 1988, it was suggested that work be initiated on developing a dimensional measurement standard.

A common industrial dimensional measurement practice is to require that gaging uncertainties be small compared with workpiece tolerances. Thus, the component of risk associated with uncertainty is small. Two points are discussed. First, some aspects of gaging uncertainty have generally been ignored, and actual uncertainty is often greater than assumed. Second, manufacturing accuracy has improved faster than measurement accuracy, so measurement uncertainty is often a larger fraction of tolerance than in the past.

A new standard under development to address these issues is discussed. A mandatory requirement of this standard, B89.3.2 Dimensional Measurement Methods, is that a plan must be prepared for each measurement. The plan is evaluated in terms of probability a feature measured as good is actually bad, and probability a feature measured as bad is actually good. This approach allows logical assessment of suitability of the plan. Formal requirements of the standard will be supplemented by extensive appendices designed to help in development and analysis of plans. To make the approach practical, most users will want to embed most of the theory in software.

The B89.3.2 standard is being developed to assess compliance with specifications which comply with Y14.5 Dimensioning and Tolerancing. Both pass-fail and process control measurements are considered, as are hard gaging and point-data-set gaging. Status and plans for completion of the standard are discussed.

BACKGROUND

A new standard, Dimensional Measurement Methods, ANSI/ASME/ B89.3.2, is being developed. The standard is designed to give direction to measuring parts that have been dimensioned and toleranced according to Y14.5. It is primarily concerned with methods for determining datums and location, form, profile, orientation, runout and size of workpiece features, including measured values and uncertainties.

The datum portion of the standard deals with the theoretically exact points, axes and planes that are the origin from which the location or geometric characteristics of part features are established. Location and position depend on the coordinate system, rectangular or polar, used to locate the features with respect to one another or as a group. Form deals with characteristics of the part, including straightness, flatness, circularity and cylindricity. Profile is used to control form or orientation, for example, and, depending on the application, size. Profile is usually specified for a line or a surface. Orientation defines the attitude of one feature or part with respect to another feature. Orientation callouts on a drawing, for instance, are usually specified as angularity, perpendicularity and parallelism. Runout, when specified, refers to either a circular or total runout condition.

RELATIONSHIP BETWEEN STANDARDS

Standard Y14.5 dictates clearly and concisely how dimensions and associated tolerances are put on a drawing. It is a dimensioning and tolerancing standard, not a measurement standard. As long as Y14.5 has existed (since the 1920's), there has never been a measurement standard. In many instances, though, Y14.5 is looked upon as a *de facto* measurement standard.

In 1969, there was an attempt to produce a gaging standard based on Y14.5, and a 25-page draft was written. For various reasons, however, the work was never completed, and the draft has lain dormant. The resurrection call came in late 1988, after Richard Walker of Westinghouse Electric's Marine Division wrote a GIDEP alert on all the data he collected on CMMs. GIDEP stands for Government-Industry Data Exchange Program. The alert points out the need for assessing alternative methods of making measurements. A decision was then made to assign the B89.3 Geometry Subcommittee the task of generating a new standard. Its working group is B89.3.2, charged with the responsibility of writing the B89.3.2 standard on Dimensional Measurement Methods.

For the gaging standard to be applicable to all gaging methods, including coordinate measuring machines (CMMs), it must be based on a mathematical description of the dimension and tolerance zones of the part to be gaged. The task of mathematizing the existing Y14.5 standard has been assigned to Y14.5.1 Mathematical Definition of Y14.5 Dimensioning and Tolerancing Principles. Its work involves establishing a clear, unambiguous mathematical definition of the tolerancing that the designer puts on a drawing.

Figure 1 shows the interdependence of these three committees. First, the designer must clearly put the dimensions of the part and the tolerance of its features on the drawing. Second, Y14.5.1 must mathematically define the meaning of those tolerances. Third, B89.3.2 must take the mathematical definition of size and tolerance zones and establish a procedure for making the measurement.

FIGURE 1: THE COMMITTEES ARE INTERDEPENDENT

MEMBERSHIP OF COMMITTEES

In general, B89 committee membership comes from three areas:

- Institutions and government, such as colleges and universities, the military and NIST, which provide an element of technical excellence.
- Users of gaging equipment, including large and small companies and government and military inspection facilities, which provide practical excellence.
- Gaging manufacturers, which provide economic expertise.

This balance leads to a standard with principles that are technically correct, practically achievable on the shop floor and economically feasible.

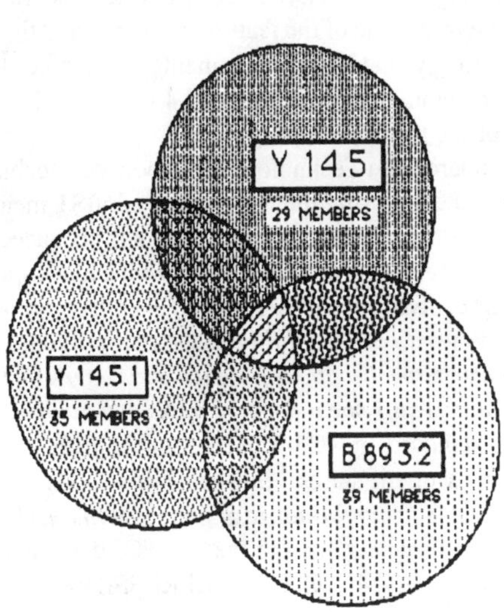

FIGURE 2: RELATIONSHIP OF COMMITTEE MEMBERSHIP

Figure 2 shows that the three committees have some members in common, there is a commonality between Y14.5 and Y14.5.1 and B89.3.2. There are several members who sit on all three committees.

PROPOSED STANDARD

The proposed new standard, Dimensional Measurement Methods, exists in draft form. An outline of the draft is presented here, although all elements are subject to change. The purpose in publishing the outline here is for information only and to show the general direction in which the committee is working. At this time, the standard has not been approved for distribution by either B89.3.2 or the ANSI Committee on Standardization.

One of the most important characteristics of this standard is that it deals with gaging uncertainty, some aspects of gaging uncertainty have generally been ignored and actual uncertainty is often greater than assumed. As an example of this the following paragraph 4.2.1 from ANSI Standard B4.4M-1981 *''no measurement can be given an absolute value because each measuring system and measuring instrument contain inherent sources of error. In addition, most measuring instruments only give diametrial sizes and, therefore, do not measure any form errors which may be present in the workpieces. Complete inspection of the workpieces to verify that no part of the surface infringes the maximum material envelope, therefore requires separate measurement of geometric form such as roundness and straightness and these measurements must be correlated with the diametrial. In the practical situation of a workshop, reliance is usually placed on the accuracy of the manufacturing process to produce workpieces having negligible errors of geometric form.*

It is assumed as the basis of this standard that the acceptability or otherwise of a workpiece will be determined by only one measurement it is also assumed that no corrections will be applied for temperature conditions of measuring instruments compression effects etc. or for any systematic errors in the measuring instrument and/or standards.''

RECENT EMPHASIS ON UNCERTAINTY

In a recent article in Quality Magazine Dr. H. S. Nielsen defined uncertainty as the interval in which the unknown difference between the true value of the feature measured and the result of that measurement can be found. He went on further to say that all measurements are subject to uncertainty. In dimensional metrology, different standards recommend ratios between 4 to 1 and 10 to 1 between the tolerance to be measured and the uncertainty of the measurement.

The National Institute of Standards and Technology has produced a technical note No. 1297 which deals with guidelines for evaluating and expressing the uncertainty of NIST measurement results.

The National Voluntary Laboratory accreditation program has produced a handbook and Section 5.7.2 states that *''well defined and documented measurement assurance techniques or uncertainty analysis may be used to verify the adequacy of a measurement process. If such techniques or analysis are not used then the uncertainty of calibration or verification shall be demonstrated to not exceed 25% of the acceptable tolerance (that is the manufacturer's specification) for each characteristic test equipment being calibrated or verified and deviation from this requirement must be approved by the customer.''*

A working draft of an ISO standard in TC3 TG1 this deals with the inspection of workpieces and measuring instruments and as its scope it states *''the scope of this standard is to define when a specific workpiece or measuring instrument is in conformance or nonconformance with a given specification taking into account the uncertainty of measurement.''* Both the ISO TC3 document and the NIST document are in accordance with the ISO document: GUIDE TO THE EXPRESSION OF CAPITAL UNCERTAINTY IN MEASUREMENT (GUM).

SCOPE OF PROPOSED STANDARD

This standard deals with the measurement of workpieces dimensioned and toleranced according to ANSI Y14.5 The document is primarily concerned with how to determine datums and location, form, profile, orientation, runout and size of workpiece features including measured values and uncertainties.

PROPOSED STANDARD INSPECTION PLAN

According to the proposed standard, the designer of the measurement process prepares a dimensional inspection plan for the design of each workpiece. The plan must include a list of measurements to be made, what gage to use for each measurement, procedure for each measurement and gaging limits for each. It is recommended that the measurement process designer document logic for the plan.

The measurement process designer assures, to the extent that it's practical, that the dimensional inspection plan is suitable for the intended purpose. Such assurance may include, for example, reference to other plans and a discussion with interested parties. Elements of suitability may include permissible probabilities of accepting a bad part, or of rejecting a good part, as well as the cost of inspection.

Background material for the steps in choosing gaging and setting gaging limits will be presented in eleven appendixes:

A. Relevant standards
B. Dimensinal measurement plan
C. Gage selection
D. Risk
E. Coordinate sampling uncertainty
F. Attributes gage uncertainty
G. Variables gage uncertainty
H.
I. Substitute geometry fitting uncertainty
J. Operator uncertainty
K. Environmental uncertainty
L. Calibration and traceability

Inspectors perform all measurements of the dimensional inspection plan, using good metrology practice as presented in what will be Appendix H, ''Good Metrology Experience.''

The inspector accepts workpieces for dimensional accuracy if all measurements are within gaging limits specified in the dimensional inspection plan. The inspector rejects workpieces when one or more measurements are outside the gaging limits specified in that plan.

It again must be emphasized that the foregoing is an unapproved outline. Arrangement and content will most likely change before the standard is issued.

WORKING TASK FORCES

- **Literature search and bibliography**
  Searches for material to help in writing the standard and develop a bibliography for inclusion in the standard.
- **Environmental effects**
  Assures that the user of the standard is aware of the environmental effects of each method of gaging.
- **Classification of gages**
  Develops an aid to the selection of a gage to establish a datum or measure a feature.
- **Specific example**
  Chooses a part or parts as a test specimen during the writing of the various phases of the standard.
- **Uncertainty/GR&R**
  Presents gage uncertainty of various gaging methods and establishes methods for measuring repeatability and reproducibility
- **Gage calibration**
  Established an outline for calibrating gages and frequency.
- **Point sample measurement**
  Investigates the difference between point or continuous measurement and its effect on accuracy.
- **Editorial and structure**
  Begins drafting and structuring the proposed standards.

OTHER CONSIDERATIONS

The gage material, contact materials, gaging force, approach speed and other characteristics of the part-gage interface must also be taken into account in writing a standard for dimensional measurement methods. When the draft of the standard is in final form it will be distributed for industry comment before it is issued.

REFERENCES

ASME Publication-Standard B4.4M-1981 (R1987), *Inspection of Workpieces*, Order No. L00062

ASME Publication-Standard Y14.5-1982 (R 1988) *Dimensioning and Tolerancing*, Order No. L00062

Francis T. Farago, *Handbook of Dimensional Measurements*, Second Edition, 1982, Industrial Press, Inc., New York, NY

Henrik S. Nielsen, *Uncertainty and Dimensional Tolerances*, Quality, Vol. 31, No. 5, pp. 25-29, Hitchcock Publishing Company, Carol Stream, Illinois, USA.

ANSI B89.1.6M-1984, *Measurement of Qualified Plain Internal Diameters for use as Master Rings and Ring Gages.*

ANSI B89.19M-1984, *Precision Inch Gage Blocks for Length Measurements (through 20 in. and 500 mm.).*

ANSI B89.1.10M-1987, *Dial Indicators (for Linear Measurements).*

ANSI B89.1.12M-1990, *Methods for Performance Evaluation of Coordinate Measuring Machines.*

ANSI B89.3.1.-1972 (R 1988). *Measurement of Out-of-Roundness.*

ANSI B89.6.2-1973 (R 1988), *Temperature and Humidity Environment for Dimensional Measurement.*

CRTD-Vol. 27, International Forum
on Dimensional Tolerancing and Metrology
ASME 1993

A HIGH LEVEL VIEW OF STEP - A FORMAL SPECIFICATION OF THE INFORMATION CONTENT OF A PRODUCT DESIGN

Neal Laurance
Research Laboratory
Ford Motor Company
Dearborn, Michigan

ABSTRACT

This talk provides a brief introduction to STEP, an emerging International Standard covering all the requirements of an enterprise for information about a product from the time the product is first conceived until it is obsolescent. This standard is intended to replace IGES as a medium of interchange for geometrical data, and to include all the other types of data about a product in an industry. The standard is a blend of CAD/CAM models and advanced database concepts. In general, STEP draws its information models from appropriate technological domains and recasts them in a common information schema.

While the early use of STEP will be for exchange between proprietary CAD systems, there is already evidence that next generation systems will have a native mode conceptually equivalent to the data base formulation prescribed by STEP. Examples of the use of STEP for relatively simple geometrical items are given. The current status and the work in progress of this standard is summarized.

STEP today handles only nominal geometry, i.e. ideal form. However, future versions of STEP will incorporate a tolerance model based on Y 14.5.1. That work is just now beginning, and the first results will not be available until next year. Standardization is similarly several years away.

INTRODUCTION

My topic today is STEP, the Standard for the Exchange of Product Data. For those of you completely unfamiliar with STEP, I hope to provide a coherent introduction; for those of you familiar with it, I hope to provide some new insights.

Unfortunately, STEP, like most modern high tech endeavors, is replete with acronyms. I will start by taking a few minutes to explain the most important ones.

PDES, which stood for Product Data Exchange Specification, is a U.S. effort that began in the 80's as a follow-on to IGES. About 1984, it was decided to move the effort into the international arena, merging it with similar efforts in Europe. The new combined program became known as STEP, the Standard for the Exchange of Product model data. However, by this time, the term PDES had become well established, and so, nine years later, to bring it into alignment with its new international flavor, the meaning of PDES has been changed to Product Data Exchange using STEP. We have arrived at an era where the acronym drives the meaning.

Other acronyms that we will come across are ISO, the International Standards Organization, and IPO, the IGES/PDES Organization. For many years, IPO was an ad hoc U.S. organization of volunteers who met to develop IGES and PDES. When IGES was promulgated as a standard, it was done by ASME. In 1992, IPO became a legal entity under a new not-for-profit corporation, U.S. Product Data, or U.S. Pro, for short. This body has become an ANSI accredited standards committee (ASC) and will issue future versions of IGES. There is also some discussion about the possibility of U.S. Pro issuing STEP as an American National Standard after it becomes an international standard.

Figure 1 illustrates how these organizations interact in the United States. The organization below each line is the administrator or secretariat of the organization above. ISO is organized into technical committees; industrial automation is technical committee 184. Each technical committee has subcommittees; subcommittee 4 is Manufacturing Languages and Data, under which STEP falls. The U.S. TAG or technical advisory group is the group responsible for formulating the U.S. position on international ballots on STEP and any other U.S. positions. As you will note, NIST acts as the administrator for SC 4 and the IPO acts as administrator for the TAG.

One other organization of interest in this diagram is PDES, Inc. This is a private consortium of about 25 companies and institutes, headquartered in Charleston, South Carolina. Its goal is to further the development of STEP, and it does this by developing application specific implementations of STEP, largely in areas of interest to its members, and by providing initial drafts of some parts of STEP. There are similar organizations in Europe.

HISTORICAL PERSPECTIVE

At this point, I would like to take a few minutes to put STEP into historical perspective. To paraphrase Richard Armour, it all started with Sutherland. Technologically, STEP is very closely linked to CAD. Like IGES, it arose to address a problem in moving information from one CAD system to another. CAD, it is generally acknowledged, began with Sutherland's thesis, Sketchpad (1980), completed at MIT in 1963. Of course, there was a lot of work going on at MIT and elsewhere investigating the use of graphics and computers, but Sutherland's thesis showed how to apply this new technology to engineering design.

Sketchpad used a doubly linked list for its data structure. Linked lists provide some primitive notion of associativity, a key component in a CAD system. However, using linked lists is a slow process, and processing speed was not a readily available commodity in 1963. Data structures were an impediment to CAD from the very beginning.

In an effort to invent a data structure better suited to graphic processing, Jerry Feldman developed a system called LEAP (Rovner & Feldman 1968). The first elements of this system were published in 1966 while he was at Lincoln Laboratory. The full LEAP language was described in a 1968 paper, published jointly with Paul Rovner. The principle innovation in LEAP was the representation of all information in the form of triples of Object, Attribute and Value. For example, the father of John is

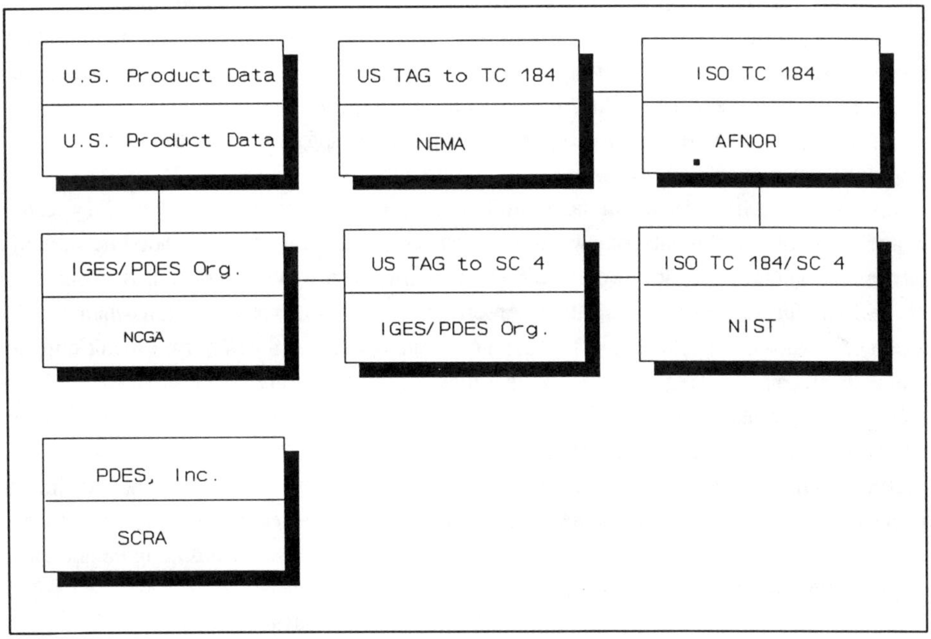

Figure 1 - STEP ORGANIZATIONS

Bill. In this relationship, John is the object, father is the attribute and Bill is the value.

One can see, even in this brief description, the seminal ideas of object oriented programming. However, those are not of so much interest here as the development of relational data base theory from these ideas. Following Feldman's work, David Childs (1968) put these ideas into the formalism of set theory and wrote several papers on a data base organized strictly on set theoretic lines. This work led directly into E.F. Codd's development of relational data bases in 1970 (Codd 1970).

Following the introduction of relational data bases and their requirement for table normalization, several techniques were developed to analyze information and provide an orderly technique for structuring the information to the relational tables. Among these were Nijssen's Information Analysis Method (NIAM) (Nijssen & Halphin 1989) and the IDEF (1992) series of languages developed as part of the Air Force's ICAM project in the mid 80's. All of these techniques use graphical depictions of elements of information and display relationships between the elements and the cardinality of these relationships. STEP represents a further development along these lines, that is, a formal method of specifying the structure and meaning of information. However, this is only one of the paths leading from Sketchpad to STEP. The other one is CAD systems themselves.

The early 1970's saw the development of several CAD systems, both commercial and private. This development coincided with the appearance of minicomputers on the market, one of the enabling as well as limiting aspects of these CAD systems. It is important to remember that CAD in 1970 stood for Computer Aided Drafting, and to some extent, it still does today. Both the A and the D of this term need to be considered.

The A for Aided means that the computer was never intended to perform the whole job; indeed much what the draftsman did (or still does) was not known in a form suitable for incorporation into a computer program. The D for Drafting meant that no new design paradigm was involved; the design was still meant to be represented in 2 dimensional drawings, usually in the form of three isographic views of the object being designed with a set of dimensional information and possibly a set of notes giving other information such as material, finish, tolerances, etc.

Thus, the information carried in early CAD systems was necessarily incomplete. This really wasn't any problem until that dramatic day when the engineering design manager decided he wanted all the drawings that had been developed on CAD system A transferred to CAD system B. After many false starts and painful attempts at translation, America responded in typical fashion when faced with a problem of unknown magnitude - we formed a committee and IGES was born. I won't bore you with all the growing pains of IGES - I feel sure that most of you are aware of them. However, I will point out that in many cases we blamed IGES for the shortcomings that were inherent in the CAD systems themselves - a case of blaming the messenger.

IGES does have severe structural limitations. The principle one is that there is no fixed meaning or semantics given for many of its syntactic elements. Each of these elements means exactly what the originating system says they mean, a severe limitation on the ability of translating engineering intent from one system to another.

It is here that STEP steps in. It brings the information modeling technology developed in the relational database world to bear on the question of prescribing semantics (i.e. meaning) of graphical entities. In this sense, STEP traces its ancestry back to Sketchpad along two parallel and very independent trajectories.

WHAT IS STEP?

So what is STEP? There are at least three answers to that question. First STEP is an IGES replacement. A large number of people who work on STEP are the same people who work or worked on IGES. Their intent is the same - to provide a translation path between dissimilar CAD systems for product design data. It is important to remember that the E in both PDES and STEP stands for exchange. A central tenant in the STEP credo is that IGES is deficient in that it does not contain the semantics of the information, only the data. Hence the job of STEP is to capture the semantics. STEP does this by focusing on the use of the information in the product design process; that is on the context in which the information is to be used. IGES, on the other hand, focuses solely on the recreation of the drawing. This is why in some cases fax has become an alternative to IGES.

The second answer is that STEP is a bottom up approach to enterprise modelling. In order to capture the semantics of the graphical elements, STEP tries to capture all the information surrounding the graphics - materials, finish, tolerance, of course, but also the context in which this is to be viewed. This includes the design release information, effectivity information, and everything else associated with product development and production. The theory is that if we know how the relationship between CAD information and all the other information in an activity area, we know its semantics. If we propagate this approach up the organization, we will ultimately arrive at a model of the information structure of the entire organization.

The third answer is the STEP is a giant information model. In an earlier version, it was thought that one STEP system might be created that could cover all of the information requirements for any corporation. Today that view has been abandoned, and the current philosophy is that a separate information model, based on a core set of elemental concepts, will be developed for each application area in each industry or business sector.

Whichever view of STEP you adopt - and they are not mutually exclusive -- STEP is a big undertaking. As it is developed, STEP will be published as a multi-part ISO standard. The conventions adopted for numbering these parts say something about the expected size of STEP. Part 1 is an overview of the entire standard. The remainder of STEP is organized into the following groups of parts.

Parts numbered 11 through 20 are used to document the tools for conveying the intent of STEP. Presently, the only part planned or completed is Part 11 - the EXPRESS language. More about that later.

Parts numbered 21 through 30 are used to document the so called implementation methods. These are really the methods by which one would access information that conforms to STEP. Part 21, the first part to be complete, describes a flat file format for expressing STEP information - the IGES replacement. Part 22 is under development. It will provide an application programmer interface into a system that supports STEP.

Parts numbered 31 through 40 are used to document test methods - methods for testing implementations for conformance to STEP. Part 31, an overview of testing, is complete. The rest are under development.

Parts numbered 41 through 200 are used to document integrated resources. Parts 41 through 100 are called generic resources; parts 101 through 200 are called application resources. These documents provide sets of related atomic concepts and their meaning. For example, Part 42 is devoted to geometry, topology, and geometric models. All the basic items used to form geometric representations are present here. They may be combined in many mix and match ways, depending on the requirements of the application.

Parts 201 through 1200 are the bulk of the standard, the so called application protocols. These documents describe a specific application area, model the information requirements of this area, and represent it in terms of the fundamental entities of the integrated resources. A key element of these documents is the scope of applicability, a precise definition of the "Universe of Discourse". The present thinking is that any implementation of STEP corresponds to one of these application protocols combined with one or more of the implementation methods. To date, only two such documents have been produced - 201 describes 2 dimensional drafting and 203 describes configuration controlled design of mechanical products. There are presently 16 other application protocols under development and more being planned.

Parts 1201 through 2200 are abstract test suites, one for each of the application protocols. None has been produced yet, so a description of these documents will have to wait.

The size of these documents is formidable. Each of the integrated resource documents is about 75 pages long. However, the first application protocols number over 600 pages, and future application protocols may well be larger. The initial release, 12 documents, total about 2400 pages. There is a real concern that STEP may fail because of weight.

Before I proceed further, I want to stress that STEP is concerned solely with product data. However, product data encompasses any process that relies on elements of the product specification, including manufacturing processes, in-service maintenance, and ultimate disposal. It covers a very broad area.

GEOMETRY IN STEP

It is now time to look at a little bit of STEP to get a flavor for it. STEP is written in a formal language called EXPRESS. EXPRESS is solely a data description language, a language for specifying the form and relationship of the data describing the product. Although some aspects of EXPRESS look procedural, it is not a procedural language. In database terminology, it is a Data Description Language (DDL), but not a Data Manipulation Language (DML).

Most other information description techniques are graphical; they document the data relationships by drawings showing relationships. Although EXPRESS has an extension that provides a graphical representation, the main part of EXPRESS is in a linear written form, like Fortran, Pascal, or other computer languages. The advantage of this approach is that it can cover any level of detail and is easily processed by computers; the disadvantage is that when reading it, one quickly looses sight of the forest for the trees.

STEP's facilities for nominal geometry are contained in part 42. This document has three principal sections, geometry, topology, and geometric models. In each of these sections, primitive concepts are defined in terms of EXPRESS data types. The definition of these entities is a combination of EXPRESS formalism and English description. The more primitive the element, the more important the English. As an example, table 1 shows the EXPRESS definition of a circle. This brief definition is accompanied by about a page of English text, providing the usual textbook definition of a circle. The EXPRESS states that a circle is a conic with one extra value, the radius. The conic itself has one parameter, its position in space which includes its orientation. Table 2 shows the EXPRESS definition of the conic.

Table 1 - STEP definition of a circle

```
EXPRESS Specification
*)
ENTITY circle
  SUBTYPE OF (conic);
  radius : positive_length_measure;
END_ENTITY;
(*
```

Table 3 shows the entities that are defined in the geometry section. For each of these elements, the EXPRESS shows the relationship to other entities and the data types of the attributes of the entity. The meaning of the entity and the attribute is almost all captured in English.

Table 2 - STEP definition of a conic

```
EXPRESS specification
*)
  ABSTRACT SUPERTYPE OF (ONEOF(circle, ellipse,
hyperbola, parabola))

  SUBTYPE OF (curve);

  position: axis2_placement;

END_ENTITY;
(*
```

Table 4 shows the similar list for the topological entities. As an example of one such entity, table 5 is the specification for a path. Here we see one example of a constraint, expressed in a WHERE rule. This constraint says that the end vertex of each edge shall be the same as the start vertex of its successor. In addition to this constraint, there are five constraints expressed only in English.

The final section of Part 42 provides a set of geometric shape models. These include solid models, both CSG and B-rep, as well as two kinds of surface models that are not solid models and two kinds of wireframe models. In addition, there is the concept of a geometric set, an arbitrary collection of points, curves, and surfaces. So you see, the answer to the question of how STEP represents geometry is anyway you like.

As I said before, the integrated resources, parts 41 through 200, are intended to supply a library of fundamental entities useful in making up an information model for an application area. The first real such application area, other than drafting, is Part 203, Configuration Controlled Design. This massive document describes the information necessary to track a design through an engineering design and release activity. It includes concepts of assembly, bill of materials, substitute parts, approvals, supplier identifications, and other such related information. This application protocol uses about 500 different entity types in its data schema to cover this application area. The document comes in six flavors, depending on the way shape is represented. Presumably, if two systems engaged in this common activity are to exchange information, they must employ the same shape representation technique.

TOLERANCING

Tolerancing will be the subject of Part 47. An initial version of such a document was written last year, but in response to many comments, it was withdrawn and the process started over. This version will be based on the concepts embodied in Y 14.5.1. It is too early to offer any specific comments on it, but based on the other 40 series parts, it is clear that it will contain only fundamental concepts, not

Table 3 - Geometry entities

GEOMETRY

- PLACEMENT
  - AXIS1 PLACEMENT
  - AXIS2 PLACEMENT 2D
  - AXIS2 PLACEMENT 3D
- CARTESIAN REPRESENTATION REPLICA
- CARTESIAN XFORM OPERATOR
  - CARTESIAN XFORM OPERATOR 2D
  - CARTESIAN XFORM OPERATOR 3D
- POINT
  - CARTESIAN POINT
  - DEGENERATE PCURCVE
  - POINT ON CURVE
  - POINT ON SURFACE
  - POINT REPLICA
- VECTOR
- DIRECTION
- CURVE
  - BOUNDED CURVE
    - B-SPLINE CURVE
      - UNIFORM CURVE
      - QUASI UNIFORM CURVE
      - BEZIER CURVE
    - COMPOSITE CURVE
      - COMPOSITE CURVE ON SURFACE
      - BOUNDARY CURVE
      - OUTER BOUNDARY CURVE
    - COMPOSITE CURVE SEGMENT
      - REPARAMETRISED COMPOSITE
  - CURVE SEGMENT
    - POLYLINE
    - TRIMMED CURVE
  - CONIC
    - CIRCLE
    - ELLIPSE
    - HYPERBOLA
    - PARABOLA
  - PCURVE
  - SURFACE CURVE
    - INTERSECTION CURVE
    - SEAM CURVE
  - LINE
  - OFFSET CURVE 2D
  - OFFSET CURVE 3D
  - CURVE REPLICA

data used in complete systems.

In February, 1993, a new project was approved on tolerancing and metrology. This project will result in a new application protocol, Part 219. That work should provide a complete description of the use of tolerancing information in metrology, but it is only in its initial stages. Both part 219 and part 47 are being coordinated at the National Institute of Standards and Technology.

In principle, STEP does not create new knowledge; it merely codifies existing technology and business practice into an unambiguous framework. In practice, of course, when gaps are found, the STEP discipline demands that they be filled with something. There is a long history of development of the mathematics of geometry, both classical and in the age of computers, and the STEP formulation can be expected to be faithful to that body of knowledge. In tolerancing and metrology, on the other hand, I perceive there are significant gaps in our knowledge - else why this conference - and the STEP formulation will have to make some assumptions.

SUMMARY

I can't really provide a summary because STEP is still in a state of flux. In a way that is ironic, since STEP's view of the world is a completely static one. STEP is concerned solely with the structure of information, not with its use. Throughout the STEP development process there is a strong philosophical

Table 4 - Topology Entities

TOPOLOGY	
CONNECTED EDGE SET	FACE BOUND
CONNECTED FACE SET	FACE OUTER BOUND
OPEN SHELL	LOOP
ORIENTED OPEN SHELL	EDGE LOOP
CLOSED SHELL	POLY LOOP
ORIENTED CLOASED SHELL	VERTEX LOOP
EDGE	PATH
EDGE CURVE	OPEN PATH
ORIENTED EDGE	EDGE LOOP
FACE	POLY LOOP
ORIENTED FACE	VERTEX SHELL
FACE SURFACE	WIRE SHELL
SUBFACE	VERTEX
	VERTEX POINT

undercurrent that today goes by the name of Semiotics. Through the centuries this philosophy has had different names, logical positivism for one, but it all can be traced back to Plato. STEP has a Platonic view of knowledge.

Let me close with some personal observations about the status of STEP and its future. Will STEP work? If by that you mean will STEP succeed where IGES has failed, the answer is probably yes. Certainly, STEP goes a long way in overcoming some of IGES' limitations, assuring a common understanding for terms contained in its files. It is not complete, however, and wherever it relies on English for communicating semantics, it is subject to misinterpretation. To be fair, part of the reason that STEP will succeed where IGES has failed is that CAD systems are much more advanced now than they were a decade ago. The problem of incomplete, ambiguous geometric representations is now well understood, the use of solid modeling systems is growing, and the user community does not naively expect CAD system A to translate to CAD system B when they know there is only a wire-frame representation underneath.

For the bigger picture, however, the outlook is less clear. We have just described two application areas that are being addressed separately, that of configuration controlled design, and that of tolerancing and metrology. Now in my company, like most companies, those activities are related at some point. However, there is no clear method for coupling multiple STEP application protocols for use within an organization. The problem is being worked on, but the solution is not obvious. As I indicated earlier, STEP uses a bottom up approach, and it is in areas such as this coupling of two application protocols that the deficiencies of bottom up design become apparent.

Table 5 - STEP definition of path

```
EXPRESS specification
*)
ENTITY path
  SUPERTYPE   OF   (ONEOF(open_path,    edge_loop,
oriented_path))
  SUBTYPE OF (topological_representation_item);
  edge_list : LIST [1:?] OF UNIQUE oriented_edge;

WHERE
  WR1: path_head_to_tail(SELF);

END_ENTITY;
(*
```

Finally, let me mention again the size of the documents. Unless STEP can find ways to organize its information better, that is, more concisely, it may well drown in its own weight. The first two application protocols are over 500 pages long, and there are about 20 such documents under development. Indeed, there are efforts to find ways to publish electronically, save a tree and all that, but regardless of how it is presented, that is formidable size for any document. I won't call it an information explosion, since I think the information content is not that high. There is a lot of redundancy in these documents.

If this overview has seemed less than conclusive, I think this is typical of trying to characterize something that is continuing to evolve. Every time I spend a week at one the STEP meetings, my understanding and outlook changes. Next week I will be in Atlanta, working over many of the documents I mentioned today. Undoubtedly, my view will be changed again following that meeting. Thank you for your attention.

REFERENCES

STEP documents can be purchased from U.S. Pro c/o NCGA, 2722 Merrilee Drive, Suite 200, Fairfax, VA 22031-4499.

Codd, E.F., *A relational model of data for large shared data banks*, Communications of the ACM, v 13, n 6, pp 377-387, June, 1970.

Childs, D.L., *Feasibility of a Set-Theoretic Data Structure: A General Structure Based on a Reconstituted Definition of Relation*, Fourth Congress of the International Federation for Information Processing. Supplement Part 10, pp 162-172, North-Holland, Amsterdam, Netherlands, 1968.

Rovner, P.D., & Feldman, J.A., *The LEAP Language and Data Structure*, Fourth Congress of the International Federation for Information Processing. Software II Part 4, pp 73-77, North-Holland, Amsterdam, Netherlands, 1968.

Sutherland, I.E., *Sketchpad: A Man-machine Graphical Communication System*, Garland Publisher, New York, 1980

Nijssen, G.M., & Halphin, T.A., *Conceptual Schema and Relational Database Design*, Prentice Hall, 1989.

Integration Definition for Information Modeling (IDEF1X), Draft Federal Information Processing Standard, September 9, 1992.

CRTD-Vol. 27, International Forum
on Dimensional Tolerancing and Metrology
ASME 1993

A CURRENT PERSPECTIVE ON TOLERANCING AND METROLOGY

Herbert B. Voelcker
Sibley School of Mechanical and Aerospace Engineering
Cornell University
Ithaca, New York

ABSTRACT

Tolerancing and metrology are primary tools for specifying, assessing, and controlling geometric variability in design and manufacturing. Both evolved from shop and drafting practice and are partially codified in standards. In recent years, the introduction of new technology – notably Coordinate Measuring Machines (CMMs) and CAD systems – exposed gaps, ambiguities, and inconsistencies in current practice. These findings triggered a wave of effort to formalize tolerancing and metrology, by 'mathematizing' the American tolerancing standard and by producing a new companion standard for measurement. These two developments are unlikely to be the end of the story, however, because other gaps and issues remain. Thus tolerancing and metrology currently stand at a watershed: the era of evolutionary practice seems to be ending, and a new era based on mathematically defined techniques and standards seems to be dawning.

This paper sets a context for the *Proceedings* papers that address these matters, by surveying the major approaches to tolerancing and metrology and the issues they engender. The paper concludes with brief discussions of issues not addressed well or at all in the *Proceedings*, and some speculations on the future.

INTRODUCTION

Manufacturing and assembly processes are inherently imprecise, and produce parts and products that vary. Variability must be accomodated in design and controlled in production, and mechanisms for doing this are woven through the entire production system. Tolerances are the primary variation-control tool in design; metrology is the primary tool for assessing conformance to toleranced specifications, and for gathering data for process control. Today tolerancing and metrology are in flux. To see why this is so, it is instructive to review some technical history.

Table 1 provides some historical highlights. Interchangeable assembly – a goal articulated in France in the late 1700s – was first realized in American Army arsenals *c*.1820. The key technology was functional gaging, which enabled parts to be made precisely enough to be assembled interchangeably. Thereafter the twin (but independent) principles of interchangeability and mechanization (loosely, the use of mechanically guided and powered tools) launched waves of change that brought, over the next 170 years, job specialization, mass production, and progressively sharper distinctions between design specifications and manufacturing methods. During this process, gages were displaced as the *de facto* medium for specifying allowable variations, and became mechanisms for testing conformance to separate, symbolic variability specs. More specifically: precise shop-floor measuring instruments (steel scales, micrometer calipers, many types of vernier instruments) became available in the mid- and late 1800s. These enabled dimensioned drawings, rather than representational drawings or models that had to be scaled with dividers, to be used as design specifications. Dimensional tolerances followed in the early 1900s to provide explicit limits on dimensional variability. The emergence of geometric tolerancing after World War-II completed the inversion: process callouts, which carry implicit variability limits, were prohibited, and tolerances were established as the sole means for specifying variability.

	< 1820 :	The artisan era
Gaging	≈ 1820 :	Gages ⇒ interchangeable assembly
↑	≈ 1930 :	First drafting standard
Limit (parametric) tolerances	≈ 1935 :	Tolerancing in the drafting standard
	≈ 1950 :	MIL-STD-8, ... MIL-STD-8A
↓	≈ 1966 :	ANSI Y14.5-1966; ... 1973,'82,'94
Geometric tolerances	≈ 1985 :	'Methods divergence'
	≈ 1990 :	Y14.5 'mathematization'; new standard for measurement
	≈ 2000 :	'Math-based' standards ?

TABLE 1: SOME HISTORICAL HIGHLIGHTS

The American Y14.5 geometric tolerancing standard was developed and is maintained by practical people versed in gaging traditions; it can be viewed as an evolving distillation of best practices codified in prose and graphics. Practice-based standards work well when technology and management undergo only evolutionary change, but radical change can destabilize them – and this happened when CMMs, and to a lesser extent CAD and CNC, entered the picture. As Dick Hook explains [Hook, 1993], CMMs led to 'methods divergence'. This brought both measuring methods and the tolerancing standard (the source of conformance criteria) into question, and engendered the current efforts aimed at Y14.5 'mathematization' [Y14.5.1: Walker and Srinivasan, 1993] and measurement codification [B89.3.2: Beckwith and Parsons, 1993]. These emerging standards are remarkable achievements that attest to the vitality of the responsible standards committees, but as I shall argue below, there is more to be done.

The next section reviews briefly the main approaches to tolerancing, current formalization efforts, and two doctrinal issues. The subsequent section sets the major issues in sampled coordinate metrology, and the last major section discusses several matters not covered elsewhere.

TOLERANCING

Two families of tolerancing schemes – parametric and geometric – have been developed for industrial use. Parametric tolerancing is based on ordinary dimensions (or vector proxies), and comes in three flavors: worst-case limit tolerancing, statistical tolerancing, and vector tolerancing. (These schemes are called 'parametric' because dimensions can be regarded as *control parameters* for an underlying mathematical representation [Requicha, 1984].) Geometric tolerancing was developed to ameliorate some intrinsic weaknesses in parametric tolerancing.

Parametric Tolerances

Figure 1a shows a conventional dimension chain for a simple part. Each dimension d_{ij} is a distance with a fixed nominal value and a variational component, i.e. $d_{ij} = D_{ij} + x_{ij}$. In worst-case limit tolerancing, x_{ij} is typically assigned ± limits. Limits for implicit dimensions (those not in the chain) may be calculated by combining ("stacking-up") worst-case conditions over the path defining the dimension. For example: d_{12} is defined by the path (d_{14}, d_{42}) and thus

$$\text{Max } d_{12} = \text{Max } d_{14} - \text{Min } d_{24}$$
$$\text{Min } d_{12} = \text{Min } d_{14} - \text{Max } d_{24}.$$

Fig. 1b shows the directly controlled variability of f_4 relative to f_1, and also the larger f_2 variability resulting from stackup. Stackups for d_{13} and d_{34} may be calculated similarly. Chains over n faces that are not "over-dimensioned" contain exactly n–1 explicit dimensions, and have ${}^{(n-1)}C_2 = (n-1)(n-2)/2$ implicit dimensions subject to stackup.

Statistical tolerancing uses the same chain rules but handles the x_{ij} components differently. The x_{ij} values are assumed to be continuous random variables, usually with zero mean value and variance σ_{ij}. If the x-components of each d_{ij} in a chain are assumed to be statistically independent (a strong assumption), they will combine on a root-mean-square basis. This means that the variance will grow only as $O(\sqrt{m})$, where m is the 'length' of the d_{ij}-defining path. Fig. 1c shows the statistical analog of Fig. 1b when x_{14} and x_{24} have equal σ's.

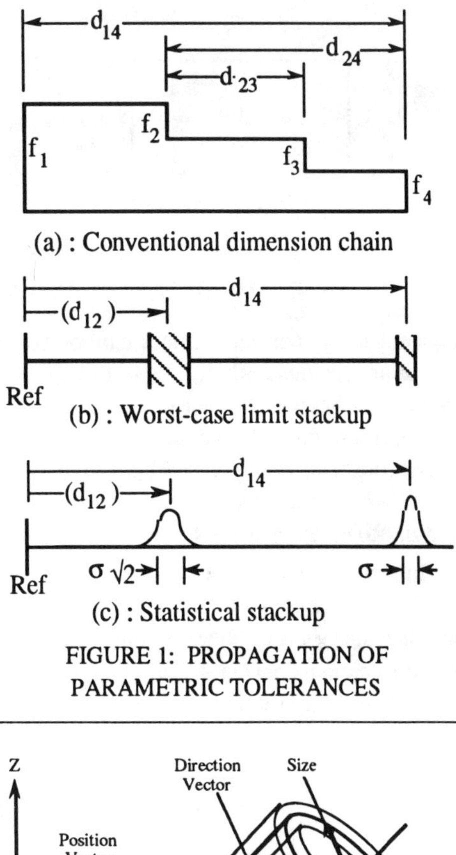

(a) : Conventional dimension chain

(b) : Worst-case limit stackup

(c) : Statistical stackup

FIGURE 1: PROPAGATION OF PARAMETRIC TOLERANCES

Figure 2 conveys the flavor of vectorial tolerancing, a relatively recent scheme proposed by Prof. Adolph Wirtz that is being considered in Europe as a possible complement to geometric tolerancing [Wirtz, 1991; Henzold, 1993]. The essential idea is to parameterize and tolerance parts in terms that relate naturally to manufacturing and inspection processes. Four descriptive entities – direction, position, size, and form – are posited as feature descriptors, with the first three defined by toleranced vectors as in Fig. 2. Wirtz *et al.* argue that these quantities are easy to inspect with CMMs, and are directly useful in machine-tool control. This argument, however, comes close to violating the process independence dictum that has governed tolerancing for several decades. Also, it

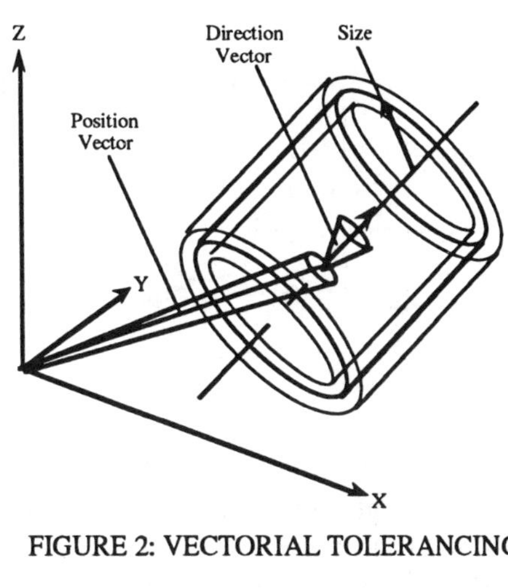

FIGURE 2: VECTORIAL TOLERANCING

is not obvious that vector tolerances will be effective for different types of processes (e.g. molding, deformation), or that they will be as effective as geometric tolerances in capturing "design intent"; more study is needed.

Figure 3 illustrates two notable weaknesses of parametric tolerancing: the lack of means for dealing with imperfect form, and ambiguous references. (Vector tolerancing handles imperfect form by requiring that perfect forms be fitted to part data.) In Fig. 3, should the 6-unit dimension be gaged as suggested in Fig. 3b, or by another of the several variants of Fig. 3b that come to mind readily? How is the indicated centerline in Fig. 3c defined *precisely?* (The three nominally

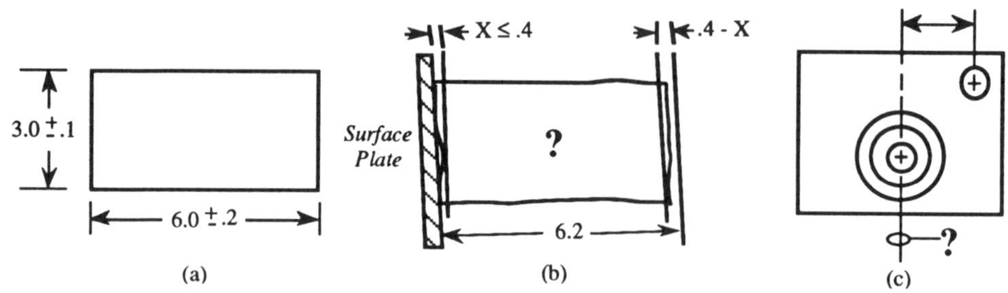

FIGURE 3: AMBIGUITIES IN PARAMETRIC DIMENSIONING AND TOLERANCING

concentric cylindrical features cannot be exactly concentric: which governs? ... and so forth.) These are serious ambiguities in the context of assembly design, and they stimulated the development of geometric tolerancing some fifty years ago. Parametric tolerancing is not yet ready for final interment, however, because it may be the most appropriate scheme available for functional tolerancing (discussed below), and it is useful when form errors can be ignored safely.

Geometric Tolerances

Figure 4 shows a part defined under the current Y14.5M-1982 geometric tolerancing standard. (Some dimensions have been omitted to avoid clutter.) The cylindrical stud has its diametral size specified by the Ø string. The blocks prefixed by symbols ⌖, ⊥, and ⌭ respectively place tolerances on (1) the stud's position relative to datum (reference) features A, B, C, (2) its perpendicularity relative to datum feature A, and (3) its cylindricity.

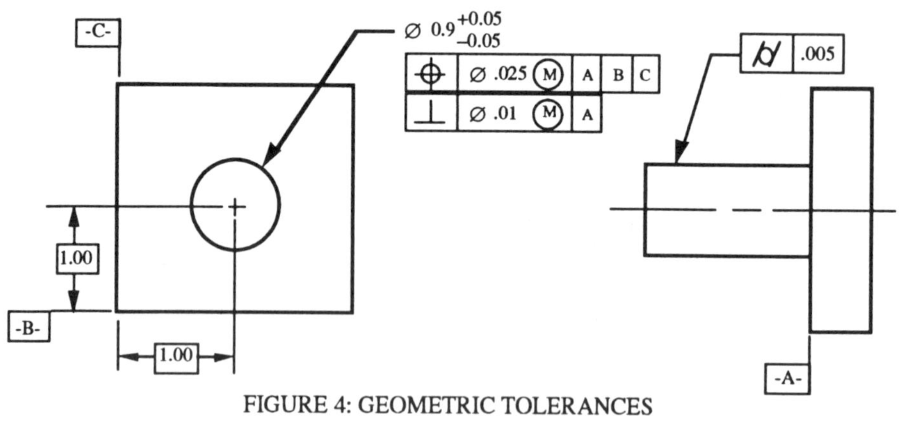

FIGURE 4: GEOMETRIC TOLERANCES

Broadly, geometric tolerancing enables tolerances to be placed on *attributes of features*, where a feature is one or more "pieces of part surface" (e.g. a hole, a slot, a planar face); feature attributes include size (for certain features), position (for certain features), form (flatness,

PACKING SLIP

1 4 3 0 5 8 5

Via

Market ID: 76901951
Order ID: 1430585
Order Date: 09/10/2026
Shipping Method:

Ship To

Alibris APEX DC 76901951-21
APEX
300 AVONDALE AVE.
GRANDVIEW HEIGHTS, OH 43212-3473, US

SKU	QTY	TITLE	TOTAL
BSM.1ALE5	1	Maternal & Child Health Nursing: Care of the Childbearing & Childrearing Family	44.19
BSM.1A8GN	1	Proceedings of the International Forum on Dimensional Tolerancing and Metrology (Papers Presented i	44.19

Order Total: $88.38

Your order has been packaged and created with great care. Enjoy! Bay State Books

#1430585

cylindricity, etc.), and relation (e.g. perpendicular-to). The semantics of geometric tolerances are established mainly by a set of rules for establishing datum systems (local coordinate systems) from physical part features, and another set of rules for constructing spatial zones in which toleranced features, or attributes thereof, must lie. (In Fig. 4, Y14.5M-1982 rcqures that the ⊥ tolerance zone contain the axis of the stud, whereas the stud's cylindrical surface must lie within a 3-D annular zone to meet the ⌭ tolerance.)

Geometric tolerancing addresses directly the weaknesses noted earlier: it handles imperfect form through containment zones, and it provides means to define reference systems. Naturally the current standard (1982) has its own weaknesses. The definitional media are prose and graphics, as noted earlier; these lead to definitions by example, and some of the examples do not have obvious generalizations. There are also some known ambiguities, e.g. "rocking datums", some known inconsistencies (e.g. between certain axial and envelope properties), and some "slippery notions" – notably *size,* which the standard handles loosely. Beyond these problems, however, the main concepts seem "essentially right" in the context of design for assembly, and the forthcoming new Y14.5 and Y14.5.1 standards will rectify many of the problems just cited.

Formalization

The *Preface* to this *Proceedings* and the paper by Walker and Srinivasan [1993] describe the motivation for the forthcoming Y14.5.1 standard. Y14.5.1 is a major step toward formalizing geometric tolerancing, because it replaces the traditional prose-and-graphics definitional style with mathematics; it also resolves some known problems and some hidden problems discovered during the mathematization process. Figure 5 illustrates the new style. The conformance zone for the surface profile tolerance on the left is defined by the vector inequality on the right; the right-side graphic merely illustrates what the vector math means. (Y14.5.1 provides more than conformance zone definitions: see Walker and Srinivasan [1993].)

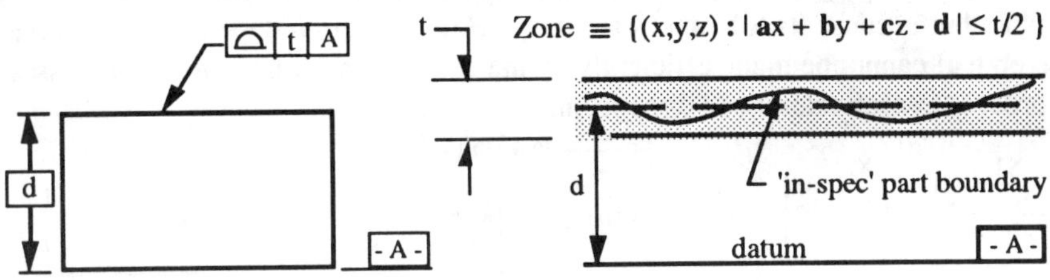

FIGURE 5: TOLERANCE SEMANTICS

The approach taken by the Y14.5.1 Committee was, in essence, to work through a draft of the new (1993/4) version of Y14.5 page by page, supplying mathematical definitions as prose-and-graphics definitions were encountered. One can imagine an alternative approach wherein the Y14.5.1 standard would contain only a sparse set of generator rules that users of the standard could instantiate to generate definitions. The alternative approach is more efficient (in principle) than the enumerative approach taken by the Committee, but it presumes the existence of a spare, regular set of underlying principles ... and these (if they exist) have yet to be enunciated precisely.

The work done to date by the Y14.5.1 Committee might be termed "cleaning-up the trenches". There is another body of work removed from industrial practice that seeks to formalize tolerancing at a higher, more holistic level; in due course it may subsume Y14.5-level tolerancing. This work is best exemplified by Requicha's early '80s research, which he summarizes in a paper below [Requicha, 1993; see also Srinivasan, 1993]. Turner and several others have also experimented with new, high level approaches to tolerancing: see Juster [1992] for a survey.

Dogma

There are two pervasive principles in tolerancing that deserve examination. The first is process independence, per Section 1.4(e) of the Y14.5 standard. This prohibits statements of the flavor, "Bore ∅3.250 x 1.500 deep", or "Finish-mill then grind ... ", which are found on old drawings in lieu of explicit tolerances. Implicit tolerancing through process specification often works well within vertically integrated companies, but it can inhibit outsourcing, and it caused enough problems in World War-II military procurement to result in its prohibition in the post-war GD&T standards. Thus for the past several decades the dictum has been, "Define the result you want, not how to get it."

The principle of process independence has come into question recently because it seems inconsistent with the doctrines of Concurrent Engineering ("Design the process with the product."). There are strong arguments on each side of the question, and perhaps we should admit two standardized classes of part and product definitions: process dependent, and process independent. Before rushing to embrace this ostensibly reasonable compromise, however, we should study closely some of the formal implications of process dependence. One such: how can equivalence be established between two part definitions based on different sets of processes?

The second issue warranting examination is the American "folk principle" and ISO default rule of tolerance independence. (A folk principle is widely believed but often undocumented.) It is easily illustrated with the hoary dictum from classical (parametric) tolerancing, "Thou shalt not over-dimension part drawings!" Consider Figure 6. The "over-dimensioned" version of 6a shown in 6b is not allowed in most organizations; it would be recast as 6c if the designer insisted on the "20.0 ± 0.1" constraint. All three versions make physical sense, however, and have the hatched acceptance regions shown. The 6b version is "illegal" not because it is physically impossible, but because its tolerances are *coupled* rather than independent. One consequence is that the 6b part cannot be made efficiently by the usual open-loop planning processes that preassign a process to each toleranced dimension.

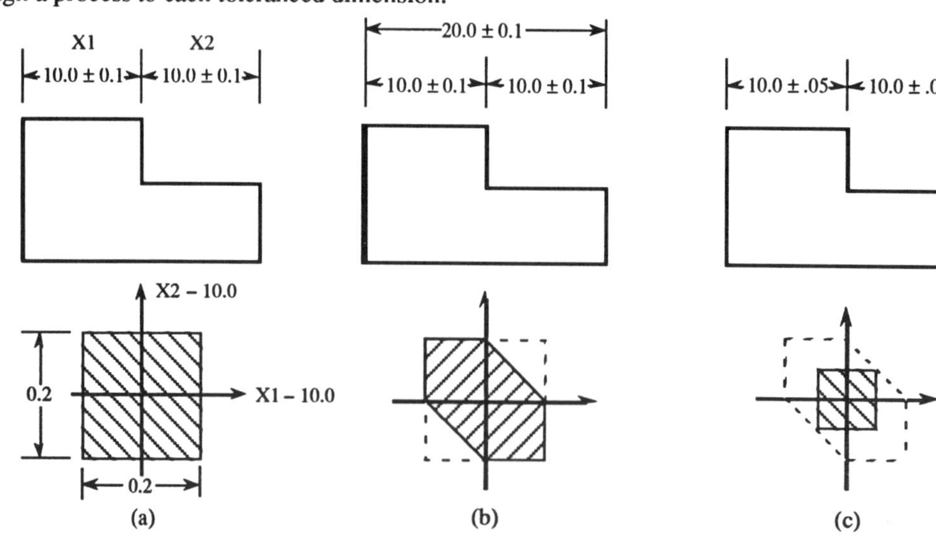

FIGURE 6: TOLERANCE COUPLING ("OVER-DIMENSIONING")

I raise this issue not to promote over-dimensioning *per se*, but to suggest that tolerance independence, if a principle at all, is best honored in the breach. It is frequently violated in geometric tolerancing (*vide* Fig. 4), and coupled tolerances may become the norm as we come to understand functional tolerancing.

METROLOGY

This section will be relatively short because the currently central issues, as I see them, are few, deep, and difficult, but not subtle. These issues, and others, are discussed in more depth in strong papers by Hocken *et al.* [1993] and Hopp [1993].

Figure 7, borrowed from Hook [1993], shows the three generic measurement technologies in use today. While manual measurement stretches back into antiquity, it was functional gaging that triggered the evolution of modern manufacturing and design. "Gage thinking" underlies some of the main principles and definitions in geometric tolerancing, and hard gages are often the conformance technology of choice when their cost can be justified. One of the major attractions of sampled coordinate metrology, as implemented currently with CMMs, is the promise of "soft gaging", i.e. the emulation of hard gages *and also* the implementation of criteria not physically gage-able (e.g. most LMC tolerances) through programmed data collection and reduction.

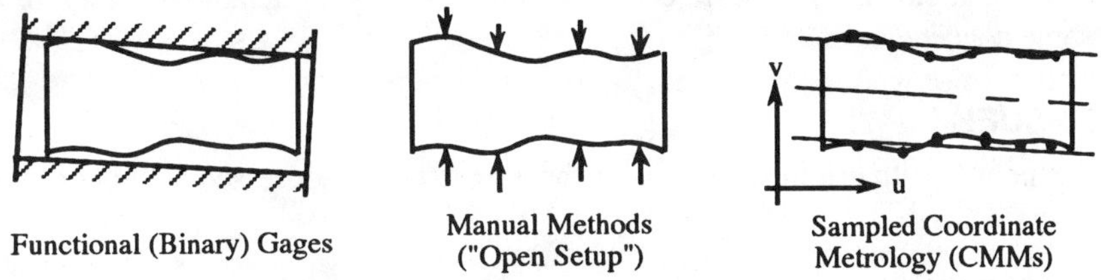

FIGURE 7: GENERIC MEASUREMENT TECHNOLOGIES

The two major issues that must be resolved for correct and efficient soft gaging are (1) sample set sufficiency, and (2) appropriate data reduction. The second, while not easy, is eas*ier* than the first. Imagine that a circularity tolerance is to be verified with the data in Figure 8a, and assume temporarily that the data are sufficient to describe the part geometry. How should the data be processed to yield a binary conformance assessment (and perhaps additional results, such as the actual circularity value)?

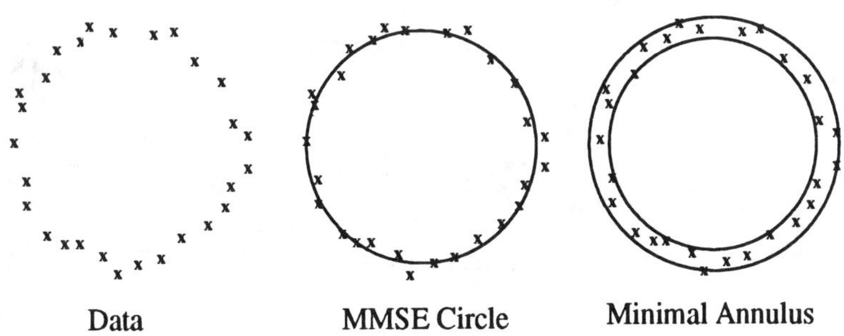

FIGURE 8: SAMPLED COORDINATE METROLOGY

Two possible answers are shown in Fig. 8: a Minimum Mean Square Error procedure, wherein maximal radial displacements are measured relative to a fitted MMSE circle, and a Minimal Containing Annulus procedure. The first is straightforward, while the second is more difficult and expensive [Roy and Zhang, 1992] but correct – because it conforms to the Y14.5 (or Y14.5.1) definition of circularity.

This little example suggests that having mathematical definitions for more complicated tolerances will be important, and that the design of correct, efficient soft gaging algorithms will provide employment for computational geometers for decades. (Circularity is conceptually easy but computationally challenging. Some difficult tolerances, such as the new interpretation of size [Srinivasan, 1993] or "rocking datums", will pose very difficult computational problems.) Thus practical difficulties lie ahead, but the work to be done seems relatively easy to define.

Let's return to the first problem, sample set sufficiency, which is the main focus of the paper by Hocken *et al.* [1993]. Criteria are needed for the distribution and density of data points, and here there are genuinely deep issues. The essential question seems to be this: can process knowledge be used, or should sampling be independent of the means used to make the part (for consistency with the process independence principle in tolerancing)? Both routes seem to require invocation of sampling principles based on spectral theory. The process-independent route requires additional criteria to enable "high frequency effects" to be rejected, but still leads to sample set sizes that are totally impractical with current CMM technology. Thus we are driven back to process-dependent sampling, where we must rely currently on limited empirical data to determine sufficiency because we do not have good process models. It is not a happy situation. It would seem that either (1) we have come upon a genuinely intractable problem, or (2) we are missing some conceptual links.

OTHER ISSUES

I shall conclude with brief discussions of three matters not addressed well or at all in the following papers.

Defining Assemblies

The Y14.5 standard is focused on part specification; virtually nothing is said about assemblies. Assembly drawings are almost always ambiguous unless covered with notes; assembly drawings generally are supplemented with (or are supplements to) an assembly process plan that specifies an assembly sequence and some feature-mating and -joining conditions. (Note that this smacks of the process dependence that is prohibited in part specification.) Some modern CAD systems do better, by providing means to declare feature matings in hierarchical graph structures [Allen, 1993]. However, Figure 9 shows that feature matings by themselves are not enough;

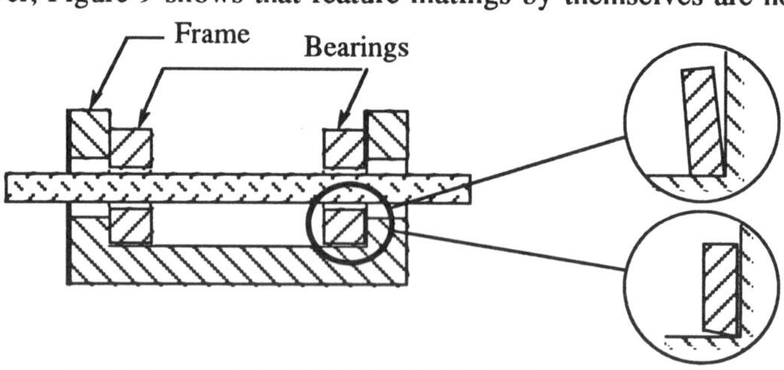

FIGURE 9: ASSEMBLY SPECIFICATION

matings often must be *ordered* in the same way that components of datum systems are ordered. No CAD systems support ordered matings insofar as I know. (Fig. 9 shows two possible face-mating orderings for assembling the bearing to the frame; which ordering is correct depends on how the parts are toleranced.) The underlying message here is that assembly specification requires more attention than it has received to date. What *is* an assembly, in precise terms [Strip and Voelcker, 1993]?

Tolerancing for Assembly

We know surprisingly little about tolerancing for assembly. We have MMC criteria for assembling simple, isolated features, and we have stackup rules for one-dimensional chains, ... and not much more. Thus assembly analyses involving complex toleranced parts almost always default to Monte Carlo simulation [Turner, 1993], and practical, systematic tolerance synthesis procedures have yet to emerge from research [Chase and Parkinson, 1991; Guildford and Turner, 1993]. Figure 10 provides a sobering example.

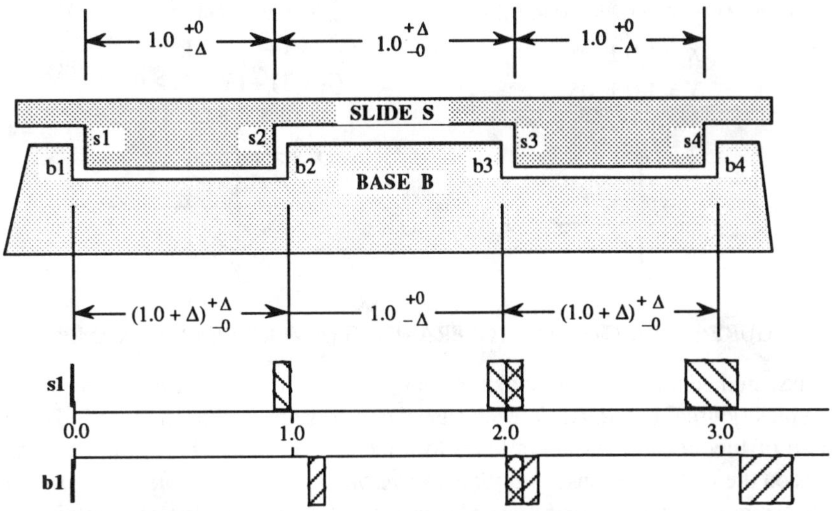

FIGURE 10: WILL THESE PARTS ASSEMBLE INTERCHANGEABLY?

Stackup analyses of the tolerances given in Fig. 10 shows interference in all four mated configurations (s1 mated to b1, s2 mated to b2, and so forth), and thus interchangeable assembly apparently cannot be guaranteed. In fact, however, it can be shown that the parts *will* assemble interchangeably; the proper conclusion to be drawn is that stackup analysis, which is one of our primary tools, is too conservative for some applications. (The example in Fig. 10 was contributed by Steffen Parratt, who has developed a new theory of one-dimensional assembly [Parratt, 1993]. Parratt's example is essentially the aircraft-door hinge problem studied less definitely, using different methods, by Altschul and Scholz [1993].) The underlying message is that we have much to learn about tolerancing for assembly.

Tolerancing for Function

Function is a much abused word in the tolerancing community. "Tolerancing for function" seems to mean something like this: "If I tolerance this part correctly, it will do what it's supposed to do, but don't ask me what that is in quantitative terms." I shall not provide a more precise definition here (it's not easy), but I shall provide an example to illustrate an important point.

Consider Figure 11, which shows the cross section of an electrical bus. The bus must fit into (assemble with) square hanger brackets of fixed size, and its resistive loss per unit length must not exceed a given value. These conditions translate into the constraints shown in Fig. 11. Two designs are given. The efficient design is a direct translation of the assembly and functional constraints (the latter is the area limit, which reflects the resistive loss limit). Note that the tolerances are coupled. The conservative design is based on independent bilateral tolerances; optimal values are found from the largest square that can be inscribed in the efficient design region.

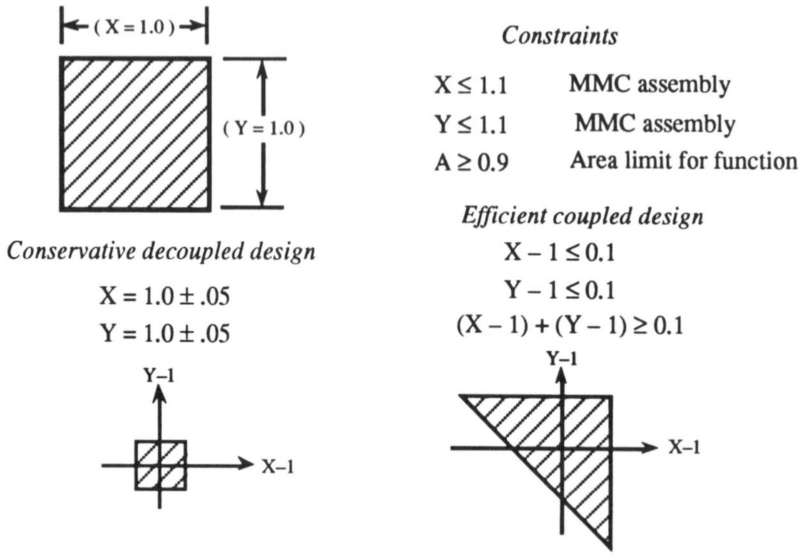

FIGURE 11: FUNCTIONAL TOLERANCING (A VERY SIMPLE EXAMPLE)

Parratt has studied more elaborate examples involving elastic deformation, heat transfer, and hydrodynamics (journal bearings). In each case the results are similar to those in Fig. 11, i.e. the introduction of functional constraints leads to tolerance coupling. This is not surprising when one reflects on the matter, because *quantitative measures of 'function' (heat transfer coefficients, eleastic deformations, and so forth) are almost never linear functions of part dimensions.* The underlying message is that we know very little about tolerancing for function.

THE FUTURE

The events of the past five years probably mark a watershed separating two eras. In the era now ending, dimensional tolerancing and metrology evolved as industrial practices without strong theoretical underpinnings. The best tolerancing practices were collected and made available to all through the evolving series of tolerancing standards. These standards played a critical role in that, in the absence of an underlying theory of tolerancing, they enforced consistency over usage in the same manner that dictionaries and thesauri enforce consistency over the use of natural languages.

The adoption of mathematics as the definitional medium for tolerances (i.e. Y14.5.1) seems to mark the opening of a new era. (It is interesting to note that the transition was triggered by a metrological 'event'.) We enter this era with only a formal means to describe tolerancing practices precisely, not a proper theory of tolerancing. Standards will continue to be critically important as long as this situation persists, but their character may change as they become 'math-based'. A proper theory of tolerancing can be expected, however, and when it appears the role of standards probably will change. They will become less important for enforcing internal consistency (theory will do that, as it does in physics and several areas of engineering), but more important for 'interfacing' with the larger world of manufacturing enterprise. The evolution of STEP [Laurance, 1993] may offer previews of the role change.

A theory of tolerancing powerful enough to subsume contemporary GD&T could appear soon, e.g. in the 1990s, based on Requicha's 1980s work. If it does, a third era may drawn a few decades later in which the 1990s theory is replaced by a more powerful theory. The reason for this prediction is that we know very little about tolerancing applications, and it will take time to learn how to tolerance systematically for assembly and for function. It is reasonable to expect that this knowledge, when we have it, will cause changes in the ways that we view, define, and use tolerances. (When pondering these speculations, bear in mind that tolerancing is *young* in terms of the time scales over which mechanical technologies develop.)

And what of metrology? CMM-based coordinate metrology is severely data-limited at present, because data are expensive when collected sequentially by moving machinery. It may be possible to finesse the currently vexing sample-set-sufficiency problem by moving to a data-rich environment based on wave phenomena rather than contact sensing. Thus the future of measurement *per se* is more likely to be paced by technology than theory.

ACKNOWLEDGEMENTS

Steffen Parratt and Dean Robinson contributed to this paper. The paper is based on studies supported by the National Science Foundation, the Alfred P. Sloan Foundation, the AT&T Bell Laboratories, and the Brown & Sharpe Manufacturing Company.

REFERENCES

Allen, G., 1993, "Tolerances and assemblies in CAD/CAM systems", this *Proceedings.*

Altschul, R. E. and Scholz, F. W., 1993, "Statistical tolerancing: A case study", this *Proceedings.*

Beckwith, W. and Parsons, F. G., 1993, "Measurement methods and the new standard B89.3.2", this *Proceedings.*

Chase, K. C. and Parkinson, A. R., 1991, "A survey of research in the application of tolerance analysis to the design of mechanical assemblies", *Research in Engineering Design,* vol. 3, pp. 23-37.

Guilford, J. and Turner, J. U., 1993, "Advanced tolerance analysis and synthesis for geometric tolerances", this *Proceedings.*

Henzold, G., 1993, "Comparison of Vectorial Tolerancing and Conventional Tolerancing", this *Proceedings.*

Hocken, R. *et al.*, 1993, "Sampling issues in coordinate metrology", this *Proceedings.*

Hook, R., 1993, "Interaction of dimensioning, tolerancing, and metrology", this *Proceedings.*

Hopp, T. H., 1993, "Computational metrology", this *Proceedings.*

Juster, N. P., 1992, "Modelling and representation of dimensions and tolerances: a survey", *Computer-Aided Design,* vol. 24, no. 1, pp. 3-17.

Laurance, N., 1993, "A high level view of STEP – A formal specification of the information content of a product design", this *Proceedings.*

Parratt, S. W., 1993, "A theory of one-dimensional tolerancing for assembly", Technical Report CPA93-2 (Ph.D dissertation nearing completion); Sibley School of Mechanical & Aerospace Engineering, Cornell University, Ithaca, NY.

Requicha, A. A. G., 1984, "Representation of tolerances in solid modeling: issues and alternative approaches", *Solid Modeling by Computers,* M. S. Pickett and J. W. Boyse, ed.; pp. 3-22, Plenum Press: New York, NY.

Requicha, A. A. G., 1993, "Mathematical meaning and computational representation of tolerance specifications", this *Proceedings.*

Roy, U. and Zhang, X., 1992, "Establishment of a pair of concentric circles with the minimum radial separation for assessing roundness error", *Computer-Aided Design,* vol. 24, no. 3, pp. 161-168.

Srinivasan, V., 1993, "Role of sweeps in tolerancing semantics", this *Proceedings.*

Strip, D. R. and Voelcker, H. B., 1993, "Geometric issues in assembly design and specification, or, What's an assembly?", *NOTES of the Workshop on Assembly & Task Planning;* IEEE International Conference on Robotics & Automation, Atlanta, GA

Turner, J. U., 1993, "Current tolerancing packages", this *Proceedings.*

Walker, R. K. and Srinivasan, V., 1993, "Creation and evolution of the ASME Y14.5.1 Standard", this *Proceedings.*

Wirtz, A., 1991, "Vectorial tolerancing for production quality control and functional analysis in design", *C.I.R.P. International Working Seminar on Computer-Aided Tolerancing*, pp. 77-84; Pennsylvania State University, University Park, PA.

CRTD-Vol. 27, International Forum
on Dimensional Tolerancing and Metrology
ASME 1993

MATHEMATICAL MEANING AND COMPUTATIONAL REPRESENTATION OF TOLERANCE SPECIFICATIONS

Aristides A. G. Requicha
Department of Computer Science
Institute for Robotics and Intelligent Systems
University of Southern California
Los Angeles, California

ABSTRACT

The meaning of tolerance specifications must be defined mathematically, to avoid ambiguous interpretations and to provide a sound basis for assessing the correctness of measurement techniques and algorithms. This paper surveys critically the various approaches to tolerancing proposed to date. It classifies these approaches at several levels. Syntactically, it distinguishes between dimensional limits and geometric tolerance specifications; semantically, it considers shape and pose (location and orientation) parameterization versus tolerance zones; and it contrasts worst-case, deterministic approaches with their statistical (stochastic) counterparts.

The principles of the offset zone theory of tolerancing, proposed over a decade ago, are presented, to show that there are general approaches, with firm mathematical foundations, which can capture the meaning of geometric tolerances.

To support automatic tolerance analysis and other important applications, tolerances must be associated with surfaces and other geometric entities within unambiguous computer representations for mechanical parts (i.e., solid models). Tolerances attached to graphic entities such as lines in a computer-generated drawing do not provide a suitable basis for supporting applications that go much beyond drafting. The paper closes by showing that relatively simple data structures can be used to computationally represent geometric tolerances. These structures can be associated with both of the two most common solid modeling schemes, boundary representation and constructive solid geometry (CSG).

TOLERANCES AND REPRESENTATIONS

Tolerance specifications are *representations*, i.e., collections of symbols that describe real-world entities. A very simple, graphic tolerance specification is shown in Figure 1. Here the symbols are graphic lines, arrows and numbers. Symbols are arranged in a representation in accordance with *syntactic rules*. Because graphic representations are intended primarily for human consumption, the corresponding syntactic rules are informal. But if representations are to be imbedded in computer programs, formal rules of syntax are required. Syntax merely determines the form of a representation. Its content, or meaning, is determined by its *semantics*, i.e., by the real-world objects that correspond to the symbols in the representation. We will have much more to say about semantics later.

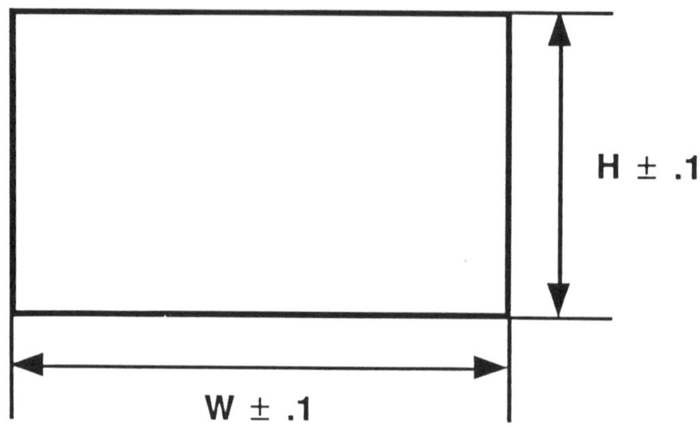

Figure 1 - A simple dimensional-limit tolerance specification.

There are two commonly-used syntactical approaches for defining tolerances. The first is shown in Figure 1, and called *dimensional limit* specification. It consists of identifying a set of parameters, usually called *dimensions*, associated with an object (or an object feature), and assigning to the parameters limits that define a range of values. (In this paper a *feature* is a portion of a solid object's boundary.) The second approach, called *geometric tolerance* specification, assigns values to certain attributes of a feature. Figure 2 shows a profile tolerance applied to the same rectangular object of Figure 1. There are several other types of geometric tolerances discussed in national and international standards.

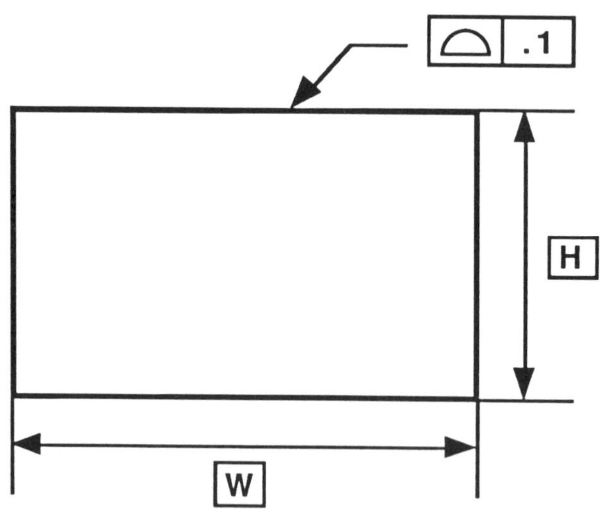

Figure 2 - A simple geometric tolerance specification

Suppose that a rectangular object is represented internally in a geometric modeling system by the triple (*Rect, W, H*), where *W* and *H* are the width and height. These two parameters provide a direct parameterization for the rectangle in the given modeling system, and are called ***natural*** or ***model*** parameters. But other parameters such as distances and angles might be more relevant to a designer's intent, and therefore other, indirect parameterizations are often useful. These user-defined parameters are called ***control*** or ***design*** parameters. Internally they must be converted by the modeler into its natural parameters. A number of interesting issues arise. For example, are the control parameters sufficient to define the object unambiguously? Are they consistent? These and related issues are discussed in Requicha (1984).

PURE PARAMETRIC TOLERANCING

The dimensional-limit tolerancing syntax is often associated with semantics based on pose (location and orientation) and shape parameterization. Figure 3 illustrates this approach, called (pure) *parametric tolerancing*. An object satisfies the specification of Figure 1, with $W = 3$ and $H = 2$, if it is a *perfect* rectangle with sides within the length limits prescribed by the specification. Thus, the rectangle shown on the left of Figure 3 meets the specification. The set of all objects that satisfy a given tolerance specification is called a *variational class* (Requicha, 1983; Requicha & Whalen 1991). All of the objects in a variational class defined by a correct tolerance specification should be interchangeable in assembly operations and functionally equivalent. Each of the objects in the variational class can be characterized by the specific values of its parameters W and H. Therefore, each element of the class can be associated with a point in the parameter or *configuration space* shown at the right on Figure 3. The entire class corresponds to the small hatched square in the figure.

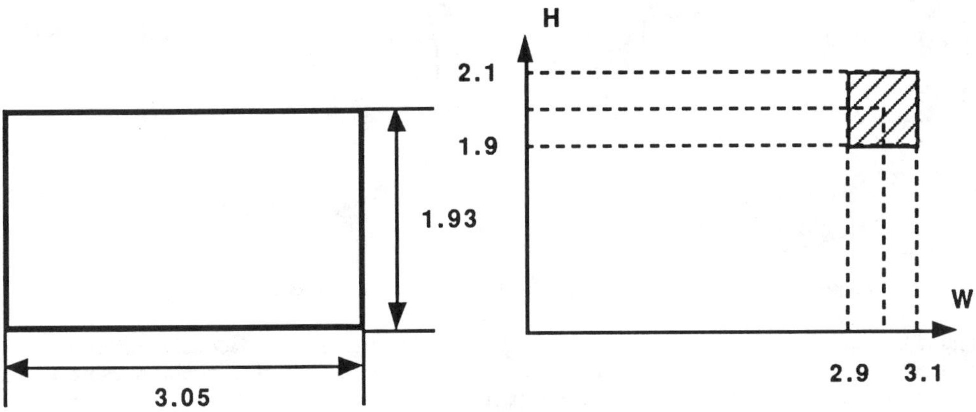

Figure 3 - An acceptable rectangle (left) and the configuration space region that corresponds to the variational class (right).

Perfect-form is an unrealistic assumption because it is impossible to manufacture perfect rectangles. (Even if it were possible we would never know, since we could not verify perfect form by physical measurements...) However, pure parametric tolerancing can be very useful. It seems to be adequate for most design calculations needed to ensure that a mechanical system fulfills its intended function (other than assemblability or fit). Also, it is easy to assign probability density functions to each of the parameters and study the corresponding variations of quantities that depend on the parameters and are important for the functionality of the part or assembly. This is the basis of statistical tolerance analysis, which is a widely used methodology. The dimensional limits are converted to parameters that characterize the probability distributions, for example, by using a Gaussian distribution with a standard deviation equal to some given constant times the dimensional limit. Most, if not all, commercial tolerance analysis programs assume pure parametric semantics, and perform either worst-case analysis, by using optimization techniques, or statistical analysis, using either closed-form or Las Vegas (the U.S. equivalent of Monte-Carlo) methods.

Pure parametric semantics ignores variations of orientation and form. But it can be extended by introducing hidden, or implicit, parameters that control orientation and form, as we will see later. It can also be extended to imperfect-form features by replacing the actual, imperfect features by perfect counterparts obtained by surface fitting. This latter approach is used in many CMM (Coordinate Measuring Machine) metrology algorithms, although it is not compatible with the ANSI or ISO tolerancing standards.

TOLERANCE ZONE SEMANTICS

Imperfect form can be acomodated by using *tolerance zones*. This approach is illustrated in Figure 4. A feature is considered acceptable if it lies within a suitably constructed zone of space (more on this later), as shown on the left of the figure for a nominally circular feature. Alternatively, one can fit a perfect circle to the imperfect actual feature and check that the center of the fitted circle is within a tolerance zone, as shown on the right. Note that the actual-geometry and the fitted-geometry methods often are not equivalent, that is, a feature may be considered acceptable by one of the methods and not by the other.

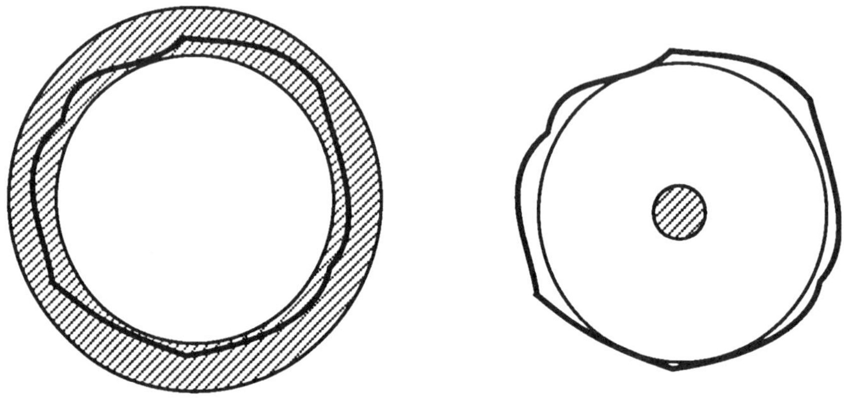

Figure 4 - Tolerance zone semantics for actual geometry (left) and fitted geometry (right).

Determining the center (or a more general characteristic) of a fitted-geometry, also called measured geometry in Requicha (1983), amounts to deriving, say, a center from an imperfect circle by using a specific mathematical rule. The derived geometry is also sometimes called *resolved* geometry. Deriving an axis for a feature that is approximately a surface of revolution is a delicate task, fraught with mathematical perils. For example, is the axis guaranteed not to self-intersect?

Parametric semantics may be extended so as to approximate tolerance zone semantics to within any desired accuracy (Turner, 1987). Figure 5 illustrates the approach. Pure parametric tolerancing, as we saw earlier, only accepts perfect rectangles within the dimensional limits. We can admit also quadrangles with imperfectly-oriented sides that lie within the hatched tolerance zone shown on the left if we proceed as follows. Introduce additional parameters, for example, consider as parameters the coordinates of the four vertices. This corresponds to an 8-dimensional configuration space, shown schematically on the right of Figure 5. For a quadrangle to lie within the tolerance zone certain constraints between the parameters must be satisfied. These constraints define a region of configuration space that corresponds precisely to the imperfect rectangles that fit in the tolerance zone. Thus, a feature is acceptable if its (extended) parameter set corresponds to a point in the appropriate configuration space region.

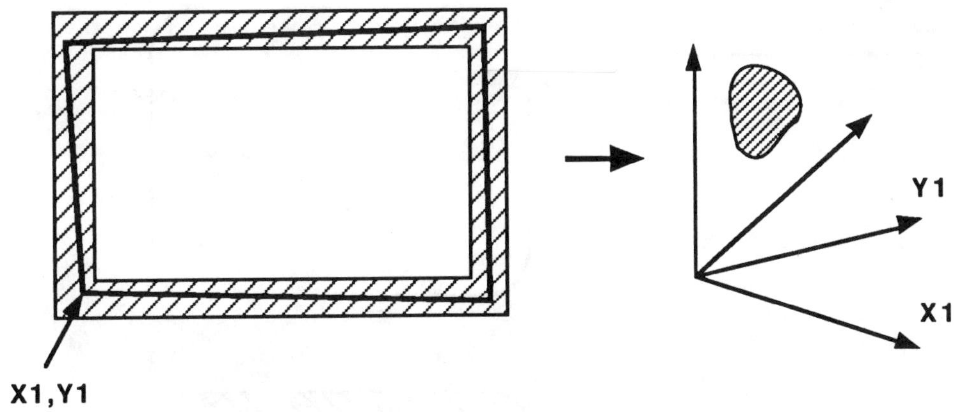

Figure 5 - A tolerance zone for imperfect rectangles (left)
and the corresponding region in configuration space (right)

Imperfect form can be accomodated by assuming a specific shape parameterization and computing configuration space regions involving the parameters. For example, a surface might be approximated by a cubic Bezier patch, defined by 16 points, and therefore with 48 parameters. The extended parametric approach for imperfect form implies large numbers of parameters, as the example indicates. Observe that the richness of the variational classes considered increases swiftly when we move from perfect-form/perfect-orientation semantics to perfect-form/imperfect-orientation and to imperfect-form semantics. The class is richest when the parameterized shape approximates closely the actual features, which typically implies that many parameters must be used. However, no finite number of parameters can describe all the objects that lie within a tolerance zone.

A theory of tolerancing based on tolerance zones must address the following issues: How is a part's surface to be decomposed into (imperfect-form) features? How are surfaces to be fitted? (Fitting is necessary for datums and certain fitted-geometry zones.) How are tolerance zones to be constructed? In the remainder of this paper we focus on this last issue, and more specifically on an approach based on offsetting operations. Note, however, that tolerance zones may be constructed in many ways. For example, one may parameterize a feature, construct its largest and smallest versions by selecting a suitable set of values for the parameters, and define the tolerance zone as the region bounded by the two extremal features (Requicha, 1984). Different approaches to tolerance-zone construction lead to different zones, and therefore whether an object is considered acceptable depends on which approach is used.

OZ THEORY

Offset zones are constructed by Minkowski operations, which are illustrated in Figure 6. Moving a circle of radius r so that its center lies within the rectangle shown on the top left of the figure sweeps a region of space shown at the top right, which is an expanded version of the rectangle. Sweeping the circle over the exterior of the rectangle erodes, or contracts, the rectangle by the same distance r, as shown at the bottom left. Subtracting the shrunk version from its grown counterpart yields the tolerance zone shown at the bottom right.

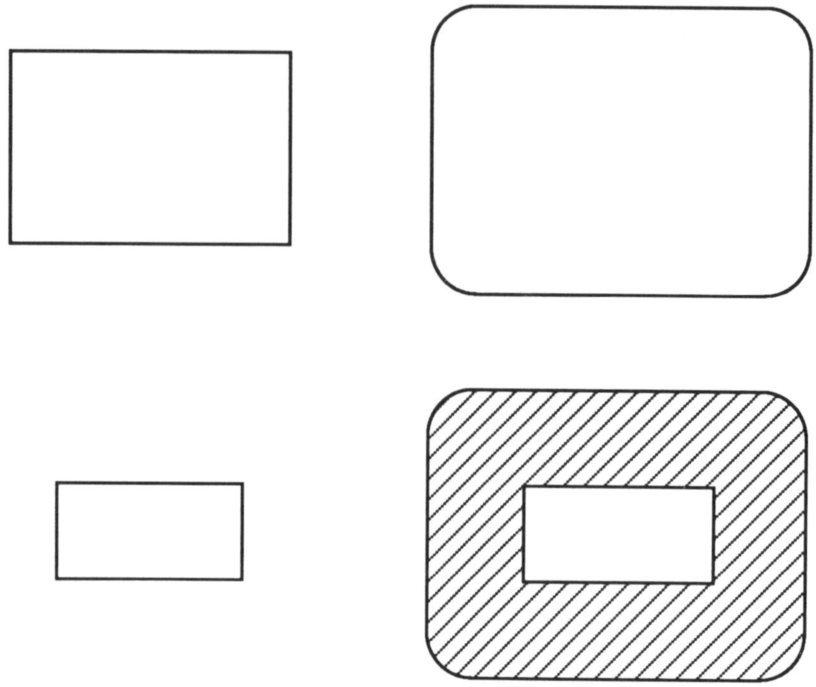

Figure 6 - A rectangle (top left), its grown version (top right), shrunk version (bottom left) and the resulting tolerance zone (bottom right)

The same basic construction is always well defined, and applies to features of *arbitrary shape*, including not only the usual cylinders, slabs, and spheres (known as "features of size"), but also more complex shapes and even objects bounded by sculptured, or free-form, surfaces. Note also that each tolerance zone is associated with a feature, and OZ theory does *not* imply a single tolerance zone for an entire mechanical part.

Armed with this basic construction procedure we can now define a variety of tolerances as in the ANSI and ISO standards. Details are given in Requicha (1983). For example, a *surface form* tolerance implies a tolerance zone of "width" (i.e., difference between the two offsets) equal to the given tolerance and unspecified size and pose. Note that it is not necessary to distinguish between cylindricity, flatness and other form tolerances—one general surface form tolerance suffices. Straightness, roundness, and similar tolerances are covered by a general *curve form* tolerance. A curve tolerance zone applies in sections of a given surface that are cut by a specified set of planes. Within each plane a tolerance zone is constructed by 2-D offsetting.

COMPUTATIONAL REPRESENTATIONS

Tolerance specifications based on the OZ theory are easy to map into computer data structures. In essence, we need means for:

1. Designating surfaces of a solid. This can be done by direct access to the faces of a boundary representation or indirectly through the primitive faces in a CSG representation.
2. Extracting subsets of surfaces. This is necessary for "local" tolerances, and, more generally, when the features to be toleranced are smaller than the faces internally represented in the modeler.
3. Grouping face subsets, to construct patterns and features such as slots, which are composed of several faces.

4. Assigning attributes to face (subset) groups. Some of the attributes represent the tolerances proper, whereas others indicate that certain feature groups constitute a datum reference frame.

When necessary, edge features can be constructed by intersecting surface features and toleranced via attributes. A system called VGraph based on the ideas just presented is described in Requicha and Chan (1986). An implementation of VGraph for the PADL-2 solid modeler and using the KnowledgeCraft AI environment is currently running on SparcStations at USC's Programmable Automation Lab.

Note that a parametric definition of tolerance zones requires a more complex implementation, because features must be described parametrically. This is not too hard to do for small sets of feature shapes. How to represent parametric features of *arbitrary shape*, and how to construct corresponding tolerance zones, however, is not trivial.

CONCLUSIONS

This paper surveyed various approaches that have been proposed for defining mathematically the meaning of tolerance specifications. It focused on the offset zone (OZ) theory, introduced a decade ago, which has interesting characteristics. It leads to rich variational classes, applies to features of arbitrary shapes, unifies several tolerance types, has well-defined semantics, does not require feature parameterization, and is easy to represent computationally.

To conclude we offer some "food for future thought", in the form of a few pointed questions directed at the broad problem of tolerancing mechanical parts and assemblies. Why are tolerance practices and standards the way they are? (A thorough description of the underlying reasoning, perhaps through case studies, would be very valuable.) Should design intent be captured more directly, perhaps at the assembly level? Should measurement (e.g. by sampling devices such as CMMs) be explicitly acknowledged in the theory (as in quantum mechanics)? Should there be tolerance specifications that can neither be verified exactly by CMMs nor by hard gages? Can current standards be simplified without significantly reducing their usefulness?

ACKNOWLEDGEMENTS

The research reported in this paper was supported in part by the National Science Foundation under grants DMC-87-96192, DDM-87-15404 and CDR-87-17322, by the industrial members of the Institute for Manufacturing and Automation Research (IMAR), and by the Industrial Associates of the Programmable Automation Laboratory, Institute for Robotics and Intelligent Systems (IRIS) of the University of Southern California.

REFERENCES

Requicha, A. A. G., 1983, "Toward a theory of geometric tolerancing", *Int. J. of Robotics Research*, Vol. 2, No. 2, pp. 45-60.

Requicha, A. A. G., 1984, "Representation of tolerances in solid modelling: issues and alternative approaches", in M. S. Pickett and J. W. Boyse, Eds., *Solid Modeling by Computers: from Theory to Applications*. New York: Plenum Press, pp. 3-22.

Requicha, A. A. G., and Chan, S. C., 1986, "Representation of geometric features, tolerances and attributes in solid modelers based on constructive geometry", *IEEE Journal of Robotics and Automation*, Vol. RA-2, No. 3, pp. 156-186.

Requicha, A. A. G., and Whalen, T. W., 1991, "Representations for assemblies", in L. S. Homem de Mello and S. Lee, Eds., *Computer-Aided Mechanical Assembly Planning*. Boston: Kluwer, pp. 15-39.

Turner, J. U., 1987, "Tolerances in computer aided design", Ph. D. Dissertation, Rensselaer Polytechnic Institute, Troy, NY.

CRTD-Vol. 27, International Forum
on Dimensional Tolerancing and Metrology
ASME 1993

ROLE OF SWEEPS IN TOLERANCING SEMANTICS

Vijay Srinivasan
Department of Manufacturing Research
IBM T. J. Watson Research Center
Yorktown Heights, New York
and
Department of Mechanical Engineering
Columbia University
New York, New York

ABSTRACT

Tolerancing semantics are defined as theoretical inspection procedures. These procedures invoke the existence of a tolerance zone which contains a geometric feature, or some geometric element extracted from it or some parameters estimated from it. Many of the tolerance zones can be defined using the offset operation. Others, however, require more general operations such as sweeps. This paper describes the role of sweeps in defining tolerance zones for size, circularity, and virtual boundaries. It also touches on the use of sweeps in manipulating tolerance zones for tolerance analysis.

INTRODUCTION

Geometric dimensioning and tolerancing (GD&T) can be viewed as a language engineers use to limit variability in their product information. This language has evolved informally over a long period of time, with a fairly rich set of symbols (syntax) and some explanation of what they mean (semantics). Figure 1 shows an example from ANSI Y14.5M-1982 that illustrates, in a particular case, both the syntax and the semantics of GD&T. Traditionally GD&T standard documents have defined the semantics using examples, such as "THIS ON THE DRAWING" and "MEANS THIS" illustrations. The ANSI Y14.5.1 subcommittee was set up to break away from this tradition and to define tolerancing semantics using sound mathematical principles.

A major part of mathematizing ANSI Y14.5M standard involves defining tolerance zones. Many of the tolerance zones can be defined using offset operations, as proposed by Requicha (1983, 1993). Others, however, were found to require more general operations such as sweeps. In this paper I will describe the direct and indirect role played by sweeps in mathematizing size and circularity tolerances. The role of sweeps in defining virtual boundaries will also be explained. Finally, I will speculate on extending the syntax of GD&T to cover sweeps and using sweeps to study tolerance accumulation for assemblies.

TOLERANCING SEMANTICS

Semantics of tolerancing is formally defined by Requicha (1983) as a *theoretical inspection procedure*:

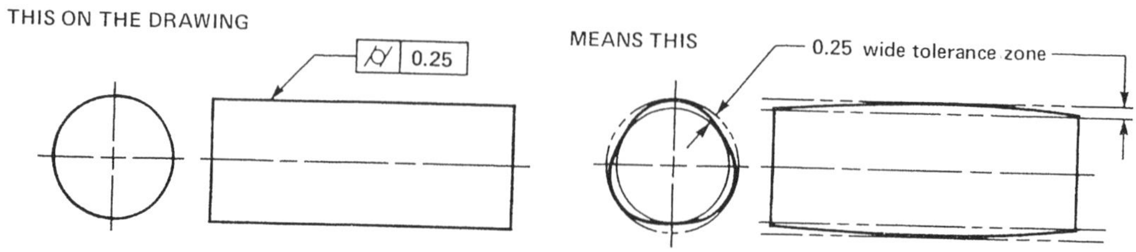

Figure 1: An example of syntax and semantics of GD&T from ANSI Y14.5M-1982.

> "A (model of a) part P satisfies a tolerance specification T if and only if there exists a decomposition of ∂P (boundary of P) into subsets G_i, called *actual surface features*, such that
>
> 1. $\cup G_i = \partial P$.
> 2. There exists a one-to-one correspondence between G_i and the nominal surface features F_i of S.
> 3. Each G_i satisfies the assertions A_{ij} associated with its corresponding F_i."

The tolerance assertions A_{ij} are usually stated as existential sentences of the form "there exists a tolerance zone ... that contains ...". The part meets the tolerance specification T if and only if the inspection procedure finds such a tolerance zone.

The tolerance assertions come in three major flavors. We can associate a tolerance zone with every one of these. For ease of further discussion, we will correspondingly classify the tolerance zones into the following three types.

1. *There exists a tolerance zone Z_i, subject to some constraints that depend on the type of tolerance (e.g., form, size, position, orientation), such that $G_i \subset Z_i$.*

   Such tolerance zones are defined in the same Euclidean space in which the part model resides. We will call them Type I tolerance zones. Figure 1 shows an example of a Type I tolerance zone. More details on Type I tolerance zones can be found in (Requicha, 1983) and (Jayaraman and Srinivasan, 1989). It is desirable to seek mathematical definition of tolerancing semantics using as many Type I tolerance zones (rather than Type II or Type III tolerance zones) as possible.

2. *Using "fitting" and "limiting" procedures derive a geometric entity $\hat{D}_i$ from G_i. Then there exists a tolerance zone Z_i, subject to some constraints, such that $\hat{D}_i \subset Z_i$.*

   These tolerance zones are also defined in the same Euclidean space in which the model resides. They will be called Type II tolerance zones. An example for Type II tolerance zone is shown in Figure 2. The distinction between the Type I and Type II tolerance zones is minor — however, it is useful to separate them for the purpose of later discussions.

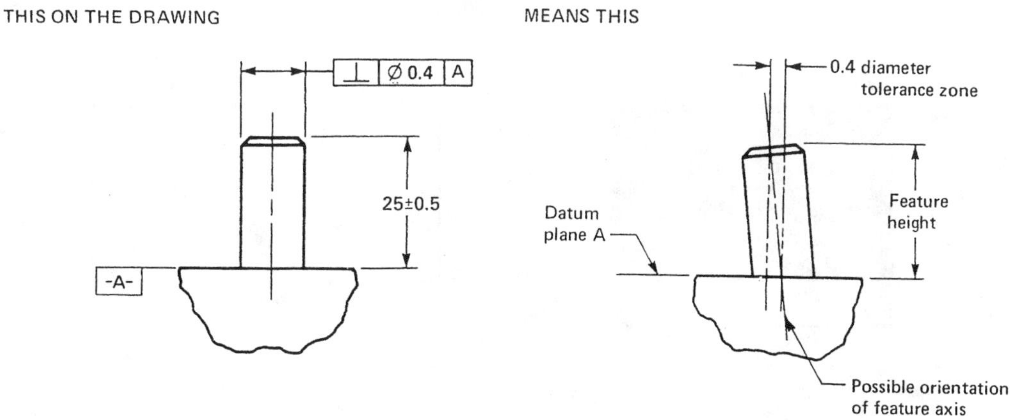

Figure 2: An example of Type II tolerance zone in ANSI Y14.5M-1982.

3. *Using a "fitting" procedure obtain a perfect-form feature $\hat{G}_i$ which acts as a substitute for G_i. Then the shape and/or pose parameters of $\hat{G}_i$ are within some parametric intervals. Such intervals define a tolerance zone in the space of the parameters.*

   Tolerance zones in the parametric space will be called Type III tolerance zones. Using the parametric intervals, we can easily construct tolerance zones in the parametric space that are bounded by hyperplanes normal to the parametric axes. It is often important to consider "non-rectilinear" tolerance zones in the parametric space. Figure 3 shows an example of a Type III tolerance zone. In general, these parametric spaces need not be vector spaces. Some times, the parametric space may contain a Euclidean subspace — in such cases, Type II tolerance zones will appear as lower dimensional subsets of Type III tolerance zones.

A study of the historic evolution of the ANSI Y14.5 standard clearly reveals a trend that favors the Type I and Type II tolerance zones over Type III zones.

OFFSETS, MINKOWSKI SUMS AND SWEEPS

Sweeps are regions obtained by moving a geometric object in space. The motion can, in general, involve both translation and rotation. In some cases, the geometric object itself is allowed to change while in motion. Though it is easy to define such general sweeps, computing useful representations of them is a daunting task.

If the motion is restricted to translations, the resulting translational sweep can also be obtained as a *Minkowski sum*. The Minkowski sum of two sets $A, B \subset \Re^d$ is defined as $A \oplus B = \{a + b : a \in A, b \in B\}$. Constant radius offsets, known simply as offsets, are a special case of Minkowski sum where one of the summands is a circular disk (in 2D) or a spherical ball (in 3D). As Minkowski sums are simply vector sums, they are applicable in any vector space. The Euclidean space and some parametric spaces are vector spaces.

Recently there has been much interest in efficient computation of Minkowski sums. For details see (Menon and Voelcker, 1991) and (Kaul, 1993). Authoritative accounts on

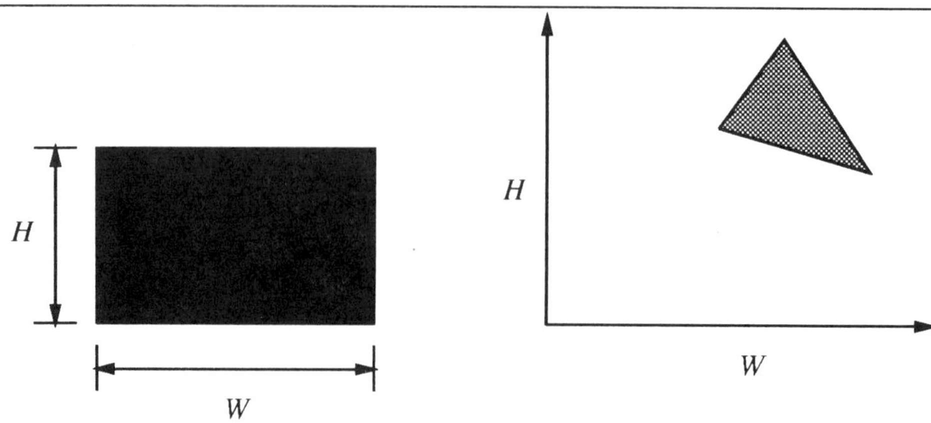

Figure 3: An example of Type III tolerance zone in the parametric space of the object on the left, after Voelcker (1993).

offsets of solids and planar curves can be found in (Rossignac and Requicha, 1986) and (Farouki and Neff, 1990), respectively.

In summary, we see that offsets are a special case of Minkowski sums which are a special case of general sweeps. The role of offsets in defining the tolerance zones has been described by Requicha (1993) elsewhere in the proceedings. Therefore, in this paper I will focus on those cases where sweeps play a role.

SIZE TOLERANCE

Size tolerancing applies only to *features of size*, which are nominally cylindrical, spherical, or composed of two opposing parallel planes. Informally, the size tolerancing semantics asserts that the "actual local size of an individual feature at each cross section shall be within the specified tolerance of size". In addition, the ANSI standard enforces a default requirement that the actual feature "shall not extend beyond a boundary of perfect form at MMC".

Historically, size evolved as a one-dimensional notion applied at every two-dimensional cross section of a three-dimensional object. The problems encountered in the search for a mathematical definition of size are described in (Srinivasan, 1993). I will now describe a solution to the size problem recently proposed by Prof. Requicha and myself using sweeps along *spines*.

Sweeping Balls Along Spines

A 0-dimensional spine is a point in space. A 1-dimensional spine is a space curve; a straight line is a special case of a 1-dimensional spine. In general, a 1–dimensional spine is an open, connected 1–manifold. A 2-dimensional spine is a surface; a plane is a special case of a 2-dimensional spine. In general, a 2–dimensional spine is an open, connected 2–manifold.

By sweeping a ball of diameter d so that its center lies in a 0-dimensional spine we get a region enclosed by a **sphere** of diameter d. A sweep of the ball with its center along a

1−dimensional spine generates a region enclosed by a canal surface. If the spine is a line, we get a region enclosed by a right circular **cylinder** of diameter d. If we sweep the the ball so that its center lies on a 2-dimensional spine we get a region enclosed by two surfaces. If the spine is a plane, we get a region bounded by **two parallel planes** that are a distance d apart. Note that such sweep operations cover all the current features of size as special cases. These sweeps are also obtainable as Minkowski sums of spines with balls.

A Tolerance Zone for Size

We are now ready to define a theoretical inspection procedure that defines the size tolerancing semantics. Let B_{MMC}, B_{LMC} be two balls with the MMC and LMC limits of size as their diameters, respectively. Consider two spines S_{MMC}, S_{LMC} associated with the two MMC and LMC balls. The entire size semantics hinges on finding these spines. The dimensionality of both the spines is $0, 1$, or 2, depending on whether we are tolerancing a spherical feature, a cylindrical feature, or two-parallel-planes feature.

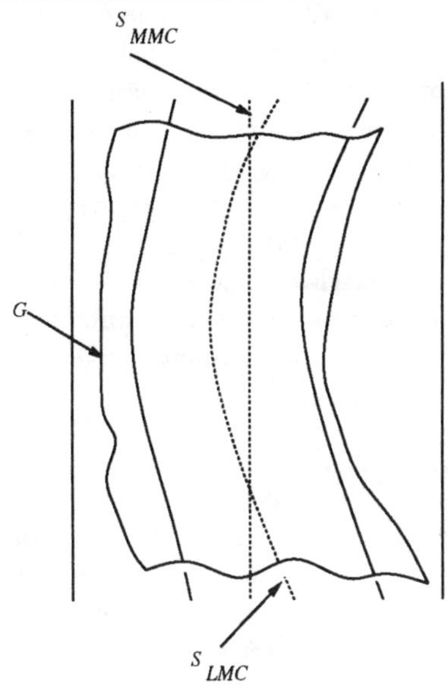

Figure 4: An example of tolerance zone for size.

In the ANSI context, if there is a note "PERFECT FORM AT MMC NOT REQD", then the shape of S_{MMC} is not restricted. Otherwise, it is restricted to be a straight line for a cylindrical feature, or a plane for two parallel-planes feature. In the ISO context, the default is not to impose such restriction on the shape of S_{MMC}. In either context, there is no restriction on the shape of S_{LMC}.

Let R_{LMC} denote the region generated by sweeping B_{LMC} so that its center lies in S_{LMC}. Define R_{MMC} similarly. For external features, ensure that R_{LMC} is contained in R_{MMC}. For internal features, ensure that R_{MMC} is contained in R_{LMC}.

An actual external feature of size G is in tolerance if we can find the spines S_{LMC} and S_{MMC} such that the above conditions are met and G is contained in $R_{MMC} - R_{LMC}$.

An actual internal feature of size G is in tolerance if we can find the spines S_{LMC} and S_{MMC} such that the above conditions are met and G is contained in $R_{LMC} - R_{MMC}$. See Figure 4 for an example involving a solid cylindrical feature. The spines as defined above must also satisfy some additional conditions to avoid accepting bad parts.

We now have a Type I tolerance zone for size. This has been adopted by the Y14.5 and Y14.5.1 subcommittees for the mathematical definition of size tolerance. Since we use balls in our sweeps simple translational sweeps will suffice to define the size semantics as described above. One may then ask whether general sweeps play any role in defining the size semantics. This question is addressed below.

General Sweeps for Size Tolerance

First consider a cylindrical feature of size. Instead of sweeping a ball along a spine, we can sweep a circular disk normal to the spine while its center lies on the spine. Since the disk is normal to the spine this will not be a purely translational sweep, and where the spine has tangent discontinuity we agree to sweep the disk over a range of angles at that point. The region swept by a ball along a spine can always be obtained by sweeping a circular disk (using a rotational motion of the disk about a diameter at the end of the spine, to round out the region), but the converse is not true.

Next consider a feature of size bounded by two parallel planes. By sweeping a line segment normal to a 2–dimensional spine while the center of the line segment lies on the spine we can generate useful regions. Such general sweeps allow us to generate a richer class of tolerance zones.

It is tempting to define a richer class of tolerance zones using general sweeps of disks and line segments. We resisted this temptation and decided to restrict ourselves to translational sweeps of balls to define size semantics due to their apparent simplicity and coverage of all current features of size.

CIRCULARITY TOLERANCE

Circularity (also known as roundness) is used to control the form of geometric elements obtained by intersecting an actual surface feature with cross sectional planes. A two-dimensional annular tolerance zone is then defined in the cross sectional plane, with the constraint that the inner and outer radii of the annular tolerance zone differ by a constant amount. Circularity can, in general, be applied to control the two-dimensional form of any surface of revolution. However, in practice, it is restricted to cylindrical, conical and spherical features.

In spite of its apparent simplicity, circularity has recently been a topic of controversy (Nasson, 1993) due to a difference in the interpretations of its semantics in Y14.5M-1982 and in B89.3.1 standards. This difference can be traced to (1) how the cross sectional planes are chosen, and (2) what, if any, restrictions are placed on the location of the center of the annular tolerance zone in that plane. An understanding of general sweeps of circular disks normal to a spine clarifies the issues involved and leads to a solution.

Sweeping Disks Along Spines

Consider the sweep of a circular disk, always orienting it normal to a straight line spine while the center of the disk is on the spine. If the the radius of the disk is kept constant during the sweep, we get a right circular cylindrical region. If the radius of the disk is varied linearly with distance along the spine as we sweep, we get a right circular conical region. In both cases, we have generated objects with perfect circularity (i.e., zero circularity tolerance).

Now consider the case where the spine is a space curve different from a straight line. If we sweep the circular disk, always orienting it normal to the spine while its center lies on the spine, we can argue that the boundary of the swept region also has perfect circularity. In fact this will be the case even if we change the radius of the circular disk as we sweep along the spine.

Tolerance Zones for Circularity

The above observation leads to a theoretical inspection procedure in which we seek a tangent continuous space curve (spine) at every point of which cross sectional planes are obtained normal to the spine. In each cross sectional plane we seek an annular tolerance zone within which the actual feature element is contained, with the constraints that (1) the center of the annular tolerance zone lies at the intersection of the spine with the cross sectional plane, and (2) the inner and outer radii of the annular tolerance zone differ by the same amount for all the cross sections.

The above solution seems to be satisfactory for current applications. The proposed semantics for circularity does not invoke general sweep explicitly — it is implicit in the underlying justification. This has been adopted by the Y14.5 and Y14.5.1 subcommittees for the mathematical definition of circularity tolerance. Note that we are still dealing with Type I tolerance zones, and by unioning all the annular tolerance zones we can obtain one composite three-dimensional tolerance zone for circularity!

VIRTUAL BOUNDARIES

Virtual boundaries were initially proposed (Jayaraman and Srinivasan, 1989) to capture functional requirements such as assembly and maintaining material bulk. Such virtual boundaries bound tolerance zones of Type I. If, however, one needs to know allowable tolerances on various parameters, these Type I tolerance zones have to be converted to Type III tolerance zones in the space of the relevant parameters. Such tolerances on parameters were called conditional tolerances (Srinivasan and Jayaraman, 1989) to reflect the fact that the resulting Type III tolerance zones usually have non-rectilinear boundaries. In any event, the original intent was that the virtual boundaries of Type I tolerance zones will be the primary tolerance representation, and the Type III tolerance zones in the parametric space will be derived, if necessary, from this primary representation.

Recently, some attention has been paid to the inverse problem. The collective effects of form, size, location and orientation tolerances on individual, as well as patterns of, features can be studied by deriving Type I tolerance zones from Type II or Type III zones. This is the motivation behind the introduction of such concepts as "virtual condition" and "resultant condition" in the current tolerancing standards. Informally, one wants to consider either the union of all parts that satisfy all the applicable tolerance assertions, or their intersections. The union will give some idea on whether parts will assemble, while the intersection will help to determine such information as minimum material bulk. Care must be exercised in positioning the various instances of the parts while computing the union or intersection, as this will have strong influence on the final result. Finally, computing the intersection can be viewed as computing the union of the complements, and then complementing the result. Therefore, the entire problem can be posed as one of determining the union.

Virtual Boundaries Using Sweeps

In some cases, the union of all instances of the part (or its complement) that satisfy all the relevant tolerance assertions can be obtained as a general sweep of one instance of the part. Such a sweep is restricted to be translational in case of MMC or LMC position tolerancing. If we have an RFS position tolerance for a feature of size, such as shown in Figure 2, then the union of all "in tolerance" parts is obtained by a general sweep of the feature of perfect form at MMC that involves both translation and rotation.

If Type II or Type III tolerance zones become the primary means for some tolerance specifications, then sweeps help us to determine corresponding Type I tolerance zones in the form of virtual boundaries. For some tolerance analysis problems it is conceivable that one may want to compute the virtual boundaries from Type II and Type III tolerance zones. There are many unanswered questions regarding the virtual and resultant conditions as proposed in the current standards. Many of these issues are still being researched and will hopefully get resolved in the near future.

EXTENSIONS USING SWEEPS

There are at least two directions in which the use of sweeps can be extended.

Syntactic Enrichment

Recently Parratt et al (1992) observed that bidirectional positional tolerancing, as described in the current standards, can be interpreted in more than one way. Different interpretations may be useful in different applications. They proposed some enrichment to the syntax of feature control frames for position tolerances to distinguish between different interpretations of current tolerancing syntax.

The last few years have seen considerable progress in the mathematical definition of the semantics of existing tolerancing syntax. It is time to consider syntactic enrichment that indicate sweeps, or their specializations such as Minkowski sums and offsets, more directly so that the tolerance zones can be defined unambiguously. I expect that this will be an active work item for the next release of the ANSI standard.

Tolerance Accumulation

The serial effects of tolerance build-up can be studied using general sweeps. But, as remarked earlier, computing such general sweeps is a daunting task. If the motion involves only small rotations, as can be expected in many tolerance applications, then the rotations can be added as vectors. This enables us to define a vector space in the space of relevant pose parameters of features. Once we have a vector space, we can use simple vector sums to add up the effects of successive motions.

The above observation has been exploited by Giordano and Duret (1993) to perform tolerance accumulation studies in general 3-dimensional cases using Minkowski sums. This approach may provide a general tool to compose Type III tolerance zones in some important problems of mechanical assembly.

CONCLUDING REMARKS

The role of offsets is well established in the mathematical theory of tolerancing. In this paper I discussed the role of sweeps, which are generalizations of offsets, in the emerging mathematical definitions of some tolerancing semantics. A strong message that I have tried to convey is that offsets, Minkowski sums and general sweeps provide powerful geometric operations to represent and manipulate tolerance zones encountered in defining the semantics of tolerances.

An unstated assumption in the development of GD&T language has been that tolerance representations can be decoupled from issues in tolerance analysis and synthesis. In the light of some recent developments, such as virtual and resultant conditions, this assumption should be questioned. As our understanding of assembly and other functional tolerances improves, there is evidence to suspect that representational issues and analysis/synthesis problems will be strongly coupled.

In this paper three types of tolerance zones were discussed. Tolerancing semantics that invoke Type II and Type III tolerance zones refer to some fitting and limiting operations that need further elaboration. Currently, our understanding of fitting is rather limited (Hopp, 1993) — further work is necessary before we can claim that the definition of tolerancing semantics is complete.

ACKNOWLEDGMENT AND A DISCLAIMER

I would like to acknowledge my colleagues in the ASME Y14.5.1 subcommittee for many of the ideas discussed in this paper. In particular Mark Nasson and Sam Thompson contributed significantly to the development of circularity semantics. The views expressed in this paper are my own, and not that of the official ASME Y14.5.1 subcommittee.

REFERENCES

Farouki,R.T., and Neff,C.A., 1990a, "Analytic Properties of Plane Offset Curves," *Computer Aided Geometric Design*, Vol.7, pp.83-99.

Farouki,R.T., and Neff,C.A., 1990b, "Algebraic Properties of Plane Offset Curves," *Computer Aided Geometric Design*, Vol.7, pp.101-127.

Giordano,M., and Duret,D., 1993, "Clearance Space and Deviation Space: Application to Three-dimensional Chain of Dimensions and Positions," *Proceedings of 3rd CIRP Seminars on Computer Aided Tolerancing*, Éditions Eyrolles, Paris, pp.179-196.

Hopp,T.H., 1993, "Computational Metrology," this proceedings.

Jayaraman,R., and Srinivasan,V., 1989, "Geometric Tolerancing: I. Virtual Boundary Requirements," *IBM Journal of Research and Development*, Vol.33, No.2, pp.90-104.

Kaul,A., 1993, "Computing Minkowski Sums," Ph.D. Thesis, Department of Mechanical Engineering, Columbia University, New York, NY.

Menon,J., and Voelcker, H.B., 1991, "Mathematical Foundations 1: Set Theoretic Properties of Ray Representations and Minkowski Operations on Solids," *Technical Report CPA91-9*, The Sibley School of Mechanical and Aerospace Engineering, Cornell University, Ithaca, NY.

Nasson,M.A, 1993, "Revelations from the Y14.5.1 Definition of Circularity and Cylindricity," *Presented at Quality Expo International*, Chicago, IL, April 20-22, also available as *Technical Report CSDL-P-3255*, The Charles Stark Draper Laboratory, Cambridge, MA.

Parratt,S.W., Robinson,D.M., and Voelcker,H.B., 1992, "Bidirectional Positional Tolerancing of Features," *Technical Report CPA92-8* The Sibley School of Mechanical and

Aerospace Engineering, Cornell University, Ithaca, NY.

Requicha,A.A.G., 1983, "Toward a Theory of Geometric Tolerancing," *International Journal of Robotics Research*, Vol.2, No.4, pp.45-60.

Requicha,A.A.G., 1993, "Mathematical Meaning and Computational Representation of Tolerance Specifications," this proceedings.

Rossignac,J.R., and Requicha,A.A.G., 1986, "Offsetting Operations in Solid Modeling," *Computer Aided Geometric Design*, Vol.3, pp.129-148.

Srinivasan,V., and Jayaraman,R., 1989, "Geometric Tolerancing: II. Conditional Tolerances," *IBM Journal of Research and Development*, Vol.33, No.2, pp.105-124.

Srinivasan,V., 1993, "Recent Efforts in Mathematization of ASME/ANSI Y14.5M Standard," *Proceedings of 3rd CIRP Seminars on Computer Aided Tolerancing*, Éditions Eyrolles, Paris, pp.223-232.

Voelcker,H.B., 1993, "A Current Perspective on Tolerancing and Metrology," this proceedings.

CRTD-Vol. 27, International Forum
on Dimensional Tolerancing and Metrology
ASME 1993

THE NEW NIST POLICY ON STATEMENTS OF UNCERTAINTY*

Barry N. Taylor and Chris E. Kuyatt
National Institute of Standards and Technology
Gaithersburg, Maryland

ABSTRACT

A measurement result is complete only when accompanied by a quantititive statement of its uncertainty. In October, 1992, then NIST director John W. Lyons announced a new NIST policy regarding statements of uncertainty associated with NIST measurement results. The new policy requires that NIST measurement results be accompanied by quantitative uncertainty statements and that a uniform approach to expressing measurement uncertainty be followed. Until the new policy was announced, there had never been a uniform approach at NIST to the expression of uncertainty. The new policy was effective immediately with a full implementation date of January 1, 1994.

The policy is based on the approach to expressing uncertainty in measurement recommended by the International Committee for Weights and Measures (CIPM) in 1981. The original recommendation was recently elaborated on by a committee composed of experts nominated by the International Bureau of Weights and Measures (BIPM), the International Electrotechnical Commission (IEC), the International Organization for Standardization (ISO), or the International Organization of Legal Metrology (OIML). The CIPM recommendation is the only recommendation concerning the expression of uncertainty in measurement adopted by an intergovernmental organization.

To aid the NIST staff in putting the policy into practice, a document containing guidelines was prepared and published.

BASIS FOR THE NEW NIST POLICY

The usefulness of measurement results is to a large extent determined by the quality of the statements of uncertainty that accompany them. Although almost all NIST measurement results are accompanied by quantitative statements of uncertainty, there has never been a uniform approach at NIST to the expression of uncertainty. In October, 1992, then NIST Director John W. Lyons announced a new NIST policy (reproduced on pages 13-15 of Taylor and Kuyatt, 1993) regarding statements of uncertainty associated with NIST measurement results. The new policy requires that NIST measurement results be accompanied by quantitative uncertainty statements and that a uniform approach to expressing measurement uncertainty be followed. The new policy was effective immediately with a full implementation date of January 1, 1994.

The new NIST policy is based on the approach to expressing uncertainty in measurement recommended by the International Committee for Weights and Measures (CIPM) in 1981 (CIPM, 1981) and the elaboration of that approach given in the *Guide to the Expression of Uncertainty in Measurement* (ISO, 1993) (hereafter called the *Guide*), which is being prepared by a committee composed of experts nominated by the International Bureau of Weights and Measures (BIPM), the International Electrotechnical Commission (IEC), the International Organization for Standardization (ISO), or the International Organization of Legal Metrology (OIML). The CIPM approach is founded on Recommendation INC-1 (1980) of the Working Group on the Statement of Uncertainties (Kaarls, 1982; Giacomo, 1982). This group was convened in 1980 by the BIPM as a consequence of a 1978 request by the CIPM that the BIPM study the question of reaching an international consensus on expressing uncertainty in measurement. The request was initiated by then CIPM member and NBS Director E. Ambler. A 1986 request by the CIPM to ISO asking it to develop a broadly applicable guidance document based on Recommendation INC-1 (1980) led to the development of the *Guide*. It is at present the most complete reference on the general application of the CIPM approach to expressing measurement uncertainty, and its development is giving further impetus to the worldwide adoption of that approach.

Although the *Guide* represents the current international view of how to express uncertainty in measurement based on the CIPM approach, it is a rather lengthy document. A succinct presentation of those aspects of the *Guide* that will be of most use to the NIST staff in implementing the new policy has been given by Taylor and Kuyatt (1993), together with some additional suggestions that are not contained in the *Guide* or policy.

COMPONENTS OF UNCERTAINTY

In general, the result of a measurement is only an approximation or estimate of the value of the specific quantity subject to measurement, that is, the **measurand**, and thus the result is complete only when accompanied by a quantitative statement of its uncertainty.

The uncertainty of the result of a measurement generally consists of several components which, in the CIPM approach, may be grouped into two categories according to the method used to estimate their numerical values:

A. those which are evaluated by statistical methods,

B. those which are evaluated by other means.

Basic to the CIPM approach is representing each component of uncertainty that contributes to the uncertainty of a measurement result by an estimated standard deviation, termed **standard uncertainty** with suggested symbol u_i, and equal to the positive square root of the estimated variance u_i^2.

An uncertainty component in category A is represented by a statistically estimated standard deviation s_i, equal to the positive square root of the statistically estimated variance s_i^2, and the associated number of degrees of freedom ν_i. For such a component the standard uncertainty is $u_i = s_i$.

The evaluation of uncertainty by the statistical analysis of series of observations is termed a **Type A evaluation (of uncertainty)**.

In a similar manner, an uncertainty component in category B is represented by a quantity u_j, which may be considered an approximation to the corresponding standard deviation; it is equal to the positive square root of u_j^2, which may be considered an approximation to the corresponding variance and which is obtained from an assumed probability distribution based on all the available information. Since the quantity u_j^2 is treated like a variance and u_j like a standard deviation, for such a component the standard uncertainty is simply u_j.

The evaluation of uncertainty by means other than the statistical analysis of series of observations is termed a **Type B evaluation (of uncertainty)**.

Further discussion of the evaluation of uncertainty components can be found in Taylor and Kuyatt (1993) and in the *Guide*.

COMBINED STANDARD UNCERTAINTY

The **combined standard uncertainty** of a measurement result, suggested symbol u_c, is taken to represent the estimated standard deviation of the result. It is obtained by combining the individual standard uncertainties (and covariances as appropriate), whether arising from a Type A evaluation or a Type B evaluation, using the usual method for combining standard deviations. This method, which is summarized in Taylor and Kuyatt (1993) and discussed in more detail in the *Guide*, is often called the *law of propagation of uncertainty* and in common parlance the "root-sum-of-squares" (square root of the sum-of-the-squares) or "RSS" method of combining uncertainty components estimated as standard deviations.

The combined standard uncertainty u_c is a widely employed measure of uncertainty. The NIST policy on expressing uncertainty states that it is commonly used for reporting results of determinations of fundamental constants, fundamental metrological research, and international comparisons of realizations of SI units.

EXPANDED UNCERTAINTY

Although the combined standard uncertainty u_c is used to express the uncertainty of many NIST measurement results, for some commercial, industrial, and regulatory applications of NIST results (e.g., when health and safety are concerned), what is often required is a measure of uncertainty that defines an interval about the measurement result y within which the value of the measurand Y can be confidently asserted to lie. The measure of uncertainty intended to meet this requirement is termed **expanded uncertainty**, suggested symbol U, and is obtained by multiplying $u_c(y)$ by a **coverage factor**, suggested symbol k. Thus $U = ku_c(y)$ and it can be confidently asserted that $y - U \leq Y \leq y + U$, which is commonly written as $Y = y \pm U$.

In general, the value of the coverage factor k is chosen on the basis of the desired level of confidence to be associated with the interval defined by $U = ku_c$. Typically, k is in the range 2 to 3. Ideally, one would like to be able to choose a specific value of k that produces an interval corresponding to a well-defined level of confidence p, such as 95 or 99 percent. This is difficult to do in practice because it requires knowing in considerable detail the probability distribution of each quantity upon which the measurand depends and combining those distributions to obtain the distribution of the measurand. The *Guide* and Taylor and Kuyatt (1993) give an approximate solution to this problem.

In view of the difficulty of determining a coverage factor for a specific level of confidence, and in keeping with the practice adopted by other national standards laboratories and several metrological organizations, the stated NIST policy is:

> Use expanded uncertainty U to report the results of all NIST measurements other than those for which u_c has traditionally been employed. To be consistent with current international practice, the value of k to be used at NIST for calculating U is, by convention, $k = 2$. Values of k other than 2 are only to be used for specific applications dictated by established and documented requirements.

REPORTING UNCERTAINTY

The stated NIST policy regarding reporting uncertainty is:

> Report U together with the coverage factor k used to obtain it, or report u_c.
>
> When reporting a measurement result and its uncertainty, include the following information in the report itself or by referring to a published document:
>
> - A list of all components of standard uncertainty, together with their degrees of freedom where appropriate, and the resulting value of u_c. The components should be identified according to the method used to estimate their numerical values:
>
>   A. those which are evaluated by statistical methods,
>
>   B. those which are evaluated by other means.
>
> - A detailed description of how each component of standard uncertainty was evaluated.
>
> - A description of how k was chosen when k is not taken equal to 2.
>
> It is often desirable to provide a probability interpretation, such as a level of confidence, for the interval defined by U or u_c. When this is done, the basis for such a statement must be given.

The NIST requirement that a full description of what was done be given is in keeping with the generally accepted view that when reporting a measurement result and its uncertainty, it is preferable to err on the side of providing too much information rather than too little. Such details are often provided to the users of NIST measurement results by referring to published documents, for example, when such results are given in calibration and test reports and certificates.

REFERENCES

*This paper is based on Taylor and Kuyatt (1993).

CIPM, *BIPM Proc.-Verb. Com. Int. Poids et Mesures* **49**, 8-9, 26 (1981) (in French); P. Giacomo, "News from the BIPM," *Metrologia* **18**, 41-44 (1982).

ISO, *Guide to the Expression of Uncertainty in Measurement*, prepared by ISO Technical Advisory Group 4 (TAG 4), Working Group 3 (WG 3), to be published (1993). ISO/TAG 4 has as its sponsors the BIPM, IEC, IFCC (International Federation of Clinical Chemistry), ISO, IUPAC (International Union of Pure and Applied Chemistry), IUPAP (International Union of Pure and Applied Physics), and OIML. Although the individual members of WG 3 were nominated by the BIPM, IEC, ISO, or OIML, it is expected that the *Guide* will be published by ISO in the name of all seven organizations. The most recent draft of the *Guide* is available from the NIST Calibration Program.

Kaarls, R., "Rapport du Groupe de Travail sur l'Expression des Incertitudes au Comité International des Poids et Mesures," *Proc.-Verb. Com. Int. Poids et Mesures* **49**, A1-A12 (1981) (in French); Giacomo, P., "News from the BIPM," *Metrologia* **17**, 69-74 (1981). (Note that the final English-language version of Recommendation INC-1 (1980), published in an internal BIPM report, differs slightly from that given in the latter reference but is consistent with the authoritative French-language version given in the former reference.)

Taylor, B. N., and Kuyatt, C. E., *Guidelines for Evaluating and Expressing the Uncertainty of NIST Measurement Results*, NIST Technical Note 1297, U.S. Government Printing Office, Washington, DC, 1993.

CRTD-Vol. 27, International Forum
on Dimensional Tolerancing and Metrology
ASME 1993

THE DETERMINISTIC APPROACH IN METROLOGY AND MANUFACTURING

James B. Bryan
Bryan Associates
Pleasanton, California

Abstract

The basic idea is that automatic machine tools and measuring machines are perfectly repeatable just like the stars and the planets. They obey cause and effect relationships that are within our ability to understand and affordably control. There is nothing random or probabilistic about their behavior. Everything happens for a reason. The list of reasons is small enough to manage by common sense, good metrology, and a reasonable investment of resources.

Some level of non-repeatability is observed in the performance of automatic machines. A determinist believes that all of this non-repeatability is caused by systematic sources and that there is no such thing as random behavior of an automatic machine. We use the term "apparent non-repeatability" to emphasize this belief since each of the sources of apparent non-repeatability is itself repeatable if examined closely enough. The magnitude of apparent non-repeatability depends on the time, money, and skill of the user in creating the proper environment for the machine.

Statistical measures of apparent non-repeatability have limited usefulness since they cannot be used to predict future machine performance if the time, money, and skill of the user changes. On line monitoring and control of the variables, such as temperature, are a better alternative. Statistical measures can be counterproductive in somehow implying that non-repeatability is the fault of the machine rather than the user. A determinist believes that any automatic machine can be made to repeat to a value that is close to its resolution. The cost of doing so is not unreasonable when compared to the benefits.

My Experience With Repeatability

Throughout my 38 year career in Precision Engineering, I have observed an amazing degree of repeatability in the behavior of the machines that I have tested. In the begining it was short term repeatability, but as we learned how to control temperature, [1], [2}, [3], [4], it became long term repeatability as well. The first machines we tested were tracer lathes, [5] followed by N.C. lathes, [6] N.C. measuring machines, [7] and diamond turning machines. [8]

My Colleagues' Experience With Repeatability

Everyone that I know or have known, who has been closely involved with the design and construction of high precision machines or has had extensive personal experience in testing machine performance, has had the same story to tell. The non-repeatabilities that we have observed have all been traceable to systematic sources that were correctable.

A partial list of these individuals includes Erwin Loewen, Floyd Hawley, Dick Clouser, Wolfgang Knapp, Don Carter, George Tlusty, Leo Sullivan, Jacques Petavel, Chuck Hendricks, Ray Kidder, and Bob Donaldson. The list also includes the late John Loxham, Fred Rolt, Cecil Timms, Bill Reason, Richard Moore, and Fred Victory.

For the past 25 years my colleagues and I have been struggling to articulate our philosophy on this subject. John Loxham has authored two papers [9, [10] I have authored one [11], and Bob Donaldson has authored one. [12] We now call it the Deterministic Point of View or the Deterministic Approach. Finding the right words is an on-going process. Everytime we speak or write on the subject it comes out a little different. The present paper is just another step toward the development of a Generalized Deterministic Theory.

John Loxham's Contributions

John Loxham deserves the credit for formally introducing the Deterministic Point of View to manufacturing engineers.

John Loxham was an astronomer, businessman, precision engineer, expert in hydrostatic bearings, metrologist, educator, and a pioneer in understanding thermal effects. He was the founder of the Cranfield Unit for Precision Engineering (CUPE) and hired Pat McKeown as its Director. He recognized the relationship between the predictability of celestial bodies and automatic machines. In a paper written in 1970, [7] Prof. Loxham made the following statement:

"A very rigorous mathematical and experimental analysis into a wide range of natural laws establishes the remarkable fact that 'there appears to be no known error in any natural law.' These laws are therefore 100% perfect, unchanging and in consequence very reliable. Industry in general has not recognized the benefits that can be obtained by exploiting these 100% perfect and very reliable natural laws as they operate in the engineering workshop. A close study of an automatic manufacturing process shows that it is operating under the influence of natural laws and is not affected by the very wonderful, complicated, hut slightly **unpredictable mechanism known as a human being**. If, as indicated above, all natural laws operate with 100% perfection, an automatic manufacturng process is always operating perfectly. It may not be doing what is required, but if that is so it is because it has not been suitably arranged. The last part of this statement is now known as:

Loxham's Principle

"An automatic manufacturng process is always operating perfectly. It may not be doing what is required, but if that is so it is because it has not been suitably arranged."

Dr. Kidder's Comments On The Perfection Of Classical Physics Laws

Some people have reacted to these quotes with some skepticism about the perfection of natural laws. I have consulted Dr. Ray Kidder, my friend and neighbor for 35 years. He is now retired from the Theoretical Physics Dept. at LLNL. In my opinion, Dr. Kidder is one of the world's greatest theoretical physicists.

According to Dr. Kidder, there are four exceptions to the deterministic nature (100 % perfection) of classical physical laws. These exceptions apply to molecular and atomic size masses and have no practical significance in the field of manufacturing. They are as follows:

1. <u>Quantum Noise</u>
   The random behavior of the elements within an atom. Sometimes called "Heisenberg Uncertainity". The product of uncertainity of momentum and the uncertainity of position is equal to Planck's constant which is equal to 10^{-27} erg-secs.

2. <u>Thermal Noise</u>
   The random exchange of energy of molecules with each other which causes a change in size of an object. The magnitude of the size change is determined by Boltzman's constant which is on the order of 10^{-23} joules/deg. C. Hocken and Teague have calculated this effect for a one meter long gage block at 20 deg. C. It amounts to one part in 10^{11} which is a change in length of a tenth of an Angstrom.

3. <u>Brownian Movement</u>
   A thermal noise effect causing the random motion of dust particles after collision with molecular particles. The magnitude of this effect is also determined by Boltzman's constant (10^{-23} Joules/deg. C.)

4. Shot Noise

The behavior of current flows of less than 10^{-19} ampere, which is one electron per second. The flow of electrons at these small currents is random.

5. Johnson Noise*

The white noise voltage generated by an open circuit resistor. The magnitude depends on Boltzman's constant, the absolute temperature, and the resistance. It can be thought of as a special case of Brownian motion. [14]

6. 1/f Noise (flicker noise)*

This noise has approximately a 1/f spectrum (equal power per decade of frequency) and is sometimes called "pink noise" Its origin is the random fluctuations in resistance of resistors and is not very well understood. It is proportional to D.C. current, and depends on the construction of the resistor. Typical values vary between .01 and 3.0 RMS microvolts per volt applied across the resistor. [14]

7. Popcorn Noise*

A characteristic of operational amplifiers that causes the offset voltage to jump at random times and durations. [14]

*The author is grateful to Prof. Bob Hocken for suggesting these additions to the original list during the Conference at Dearborn.

Assuming that the above list is correct, we can see that the probabilistic approach to a manufacturing problem involving an automatic machine is simply not necessary. It can also be a distraction. It is more important for the user to realize that the apparent non-repeatability that is observed is **his** fault and not the fault of the machine. This realization is very important in creating an aggressive, optimistic, willingness-to-interfere, positive attitude on the part of the user.

History tells us that it takes time to become completely comfortable with any new theory and particularly with the Deterministic Point of View. It took the astronomers and navigators 150 years to become dedicated determinists and finally believe in Newton's laws. John Loxham [11] told the following story about the discovery of the planet Neptune:

Discovery of the Planet Neptune

"The story of the discovery of the planet Neptune in 1846 illustrates the ever present temptation of researchers to disregard fundamentals and make up wild theories to explain experimental evidence that is unusual."

"It all arose from the non-predictability of the motion of the planet Uranus which was first discovered in 1781. After 156 years of observations no one could calculate an orbit based on the pulls of the known planets that would fit the data. In 1837 G. B. Airy, the new Astronomer Royal, wrote: 'The errors of longitude are increasing with fearful rapidity.' Wild theories were proposed at this time: a resisting medium; an undetected satellite; or a collision with a c:omet. No one could explain, however, why a resisting medium should afflict Uranus and no other planet."

"By 1845 Uranus was out of place by the 'intolerable quantity' of two minutes of arc, enough to darken the lives of astronomers and mathematicians. Yet the dawn was at hand. The solution to the mystery was based on the calculations of two young mathematicians, John Adams, an Englishman, and J. LeVerrier, a Frenchman."

"Working independently, but with faith in Newton 's law of gravitation, they calculated the mass, orbit and position of an undiscovered planet whose influence would explain the strange behavior of Uranus. The calculations took three years, but when a telescope at the Berlin observatory was finally pointed at the predicted location in the sky on the night of September 23, 1846, **the planet Neptune was there.**"

Repeatability of the Moon

Early navigators regarded the Moon as being unreliable and non-repeatable. They were right, considering the accuracy of the tables that were then available. The tables were in error by tens of miles because the Moon is affected by the varying gravitational field of the earth. Ship captains who used the Moon went on the rocks.

We now know that the Moon is just as reliable and repeatable as the Sun and the Stars. The equations that determine the Moon's position are more complex. The Nautical Almanac, generated on a large computer, is available for $13.00, and gives the position of the Moon to the same 0.1 mile accuracy as the Sun and the Stars.

Analogy Between The Moon and a Friction Way Measuring Machine

There is an analogy between the complicated repeatability of the Moon and a Moore Special Tool Co. friction way measuring machine. At one time, we believed that the Moore friction way machine was not as repeatable as the roller way machine. We now know that they will both repeat to the same magnitude (one microinch under N.C. control.)

The procedure to exploit the repeatability of the friction way machine is more complicated than it is for the roller way machine. The machine must be programmed to go to the desired position, back up a distance of .020 inches, go to the desired position, and back up, repeating the cycle three times to eliminate the oil wedges. The solenoid controlled air brakes are applied after the final approach. The screw is then backed off 100 microinches to relieve the stress in the screw and the thrust bearing. It required years of hard work to finally understand the additional complexity of dealing with the oil films and larger lead screw drive forces and to work out the solution to the problem. The story of the Moon is quite similar.

The Portas Principle

John Loxham and I loved to tell each other stories about repeatability. Jeff Portas, now Managing Director of Cranfield Precision Industries, was having lunch with us one day at Cranfield while we were discussing our favorite topic. We totally ignored him, which was not very polite, but at the end of the lunch he made a rather profound statement, which is now known as the Portas Principle:

"Random results are the consequence of random procedures."

The Portas Principle says everything that needs to be said if interpreted broadly enough. The following list indicates the typical sources of apparent non-repeatability of machine tools operating at zero load. It is important to realize that the list has a finite length. When these sources are brought under control, the machine is deterministic. The sources are as follows:

1. Random procedures
2. Thermal effects
3. Hysteresis from loose joints, cables and hoses
4. Vibration
5. Inadequate resolution
6. Dirt
7. Coulomb friction
8. Sliding oil films
9. Change of position of rolling elements
10. Variations in the supply of electricity, compressed air, vacuum, and electrical ground loops

If we speak about the repeatability of the machined workpiece, the list is longer. We must add tool wear, tool geometry, tool deflection, material variability, and workpiece clamping. We have been able to control these sources as well as the ones in the above list. The example of the X-Ray microscope described below, demonstrates this capability.

Controlling the Sources of Non-Repeatability in a Diamond Turning Machine

Random procedures must be corrected as the first step in controlling apparent non-repeatability. Humans are involved in this step since they are responsible for setting up the automatic machine. Controlling the behavior of humans can be done by using carefully written procedures and by performing critical tests to determine that each step in the setup has been correctly performed..

Thermal effects are by far the largest source of apparent non-repeatability after random procedures have been corrected. We were able to bring thermal effects under control in 1972 [10]. We built a complete box around the machine and used 40 gals.per min. of temperature controlled oil held to +/- .01 deg. F. to shower the entire machine. The drift of the structural loop between the tool and the workpiece was held to one microinch for weeks at a time.

We were able to bring vibration under control by the use of self-leveling pneumatic vibration isolators.

The problem of dirt was solved by super-fine filtering of the shower oil. These filters are able to filter the oil so that it is cleaner after a few days in the machine that it is brand new.

The problem of variations in utility supplies was solved step by step using commercial pressure regulators and power supplies.

Benefits of The Deterministic Approach in Diamond Turning an X-Ray Microscope

This machine uses a Moore, roller way, continuous path jig grinding machine for its X, Z slideway system. Leadweights are used to eliminate backlash by pulling the X and Z slideway nuts against the screw. It uses the 40 GPM liquid shower system for thermal stability. A 1000 count encoder mounted to a 100 to one gearbox on the end of the screw gives a resolution of one microinch. The slideway system was designed by Moore more than 50 years ago.

This machine has demonstrated its ability to repeatably diamond bore the same contour on an electroless nickel plated X-Ray microscope within 1.5 microinches. The feedback resolution from a rotary encoder on the acme thread leadscrew is one microinch. This repeatability (in the metal) was achieved with different operators, on different shifts, on different days of the week, making new setups for every trial.

Benefits of the Deterministic Approach in Manufacturing Microballoons

Background

Dr. Kidder was the first person (1965) to understand the possibilities of using lasers to generate temperatures high enough to ignite deuterium-tritium gas in a pure fusion reaction. According to Dr. Kidder, it will soon be possible to ignite a deuterium capsule and achieve a gain of 100 over the laser energy striking the target for an equivalent yield of 400 pounds of high explosive. This yield is obtained for a microballoon having a typical diameter of 120 micrometers, a wall thickness of 3 micrometers and a fill pressure of 200 atmospheres of deuterium-tritium. (The microballoons are filled with gas at a temperature which allows the gas to pass through the solid glass wall by diffusion.)

The Hendricks Process For Blowing Microballoons

In a recent interview, Dr. Charles Hendricks, a retired LLNL physicist, told the following story about the development of a deterministic process for blowing microballoons:

In the early days of laser fusion research, large quantities of randomly manufactured microballoons were purchased from commercial sources. The ones that met our requirements for roundness, concentricity, wall thickness, and diameter had to be selected. The **acceptance rate was one in 10^{11}**. Needless to say, this situation was not very satisfactory!

Dr. Hendricks believed that a deterministically designed blowing tower could be built to solve this problem. Most of his colleagues said that it couldn't be done, because, they said, "the process of blowing microballoons is inherently random."

Dr. Hendricks had extensive experience with cloud chamber physics, glass chemistry, glass blowing, and pottery making. He called on this experience to design and build a microballoon blowing process that was deterministic and could make microballoons to order. After three years of hard work and dedication to the laws of physics, the **reject rate was one in 10^3**.

Fourteen variables have to be controlled to achieve this performance. The process cycle time is about three seconds from raw material to finished product! In my opinion, this accomplishment ranks with the discovery of Neptune.

The process begins by mixing the glass ingredients such as, silicate, sodium, potassium, boron, etc. with water. The mixture is forced through a capillary tube and emerges as a continuous jet. The capillary tube has a piezo- electric actuator which induces an ultrasonic vibration into the jet. As the jet flows vertically downward through the 15 foot long tower, it breaks up into droplets whose mass is determined by the wavelength of the vibration and the diameter of the jet. The silicate and water droplets are small since they are destined to provide the right amount of material to make hollow microballoons having a 3 micron wall thickness and 120 micron diameter.

Surface tension causes the solid droplets to become spherical as they fall downward through the tower which is filled with clean, dry air at atmospheric pressure. Electric heaters in the tower raise the temperature of the droplets so the water can evaporate. The temperature in the next section of the tower is rapidly increased so the ingredients in the outer surface of the droplet melt and become molten glass. The inner material is now sealed. As it outgasses, small bubbles are formed which attach themselves to each other and finally make one large bubble. At this point all the material has melted and the hollow molten sphere continues to fall through the tower which has sections of higher and lower temperatures. The higher temperatures cause the balloon to expand, while the lower temperatures cause it to contract. This expansion and contraction creates an effect which causes the wall thickness to become uniform. The final temperature in the tower determines the final diameter just before it is quenched."

The microballoons then have to be inspected. Reporting the results of the inspection in statistical terms is not a problem for a determinist. Calculation of statistical parameters such as the average diameter, the dispersion of diameters about their mean, the range, etc.are not controversial. Its the same as reporting baseball players's statistics. Dr. Hendricks reports, however, that inspecting one microballoon in a batch gives essentially the same information as inspecting all of them.

After inspection, the microballoons are pressurized and leak tested. Their burst safety factor must be certified just like any other pressure vessel. This is done by filling different vessels to different pressure levels and recording which ones burst at atmospheric pressure and temperature.

The Hendricks story tends to confirm The Loxham Principle which says that an "automatic manufacturing process" is always operating perfectly etc. etc. My own experience has been limited to automatic machine tools and measuring machines. After hearing the Hendricks story, I am much more comfortable with Loxham's choice of the term "manufacturing process."

The Deterministic Position

I would like to conclude this paper by stating the deterministic position (latest version) as clearly as possible in the hope that our differences with the probabilists can be narrowed or possibly resolved:

1) A determinist will never agree that a fixed value of non-repeatability can be assigned to a given machine. Such a value does not exist. Non-repeatability depends on the time, money, and skill (culture) of the user.

2) A determinist will be happy to agree that some level of apparent non-repeatability does exist for a given machine, on a given day, for a given series of tests, programmed in a given way, conducted by a given person, in a given environment consisting of a given level of temperature variation, vibration, and dirt, using a given set of instruments with a given limitation on time and money.

3) A determinist will have great doubt and anguish in ever agreeing that some level of non-repeatability is inevitable (Bryan, Loewen, [13] 1974) regardless of the time, money, and skill available. This issue is negotiable, however.

4) If humans are continuously involved in the process, the deterministic point of view doesn't work. Dice throwing is a good example. Humans are random. Gaussian statistics work just fine.

5) If human beings are involved, and there is justifiable ignorance about the cause and effect relationships, the probabalistic approach is the only alternative. The field of medicine is a good example. Every advancement in medical research depends on probability.

What about Chaos?

In my opinion, the advent of Chaotic Theory tends to reinforce and support the Deterministic Approach. In any case, I am happy to see that the assumption of random behavior for everything we don't understand is rapidly going out of style. We need the input of the Chaotic Theorists to help us toward our goal of establishing a Generalised Deterministic Theory.

Conclusion

Writing this paper has forced me to review every idea that has come to mind over the years and to try to find some better ways to express these ideas. I hope the discussions at this conference will lead to some new proposals that will narrow our differences with the probabalists.

References

1. Bryan, McClure, Brewer, Pearson, "Thermal Effects in Dimensional Metrology," ASME Production Engineering Conference, Berkeley, Calif. Report #65-PROD-13 (1965).

2. Bryan, J. and McClure, E. , "Heat vs Tolerances" American Machinist Special Report #605 June 5, 1967

3. Bryan, McClure, Clouser, Donaldson, ' A Practical Solution to the Problem of Thermal Stability in Machine Tools" Society of Mfg. Engrs. Int'l Conf. , Chicago, Paper # MR72-138, April 1972

4. Bryan, J. , "International Status of Thermal Error Research" Annals of CIRP, Vol. 2 Berlin, 1990

5. Bryan, Bowerbank, Holland, Mohl, "Upgrading Tracer Lathe Accuracy" American Society of Tool and Mfg. Engrs. paper no. 362 Vol. 61, Book 1, 1962

6. Bryan, J. Pearson, J., "Machine Tool Metrology" American Society of Tool and Mfg. Engrs. Golden Gate Conference, San Francisco Sept. 24, 1968

7. Bryan, J. 'Interferometrically Controlled Continuous Path Tape Control for Rotary Contour Gage" Lawrence Livermore National Laboratory Report # UCRL- 7072, 1962

8 Bryan, J. "Design and Construction of an 84 inch Diamond Turning Machine" Precision Engineering, Vol I, Jan. 1979

9. J. Loxham, "The Commercial Value of Investigations into Repeatability," Cranfield Unit for Precision Engineering, Bedford, England, (1970)

10. Loxham, J. , "The Presidents Evening" Quality Engineer, May 1972, Vol 36, No. 5

11. Bryan, J. , "The Power of Deterministic Thinking in Machine Tool Accuracy" The First International Machine Tool Engineers Conference, Tokyo, Japan Nov. 7, 1984 , Lawrence Livermore National Laboratory Report #UCRL- 91531

12. Donaldson, R. "The Deterministic Approach to Machining Accuracy," Society of Manufacturing Engineers Symposium, Golden, Colorado, Nov. 1972.

13. Bryan, J. , Loewen, E. ,"The Limit of Accuracy of Machine Tools," International Conference on Production Engineering, Aug. 26-29, 1974, Tokyo, Japan. Lawrence Livermore National Laboratory Report # UCRL 75645

14. Horowitz, P. , Hill, W. "The Art of Electronics", Cambridge University Press , Second Edition 1989

CRTD-Vol. 27, International Forum
on Dimensional Tolerancing and Metrology
ASME 1993

SAMPLING ISSUES IN COORDINATE METROLOGY

**Robert J. Hocken, (Jay)araman Raja,
and Uppliappan Babu**
Department of Mechanical Engineering
and Engineering Science
Precision Engineering Laboratory
University of North Carolina–Charlotte
Charlotte, North Carolina

ABSTRACT

Coordinate measuring machines acquire data one point at a time on the surfaces of the parts being measured. These data must be analyzed in order to produce a substitute geometry to be compared to the design intent. Algorithms are used to analyze these data. The results obtained are strongly dependent upon measuring errors, the form error of the part, the sampling strategy, and the density of points sampled. In this paper we review these problems and discuss some of the progress addressing them during the past few years.

BACKGROUND

The computer-controlled coordinate measuring machine has become the inspection instrument of choice for the dimensional measurement of mechanical parts. Coordinate measuring machines allow the acquisition of accurate dimensional data on most part surfaces and readily give the relationship(s) between various part features. Coordinate measuring machines are easily reprogrammed for changes in product and product design. The results of a coordinate measuring machine inspection can also be used readily for process control [Wirtz, 1986].

Coordinate metrology has some drawbacks. A coordinate measuring machine acquires data on a point-by-point basis. These data need to be analyzed to create a geometric model, usually called the substitute geometry, of the feature being measured. This analysis is performed using algorithms whose results are strongly dependent upon many factors including systematic and pseudo-random machine error, form error (systematic deviations of the measured surface from ideal), surface finish, algorithm correctness and robustness, the sampling method, and the sampling density. Because of these effects the results reported after analysis may contain large errors even on high accuracy machines. In real inspection this problem is compounded by issues of tolerance definition and algorithm selection which are discussed elsewhere in this issue.

A few years ago some of these problems became the subject of articles in the popular press [Pond, 1988; Plazek, 1988] due to a Government-Industry Data Exchange Program (GIDEP) alert [GIDEP, 1988]. This alert sparked industrial and academic interest [Tipnis, 1989]. The problem is fairly complex but can be divided into three broad areas: establishing the relationship between the design intent and the data acquired by the inspection procedure; controlling the influence of the measuring system and the sampling algorithm on the inspection results; and identifying systematic biases in the results, due to the inspection procedures and algorithms chosen for analysis. In this paper we will concentrate upon the last two issues and report on some of the progress that has been made in this area since 1988. First we will explain the form and magnitude of the normally encountered machine and part errors. The concern here is that measurement data contain systematic and pseudo-random errors that ultimately must affect the sampling strategy. Next we will briefly discuss the algorithms in common use and their sensitivity to parameterization. This will be followed by a discussion on sampling where we report the results of some of the computer and real experiments that have been performed.

SYSTEMATIC AND PSEUDO-RANDOM MACHINE ERRORS

Geometric/Parametric Errors

Nearly all precision coordinate metrology instruments/machines are composed of one or more moving elements designed to position a measurement probe with respect to a workpiece [Hocken, 1980]. It is traditional to design each positioning element such that it behaves as a rigid body with five of its six degrees of freedom eliminated, then to drive the element in the remaining direction and measure its motion in that direction as accurately as possible. Normally the motion desired is a pure linear or rotary motion. Complicated forms, like gears, threads, turbine blades, and the like, are measured by coordinated action of the different elements. Errors in the motion of an individual element are referred to as "parametric errors," and there are Standards dealing with the parametric errors of linear axes and rotary axes; most specifically, ANSI/ASME B5.54 [ANSI/ASME, 1992] and ANSI/ASME B89.3.4M [ANSI/ASME, 1985], as well as technical papers [Beckwith, 1991; Bryan, 1968a, 1979a, 1989; Hocken, 1977; Lane, 1969; Tlusty, 1959, 1980b]. For example, a linear carriage has six degrees of freedom, five of which are constrained. Motion in the unconstrained direction may have an error which is normally referred to as a "scale error," since this motion is measured by some type of scale. Linear motions orthogonal to the intended motion direction are called "straightness errors" or "straightness motions," and rotations of the carriage are called "angular errors," usually yaw, pitch, and roll, analogous to terms used in aircraft. Rotary axes have similar errors but different nomenclature [Bryan, 1967; ANSI/ASME, 1985]. Besides parametric errors of individual axes, there are also errors in the squareness or parallelism required between axes. For a typical three-axis machine, there are 21 degrees of freedom, six for each axis plus three squareness terms. These "geometric errors" or "parametric errors" are highly systematic.

Machine error characterization is now well established. First a model of the machine is created using either matrix algebra [Hocken, 1977; Zhang, 1985] or homogeneous transformation matrices [Donmez, 1986a; Soons, 1992]. Next each of the 21 error terms of the machine is measured using laser interferometry and other instrumentation [Caskey, 1991, 1992a, 1992b]. Typical "shop-floor" type measuring machines have systematic errors of many microns which approach current tolerances on a wide variety of manufactured parts [Swyt, 1992]. Thus, in a great many cases, either these errors must be reduced or sampling strategies devised that allow error cancellation.

Probe Errors

Coordinate measuring machine probes also introduce systematic error. This is true for all probe types, including optical probes. In spite of some advances, the most common probes in use are switching probes. Switching probe characterization is normally accomplished by measurements on a precision test sphere [Caskey, 1991; Wilson, 1992; ANSI/ASME, 1990; Krejci, 1990] which is round to better than 0.1 micrometer. The ball is probed using the manufacturer's suggested probe approach rate and probe approach distance over the complete vertically accessible hemisphere. Typically, switching probes yields a three-lobed pattern with several microns of out-of-roundness [Harary, 1989; Wilson, 1992]. Note that such lobing is also of the order of magnitude of part tolerances. Studies have been conducted varying other parameters such as probe approach rate and distance, stylus length, and stylus orientation. They demonstrate that these effects can be many times the tolerance desired for the measurement [Phillips, 1993].

Thermal Errors

A third broad category of errors which affect CMM performance are "thermal errors." Thermal effects which cause changes in the relative position between probe and workpiece can be comparable in size and type to those that result from geometrical imperfections of the machine. They arise from a wide variety of sources, including sources not directly related to the measuring processes at all, such as room temperature, lights, pneumatics, etc. A very comprehensive description of these effects was given some years ago by Bryan [Bryan, 1968b] and thermal effects in metrology are a subject of national standards [ANSI/ASME, 1973].

There have been many efforts to understand the effects of thermal errors on coordinate measurement. A good review is available in [Bryan, 1990]. Many researchers have also contributed [Balsamo, 1990; Sartori, 1989; Lingard, 1991]. Thermal effects distort the machine geometry and lead to changes in the parametric errors discussed above. In real shops these errors often dominate. Since they are time varying, they create measurement "noise" that pseudo-randomizes the measurement results. It is rare that errors produced yield normal distributions, thus making analysis less tractable.

Other Errors

Many other errors are also present in measuring machines. They all yield systematic or pseudo-random type errors. A good overview of these errors was given recently in the literature [Phillips, 1993]. Besides the errors above, Phillips, et al., include (among others) effective stylus diameter (set by probe calibration), offset vectors between different styli, probe repeatability, stylus bending, indexable probe head repeatability, stylus changer repeatability, probe spatial frequency response and fictional effects, non-rigid-body geometric errors, CMM part loading, CMM dynamic behavior, CMM repeatability, other environmental factors, variation of utility services, part dynamics, part fixturing, and part thermal properties.

Machine Errors - Conclusions

The above discussion points out the fact that the errors in the measurement system are sufficiently large that they can appreciably affect the measurement results. The fact that most of the errors are systematic or pseudo-random places particularly difficult constraints upon the appropriate sampling strategy. This will be discussed more fully in Section 5.

SURFACE AND FORM ERROR IN MANUFACTURED PARTS

The manufacturing processes used to generate mechanical parts leave a fingerprint on the geometrical surface of the part which is a combination usually expressed as a range of frequencies and amplitudes. The high-frequency component is usually referred to as the surface roughness, the medium-frequency component as the waviness, and the low-frequency component as the form error. The surface roughness parameters, such as R_a (average roughness) and S_m (average wavelength), together with the roughness cut-off value, can be used to establish an upper limit on the number of points required for an accurate form error measurement. (This assumes that the radius of the CMM probe is comparable in size to the cut-off wavelength used for the measurement of surface roughness. This prevents inclusion of high-frequency components in the form error measurement [Anbari, 1989].) The use of surface finish as a process characterization parameter has led to a reasonable understanding of what surface finish can be expected from different manufacturing processes. Bar graphs showing the ranges of these expectations are available in textbooks [Kalpakjian, 1989].

There is less guidance regarding the form error generated by manufacturing processes. There is extensive literature on the errors present in machine tools of all kinds (for example, see [Hocken, 1980; Bryan, 1968a; Burdekin, 1988]); however, only in certain special cases has a detailed relationship between manufacturing process and part form error been developed. For the most part, these relationships have been restricted to simple turning and boring [Kumaran, 1992]. More complex relationships are just being examined [Salsbury, 1991]. Experienced process engineers, however, surely know the capabilities of different processes. That is, specific processes certainly more commonly produce particular form errors and these form errors should have predictable magnitudes. In order to ascertain the breadth of knowledge in this area we conducted a literature search [Caskey, 1991]. The results were disappointing, though data were found on the influence of surface roughness [Anbari, 1989], form errors in roundness from different processes [Farago, 1982], errors in measuring curved surfaces [Morgan, 1984], and generalized discussions of geometrical errors in certain types of machining processes [Kang, 1976, 1978; Hocken, 1980]. Considering the number of manufacturing processes and the variety of forms produced, this search was not as successful as we had hoped. Results of an international survey were equally sparse. Over 200 "experts" were contacted. Solid data were supplied [Osanna, 1990], but the data reduction had been such that only error magnitudes could be assessed. As with the machine measurement errors these error magnitudes are of the order of many manufacturing tolerances.

SUBSTITUTE GEOMETRY AND ALGORITHMS

Two types of algorithms are in use by coordinate measuring machine manufacturers and users. The first type, and the most common one, is simply the maximum likelihood (Gaussian, least-squares, best fit, etc.), that is, the minimization of the sum of the squares of the deviations of the measured values from the mathematical expression of the "ideal" geometry. The second type, normally called Mini-Max, extrema, or Tschebycheff, minimizes the perpendicular distance between the measured points and the "ideal" geometry. These Mini-Max algorithms are really only used for very specific cases, such as circular geometry, and are not normally supplied with measuring machines. A subset of the Mini-Max algorithms, the maximum inscribed feature and the minimum circumscribed feature, are used for particular forms (circles, cylinders, spheres). Many other methods of analyzing the data are possible (for example, see [Daniel, 1980]). Most metrologists concentrate on the Gaussian and the Tschebycheff methodologies. Least-squares algorithms have been developed by the National Institute for Standards and Technology (NIST) [Hopp, 1993], the National Physical Laboratory (NPL, UK) [Chetwynd, 1985; Cox, 1983, 1985; Forbes, 1989], and the Physikalisch Technische Bundesanstalt (PTB, Germany) [Porta, 1986]. At UNCC we wrote algorithms, both least-squares and Mini-Max, for the basic features which are the 3D line, the 2D circle, the 3D circle, the 3D plane, the cylinder, the sphere, and the cone [Caskey, 1992]. For the least-squares we used the methods suggested by Forbes [Forbes, 1989] and for the mini-max the methods of Anderson, Osborne, and Watson [Osborne, 1967]. Parameterization of the elements followed the suggestions of British Standard 7172 [BSI, 1989]. The algorithms were tested using data sets provided by PTB [Porta, 1986]. The least-squares algorithms were compared to those of NIST and after some alterations good agreement was obtained.

Algorithm Testing

Algorithm testing is a difficult problem in itself. In order to test algorithms, researchers have developed test procedures which normally involve generating test data sets, running these data sets through the reference algorithms, comparing the results of those algorithms to the "correct" answer, and graphically displaying the results [Porta, 1986; Diaz, 1992]. Similar generators are also used to estimate the effects of sampling. In those types of generators form error and measurement uncertainty are added and the sampling methodology is altered. This is discussed more fully in Section 5.

Algorithm Stability/Errors

There are two important issues regarding algorithmic errors. The first simply deals with algorithm selection. Most current tolerancing schema, particularly those that are standardized, specify a particular geometrical interpretation for a tolerance call-out. For example, in the case of the cylinder the desired representation might be the minimum circumscribed cylinder, or, in the case of the hole, the maximum inscribed cylinder. Since most coordinate measuring machines only perform least-squares fits, the algorithm selected would not give the correct feature size. The other issue is simply one of implementation. Most of the problems encountered in fitting features are non-linear, and recursive techniques need to be used to establish the correct fit parameters. Such recursive techniques are generally slow, and algorithm designers often attempt to speed up the operation by linear approximations. This can lead to algorithms that give incorrect results [Porta, 1986; Phillips, 1993].

SAMPLING

The above discussion was intended to set the stage for the discussion of sampling issues. In the general case we are faced with attempting to measure a part feature or true position with a machine measurement system containing systematic and pseudo-random error. Further, the form of the feature deviates from ideal over a range of wavelengths and amplitudes that are representative of a manufacturing process (often unknown). We are going to use algorithms to fit these measurement data that never can be completely tested and, in some cases, are quite sensitive to "outliers," a term that is more subjective than quantitative. Finally, in production, we want to measure this part as quickly as possible, i.e., with the minimum number of points.

Before addressing this issue, it should be pointed out that in the discipline of metrology a great deal of effort has been expended in developing measurement methodologies which are quite successful at reducing errors due to sampling. We refer here to what are commonly called "reversal techniques" or "multiple redundancy." These methods effectively over-sample to obtain the required measurements. We discuss these briefly before discussing production sampling strategies.

Metrological Sampling Strategies

The problem of sampling to obtain true position and or form in the presence of machine error has been often called self-calibration or self-correction. Some of the earliest examples are lost in antiquity; for example, the so-called "three flat" method by which a perfect flat surface may be generated or measured by intercomparison of three non-flat surfaces [Moore, 1970]. Another very old method, called "straightedge reversal," was so common to the craftsman that only recently has it been discussed in the technical literature [Estler, 1985a; Bryan, 1989]. This reversal principle was applied to roundness measurement two decades ago

[Donaldson, 1972]. Other authors have examined multi-position measurement [Reeve, 1979; Reason, 1966] and multi-probe measurements [Whitehouse, 1976; Mitsui, 1982] for the same purposes. Opticians traditionally have used rotation and shear tests [Reid, 1992] in order to measure optical form. In all these cases, the concept is similar; that is, given a stable artifact and a measuring machine or method with high repeatability, the exact geometry of the machine or the artifact may be deduced by cleverly arranged multiple measurements of the same artifact. Although the concept has been widely used, no generalized theory has ever been developed.

In the 1970s, this concept of multiple measurements was applied to coordinate measuring machines [Reeve, 1974; Hocken, 1979]. The procedures were drawn from those used in photogrammetry to deduce camera locations and to calibrate camera lens distortions [Brown, 1958, 1967]. These procedures are still in use calibrating ball plates at NIST and in camera calibration for vision systems [Doiron, 1986]. Other researchers applied these techniques to machine tools, using curvilinear coordinate systems to represent the machine errors, and then measuring artifacts on the machine to deduce the curvature parameters [Clément, 1981; Juoy, 1986]. Simultaneously to this work, researchers were addressing the mapping of the imperfect geometry of these machines using parametric error calibrations [Hocken, 1977; Zhang, 1985, 1988; Estler, 1985b; Beckwith, 1991]. Sometimes the redundant artifact measurements were used to check the imperfect geometry deduced from parametric calibration [Hocken, 1977] or procedures devised which combine both techniques [Soons, 1992]. This method is also used in the lithography community for mapping errors in e-beam stages [Raugh, 1985] and was used to measure ball plates as part of an international round robin [Caskey, 1992b]. It is important to realize that these methods could be applied to the problems discussed in this paper, except that they require considerable education of the machine operator and also double or triple the measurement time.

Production Sampling Strategies

The task here is to develop strategies for the distribution of data points to cover the range of characteristic shapes expected from the results of machine and part error analysis, while using the minimum measurement time. Sampling theory is well developed in electrical engineering [Cochran, 1976] and several statisticians have provided textbooks for the more general cases [Rabiner, 1975]. Redundant sampling techniques were discussed in Section 5.1. In production coordinate measurement, however, due to the limitations in speed of most machines, much more limited sampling, i.e., under-sampling, is desirable. Several possible sampling strategies have been addressed in the literature. These include uniform sampling, random sampling, stratified sampling and other more esoteric schema (such as Hammersley sampling [Woo, 1993]). Stratified sampling is a variation on uniform sampling, where the surface is divided into uniform length intervals (for a line) or area elements (for a surface) and the point chosen for each interval or area is randomly selected within the length interval or area element [BSI, 1989; Cochran, 1976].

It should be pointed out that on real coordinate measuring machines the software provided by the manufacturer is somewhat limited. It is much easier to program a measuring machine to do equal intervals in angle or space than it is to use any of the other sampling techniques. In our discussion below we will concentrate primarily on uniform sampling, since it has proven to be most practical for normal users.

"Experimental" Methodology

There are several possible approaches to examine the effects of sampling in the presence of form and machine error. The most obvious is to simply acquire a large number of parts, measure them using different sampling strategies, and compare the results. This has been done for cylindrical geometries [Wilson, 1992]. Another approach is to attempt analytic solutions [Phillips, 1994] as was done for spheres. A more common approach is to examine these issues using computer experiments. This requires creating a data generator to produce data on imperfect features, fitting these features to tested algorithms, and comparing the results using the output parameters from the fits. Results from these types of experiments are discussed below. These experiments were conducted using machine errors and form errors commensurate with those described in Sections 2 and 3, using the algorithms discussed in Section 4.

Data generator for computer simulations. The generator normally begins by constructing a perfect feature and then adding systematic and random error to the feature. The feature is then sampled according to different strategies and with different data densities in order to create the data set to be used for the computer experiment. Randomness is added by varying the positions of the sampling points with respect to the form error. The procedure is illustrated in Fig. 1 for circles and cylinders.

Feature parameterization and representation of results. Because of the large number of different test cases and features involved, it is necessary to develop standard means of parameterizing each of the features and physically meaningful, graphical methods for displaying the results obtained from a fitting procedure. Most commonly the representations use a point and direction cosines and in some cases extra parameters following national and international standards such as IGES and DMIS. For example, the cone is specified by the coordinates of a point on the cone axis, the direction cosines of the axis, and the cone angle. A complete description of these parameterizations is part of the working documents of the ANSI/ASME B89.4.12 Committee.

Representing the results in a concise and physically meaningful manner also presents some difficulties. The first of these is deciding upon the "correct" answer. We define the correct answer as the answer obtained in the presence of systematic error only, with "infinite" sampling density. Other definitions are possible. Using our definition, the correct answer is not always the original "ideal" feature. For example, consider the slope of the least-squares best-fit line to a line with one complete cycle of sinusoidal form error. In some cases, such as lines and spheres fit with least-squares algorithms, analytical solutions can be found [Phillips, 1994], but for the other cases we had to use numerical methods.

In order to display the results of many simulations, graphical representations that would be physically meaningful were developed. Some obvious ones were quickly discarded. For example, on the 3D line an obvious one was simply differences between the true mean coordinates of the line and the fit mean coordinates. However, this is not really physically meaningful as, if the true line and the substitute line had any point in common and had the same direction cosines, the fit would be perfect. We therefore chose sets of parameters that were physically meaningful (see [Caskey, 1991] for a complete list).

Current Results

Lines. Some years ago we reported on our results for the 3D line [Caskey, 1991; Palanvelu, 1990]. We used a line of 1,000 mm and varied the size of the sample, the amplitude and spatial frequency of a sinusoidal form error, and amplitudes of random errors drawn from a uniform distribution. The line was rotated to an arbitrary position in 3D space. Fits were performed varying the sample size, the number of cycles of form error, the form error magnitude, and the random error magnitude. Even in this simple case, the parameters described in the fit line did not converge to stable values until a large number of points (over 50) were sampled. **In order to establish a datum line, most inspectors use two points, not 50.** This theme will be repeated throughout the Results section of this paper.

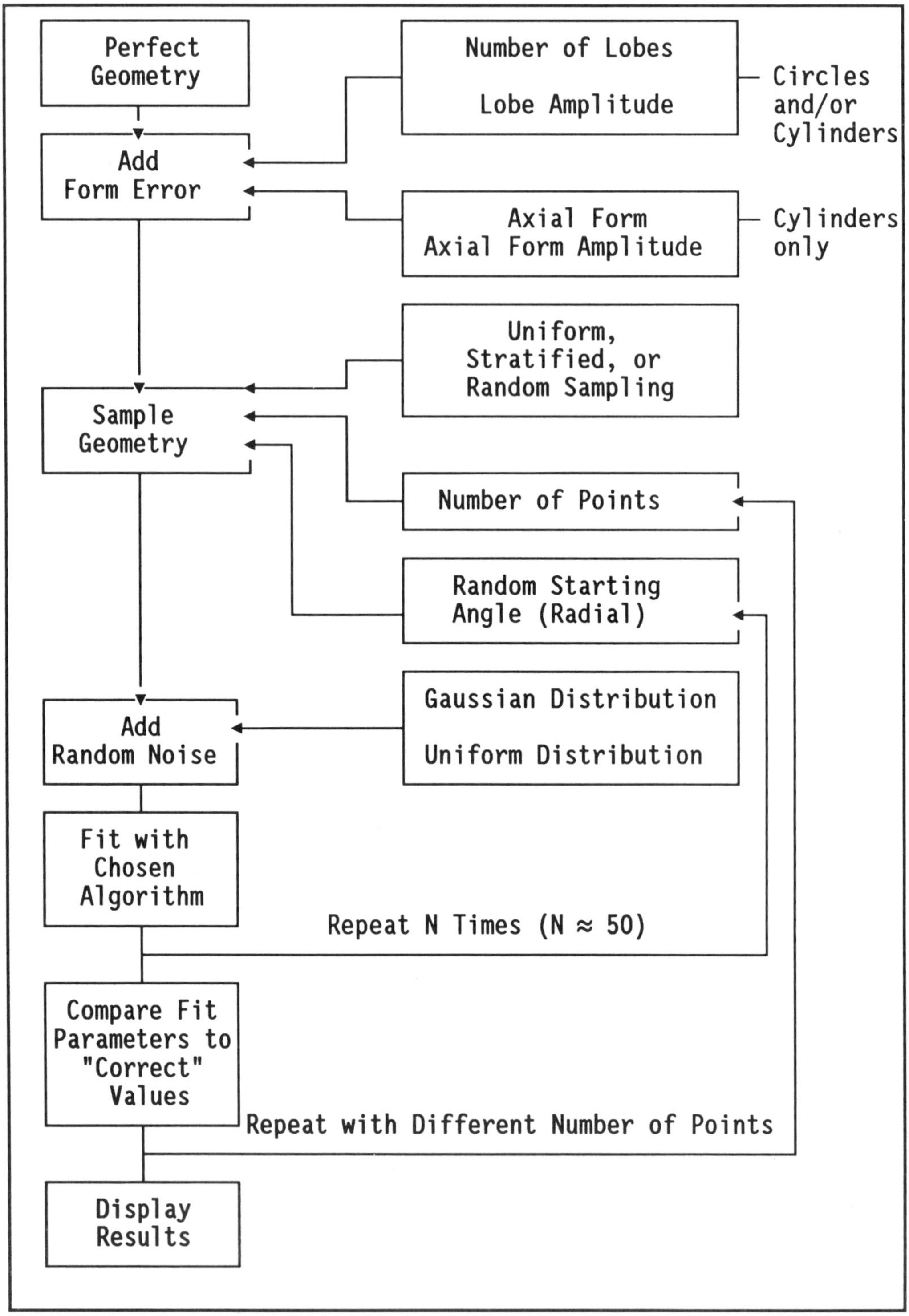

FIGURE 1. A BLOCK DIAGRAM OF A TYPICAL PROCEDURE FOR PERFORMING COMPUTER SIMULATIONS. THIS DIAGRAM IS FOR A CIRCLE OR CYLINDER MEASUREMENT.

Planes. Similar experiments were conducted on planes [Machireddy, 1993]. Sinusoidal form errors and random errors from a uniform distribution were added to the true plane, and the results were sampled and fit to both least-squares and Mini-Max algorithms. Both uniform sampling and stratified sampling were performed, and sample sizes ranged from four to 49 points. Parameters describing a typical plane were form error of amplitude 0.01 mm in the X and Y directions, with random error of the same order. The form error periodicity was varied from one to three cycles.

For planes we were more concerned with the angular deviation between the normal of the fit plane and the original plane since planes are used for datum surfaces and any angular errors resulting from this fit would be magnified by offsets when other features on a part were measured. In these experiments we found that the range of the angular deviations was also quite large until nearly 50 data points were acquired. **Again, inspectors often use only the minimum number of points (three) to establish such a plane.** Form errors of 10 micrometers (such as used in the simulation) would be common on milled surfaces.

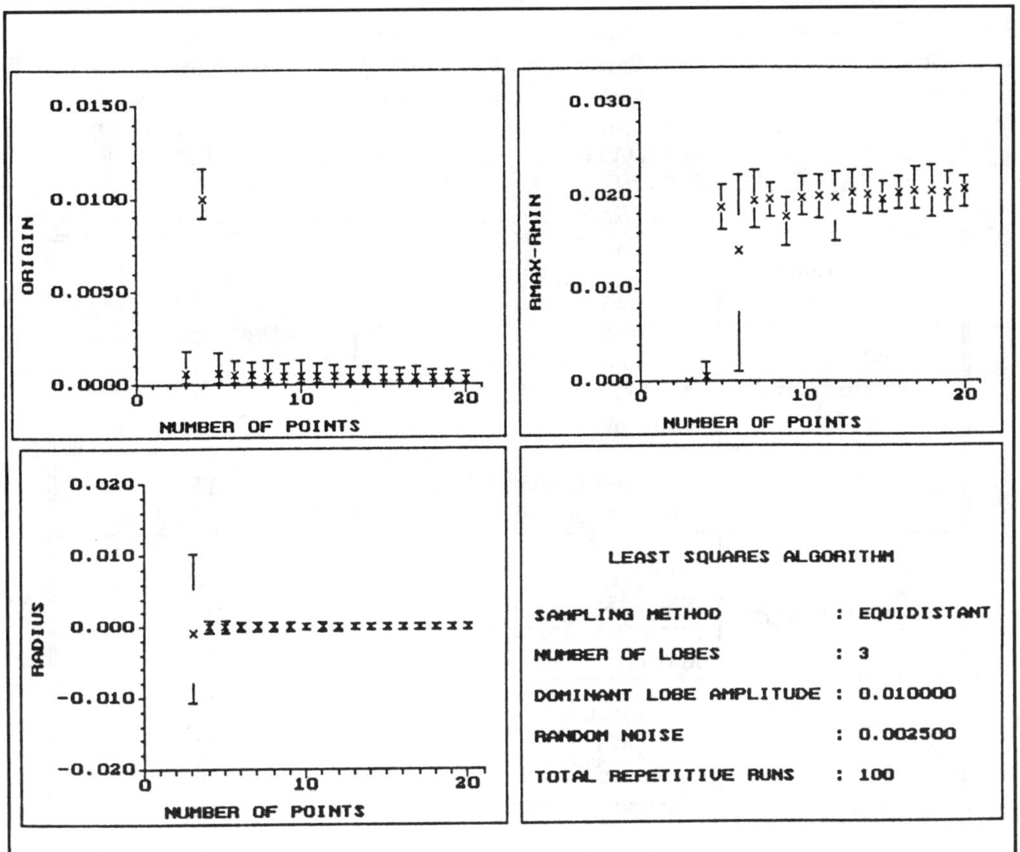

FIGURE 2. RANGES OF RESULTS WHEN FITTING A THREE-LOBED CIRCLE WITH RANDOM NOISE. THE STARTING ANGLE OF THE UNIFORM SAMPLING IS VARIED RANDOMLY 50 TIMES FOR EACH NUMBER OF POINTS.

Circles. Circles have been more extensively studied than any other geometry. A particular concern has been the three-lobed circle since it is commonly produced in cylindrical grinding and turning using a three-jawed chuck. An analysis showing the results obtained from different algorithms on the three-lobed circle is given in the literature [Weckenmann, 1991]. Extensive computer experiments were also performed at Charlotte covering circles with many different lobe patterns of different amplitudes. Different levels of random noise were also added [Odayappan, 1992]. This work was funded in part by the DITS Program of CAM-I [DITS, 1990]. A typical result is shown in Fig. 2 which consists of plots showing the range in the origin, the out-of-roundness, and the radius in a three-lobed circle with equidistant sampling. The starting point was drawn randomly. Note that certain patterns emerge. These patterns are caused by aliasing-type effects. For example, one should never use four points to measure a three-lobed circle, as the correct origin will not be ascertained. These results were confirmed by CMM measurements on actual parts [Wilson, 1992].

Due to the symmetry present in the problem of the lobed circle, it was possible in this one case to make some fairly concrete recommendations regarding sampling. Table 1, reproduced from Odayappan [Odayappan, 1992], shows his recommendations for circles with different numbers of lobes and different algorithms. Note that the maximum inscribed circle is unstable for two lobes, which can be easily seen by examination of the geometry.

TABLE 1. RECOMMENDATIONS REGARDING THE ACCEPTABLE SAMPLING DENSITIES FOR CIRCLES IN THE PRESENCE OF FORM ERROR [ODAYAPPAN, 1992]

No. of Lobes	Algorithm	No. Points to Limit Sampling Uncertainty to 10% of Form Error Amplitude
2	Least-squares Mini-Max Minimum Circumscribed Maximum Inscribed	7, 9, 11 13 13 (Unstable)
3	Least-squares Mini-Max Minimum Circumscribed Maximum Inscribed	7, 11 11, 13 11, 13, 14 11, 13, 14
4	Least-squares Mini-Max Minimum Circumscribed Maximum Inscribed	7, 9, 11, 13 14 14 19
5	Least-squares Mini-Max Minimum Circumscribed Maximum Inscribed	7, 9, 11 17 12, 13 17

Spheres. A study on the interaction of form error in sampling strategy for spheres has been conducted at the National Institute of Standards and Technology (NIST) [Phillips, 1994]. In this paper they examine the interaction between the sampling strategy and the form error and introduce a formalism which expresses these interactions. The major result is that sampling strategies can be found which completely suppress the form error in specific situations where the coordinates of the center of the sphere are the desired parameters. Radius variations can also be suppressed in particular situations. The situation is somewhat analogous to that of circles, although they obtain their results primarily analytically, rather than through the use of computer simulation.

Cylinders. In cylindrical parts, form errors are present in both axial and radial (lobing) directions. For most manufacturing processes, the dominant lobing is of low order (two to seven) and the lobe amplitude reduces dramatically with the number of undulations per revolution [Wilson, 1992]. The axial form errors are also low order and can be classified as barrel, hourglass, banana, undulated, and bell mouth. As with the circles, we conducted a computer simulation to assess sampling effects [Babu, 1993]. Some typical results (reprinted from [Babu, 1993]) are shown in Fig. 3. Figure 3 shows the simulation results on a three-lobed, barrel-shaped cylinder where different numbers of points per level were used and the number of levels was varied up to 1,000. The approach to the "correct" answer is discouragingly slow.

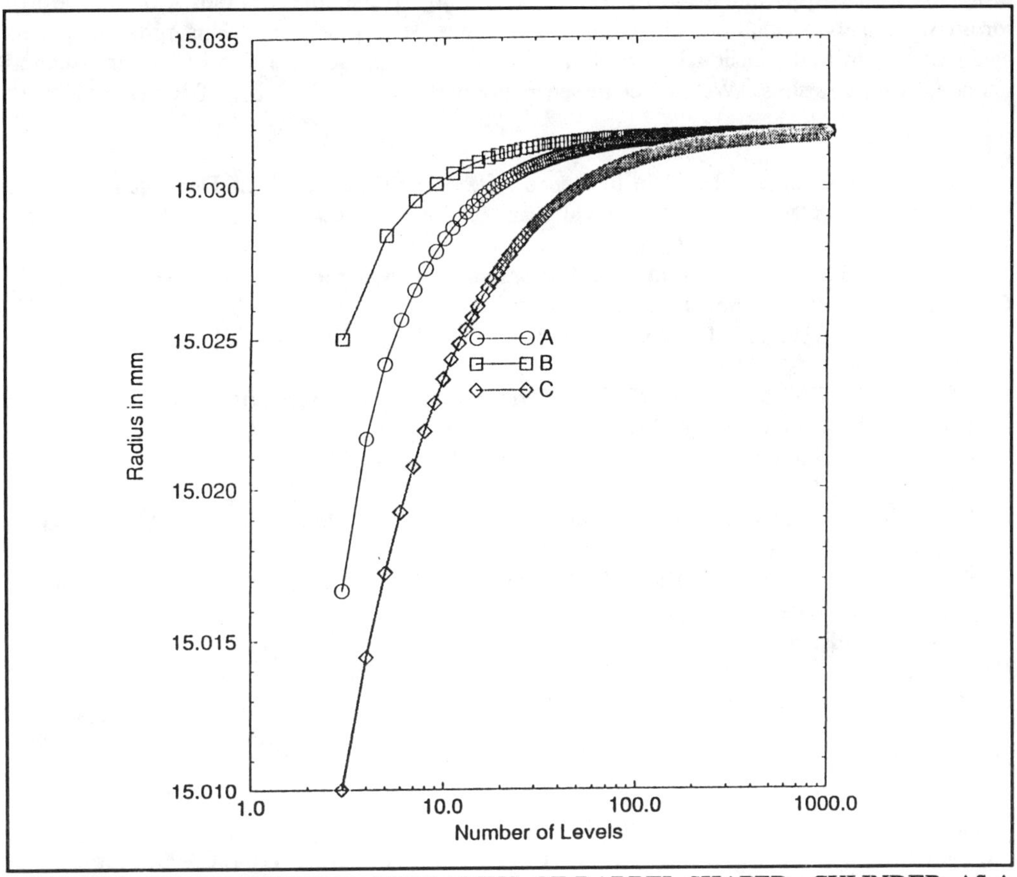

FIGURE 3. LEAST-SQUARES FIT RADIUS OF BARREL-SHAPED CYLINDER AS A FUNCTION OF NO. OF LEVELS, NO. OF PTS./LEVEL. A: 5 PTS./LEVEL; B: 10 PTS. MID LEVEL, ALL OTHERS 5; C: 10 PTS. TOP/BOTTOM LEVELS, OTHERS 5.

CONCLUSIONS

Compared to 1988 there are now a large number of workers concerned with the limited sampling capabilities of CMMs and their effects on part measurement. From what we have seen so far the results of this work can be summarized succinctly (if slightly over-simplified). Current inspection techniques, used daily in manufacturing, drastically under-sample geometric features in the presence of unknown part form and measuring machine errors. This leads to two corollaries:

- In order to accurately measure part geometry in the presence of unknown normal form error, much higher sampling densities than currently used must be incorporated. This will require, if inspection times are not to be drastically increased, a new generation of machines capable of high-speed scanning of part surfaces.
- Decisions regarding how to measure a part and the choice of algorithm(s) for data analysis are often counter-intuitive [Phillips, 1994]. This leads to the conclusion that intelligent decision systems or procedures need to be used to control the inspection and analysis process.

There is also a great deal of work that still needs to be done. One of the most pressing problems is that of partial features; that is, attempting to measure the radius of a sphere or toroid when only a small section of the surface has been produced. This problem occurs again and again in dimensional inspection. Interrupted features are also of interest, such as cylinders with key slots. We expect to see many new results in the next few years.

REFERENCES

Anbari, N., Trumpold, H., "The Influence of Surface Roughness in Dependence of the Probe Ball Radius Measuring the Actual Size," CIRP Sci. Tech. Comm. Rpt. (Surf. Met.), 1989.

ANSI B5.54-1992, "Methods for the Performance Evaluation of Computer Numerically Controlled Machining Centers," ASME, New York, 1992.

ANSI/ASME B89.1.12M-1990, "Methods for Performance Evaluation of Coordinate Measuring Machines," ASME, New York, 1990.

ANSI/ASME B89.6.2-1973(R1979), "Temperature and Humidity Environment for Dimensional Measurement," ASME, New York, 1973.

ANSI/ASME B89.3.4M-1985,"Axes of Rotation," ASME, New York, 1985.

Babu, U., Raja, J., Hocken, R. J., "Sampling Methods and Substitute Geometry Algorithms for Measuring Cylinders in Coordinate Measuring Machines," Proc. Ann. Mtg. ASPE, Seattle, WA, Nov. 1993.

Balsamo, A., Marques, D., Sartori, S., "A Method for Thermal-deformation Corrections of CMMs," CIRP Annals, 1990.

Beckwith, W., Bryan, J., Callaghan, R., Charlton, T., Taylor, B., "Parametric Calibration of Coordinate Measuring Machines," ASME Tech. Paper, 1991.

Brown, Duane C., "Computation Tradeoffs in the Design of a One Micron Plate Comparator," pres. at the 1967 Semi-ann. Conv. of Amer. Soc. of Photogram. and The Amer. Cong. on Survey. & Mapping, St. Louis, Oct. 2-5, 1967.

Brown, Duane C., "A Solution to the General Problem of Multiple Station Analytical Stereotriangulation," AFMTC TR 58-8, Feb. 1958.

Bryan, J. B., Pearson, J. W., "Machine Tool Metrology," ASME TP 1068-753, 1968a.

Bryan, J. B., "International Status of Thermal Error Research," CIRP Annals, 1968b.

Bryan, J., "The Abbe Principle Revisited--An Updated Interpretation," Prec. Eng. 1, 3, 1979a.

Bryan, J., "International Status of Thermal Error Research," CIRP Annals, 39/2, 1990.

Bryan, J., Carter, D., "How Straight is Straight," Am. Mach., Oct. 1989.

Bryan, J., Clouser, R., Holland, E., "Spindle Accuracy," Am. Mach. Spec. Rpt. No. 612, 1967.

BSI, "Assessment of Position, Size, and Departure from Nominal Form of Geometric Features," British Standards Inst., BS 7172, 1989.

Burdekin, M., Park, J., "Contisure - A Computer Aided System for Assessing the Contouring Accuracy of NC Machine Tools," Proc. 27th Int'l. MATADOR Conf., April 1988.

Caskey, G., Hari, Y., Hocken, R., Machireddy, R., Raja, J., Wilson, R., Zhang, G., Chen, K., and Yang, J., "Sampling Techniques for Coordinate Measuring Machines," Proc. 1992 NSF Design & Mfg. System. Conf., Atlanta, GA, Jan. 1992a.

Caskey, G. W., "A Streamlined Method for the Parametric Calibration and Compensation of a Coordinate Measuring Machine," Masters Thesis, UNC Charlotte, Charlotte, NC, 1992b.

Caskey, G., Hari, Y., Hocken, R., Palanvelu, D., Raja, J., Wilson, R., Chen, K., and Yang, J., "Sampling Techniques for Coordinate Measuring Machines," Proc. of the NSF Conf., 1991.

Chetwynd, D. G., "Applications of linear programming to engineering metrology," Proc. Instr. Mech. Engrs. 199, B2, 1985.

Clément, A., and Bourdet, P. (submitted by Weill, R.), "Commande adaptive dimensionelle d'une machine a mesurer tridimensionelle," Annals of the CIRP 30, 1, 1981.

Cochran, W. G., Sampling Techniques, Wiley & Sons, New York, NY, 1976.

Cox, M. G., and Jackson, K., "Algorithm and software for engineering metrology: A statement of need," National Physical Laboratory, UK, Report MOM 65, 1983.

Cox, M. G., "Software for computer aided measurement," Laboratory Practice, National Physical Laboratory, UK, June 1985.

Daniel, Cutchbert, and Wood, Fred S., Fitting Equations to Data: Computer Analysis of Multi-factor Data, 2nd Edition, John Wiley & Sons, New York, 1980.

Diaz, C., "User's Guide for the Algorithm Testing System/Version 1.1,"NISTIR 5137, Nat'l. Inst. of Stds. and Technol., Gaithersburg, MD, Nov. 1992.

DITS: "Quality Assurance Program: Dimensional Inspection Techniques Specification (DITS)," Proposal, CAM-I, Arlington, TX, 1990.

Doiron, T., "Computer Vision in Dimensional Metrology at NBS," Proc. Meas. Sci. Conf., Irvine, CA, Jan., 1986.

Donaldson, R. R., "A Simple Method for Separating Spindle Error from Test Ball Roundness Error, UCRL-73805 Preprint, LLNL, Livermore, CA, 1972.

Donmez, A., Blomquist, D. S., Hocken, R. J., Liu, C. R., Barash, M. M., "A general methodology for machine tool accuracy enhancement by error compensation," Prec. Eng., 1986a.

Estler, W. T., "Calibration and use of optical straightedges in the metrology of precision machines," Optical Engineering 24, No. 3, May/June 1985a.

Estler, W. Tyler, Magrab, Edward B., "Validation Metrology of the Large Optics Diamond Turning Machine," U.S. Department of Commerce, NBSIR 85-3182(R), Final Report, 1985b.

Farago, F. T., Handbook of Dimensional Measurement, Industrial Press, Inc., NY, 1982.

Forbes, B. A., "Least-squares best-fit geometric elements," National Physical Laboratory, UK, Report DITC 140/89, 1989.

GIDEP: Government and Industry Data Exchange Program (GIDEP) Alert No. X1-A-88-01, August 22, 1988, and its Addendum, GIDEP Alert No. X1-A-88-01A, September 20, 1988.

Harary, H., Creismeas, P., Chollet, P., and Clément, A., "A New Method for the Characterization of Analog Probes," Presented at the IPES '89, Monterey, CA, September 1989.

Hocken, Robert J., et al., "Technology of Machine Tools - A Survey of the State-of-the-art by the Machine Tool Task Force, Volume 5: Machine Tool Accuracy," NTIS Document No. UCRL-52960-5, October 1980, plus references therein.

Hocken, R. J., and Borchardt, B. R., "On Characterizing Measuring Machine Geometry," U.S. Dept. of Comm., NBS, NBSIR 79-1752, June 1979.

Hocken, R., Simpson, J. A., Borchardt, B., Lazar, J., Reeve, C., and Stein, P., "Three Dimensional Metrology," CIRP Annals 26/2, 1977.

Hopp, T., "Computational Metrology," spec. sec. of ASME Mfg. Rev., Part 1, Dec. 1993.

Juoy, F., et al., "Theoretical Modelisation and Experimental Identification of the Geometrical Parameters of Coordinate Machines by Measuring a Multi-directed Bar," Annals of CIRP 35/1, 1986.

Kalpakjian, S., Manufacturing Engineering and Technology, Addison-Wesley, MA, 1989.

Kang, T. H., "Sources of geometrical error in machining processes," SME TP MR71-263, 1978.

Kang, T. H., "The effect of chucking methods on roundness error in the boring process," J. Eng. for Indus., Feb. 1976, pp. 233-238.

Krejci, James V., "CMM Measurement Enhancement Using Probe Compensation Algorithms," SME, 1990.

Kumaran, S., "The Analysis of Spindle Errors," Masters Thesis, UNC Charlotte, Charlotte, NC, 1992.

Lane, D. L., "Method for Evaluation and Certification of the Moore Model No. 3 Universal Measuring Machine," Lawrence Livermore National Laboratory, Livermore, CA, UCRL-50679, 1969.

Lingard, P. S., Purss, M. E., Sona, C. M., Thwaite, E. G., Mariasson, G. H., "Temperature Perturbation Effects in a High Precision CMM," Prec. Engrg., 1991.

Machireddy, R. K., Hari, Y., and Hocken, R., "Sampling Techniques for Measurement of a Plane Surface Using a Coordinate Measurement Machine," Nat'l. Design Engrg. Conf., Chicago, IL, ASME, 1993.

Mitsui, K., "Development of a New Measurement Method for Spindle Rotation Accuracy by Three Points Method," Proc. 23rd International MTDR Conf., 1982.

Moore, W., Foundations of Mechanical Accuracy, Moore Special Tool, Bridgept., CT, 1970.

Morgan, D. E., Piazza, J. E., and Plyler, R. E., "Inherent Errors on X, Y, Z Coordinate Measuring Machines on Curved Surfaces," Proceedings of the 1984 Annual International Industrial Engineering Conference, Institute of Industrial Engineers, p. 400, 1984.

Odayappan, O., "Development of Methodologies for Measurement of Circles on a Coordinate Measuring Machine," Masters Thesis, UNC Charlotte, Charlotte, NC, 1992.

Osanna, P. H., "Workpiece accuracy and geometrical deviation data for cylindrical parts," Report from Dept. of Production Engrg., Tech. Univ. Vienna, Austria, 1990.

Osborne, M. R., Watson, G. A., "On the Best Linear Tschebycheff Approximation," Computer J., 1967.

Palanvelu, D., "Algorithm Evaluation for Coordinate Measuring Machines," M.Sc. Thesis, University of North Carolina at Charlotte, 1990.

Phillips, S. D., Borchardt, B., Estler, W. T., and Henry, J., "A Study on the Interaction of Form Error and Sampling Strategy for Spheres," NIST J. of Res., 1994 (to be published).

Phillips, S. D., Borchardt, B., Caskey, G., "Measurement Uncertainty Considerations for Coordinate Measuring Machines," NISTIR 5170, NIST, Gaithersburg, MD, 1993.

Plazek, Chester, "Efforts to Relate CMM Data to Design Specifications Underway," Quality Magazine (Plazek, Editor), December 1988.

Pond, James B., "Answers to CMM Problems Sought After GIDEP 'Alert'," Metalworking News, December 12, 1988, p. 23.

Porta, C., and Waeldele, F., "Field Testing of Three Coordinate Measuring Machine Evaluation Algorithms," Physikalisch Technische Bundesanstalt, March 1986.

Rabiner, L. R., and Gold, B., Theory and Application of Signal Processing, Prentice-Hall, New York, NY, 1975.

Raugh, M. R., "Absolute two-dimensional sub-micron metrology for electron beam lithography: a theory of calibration with applications," Prec. Engrg. 7, 1, Jan. 1985.

Reason, R. E., 1966 Report on the Measurement of Roundness, The Rank Organization, Rank Taylor Hobson (1966).

Reeve, Charles P., "The Calibration of a Roundness Standard," National Bureau of Standards Report NBSIR 79-1758, Washington, DC, 1979.

Reeve, Charles P., "A Method of Calibrating Two-dimensional Reference Plates," National Bureau of Standards Report NBSIR 74-532, Washington, DC, July 1974.

Reid, Paul, Chief Scientist, AXAF Program, Hughes Danbury Optical Systems, Danbury, CT, private communication, 1992.

Salsbury, J., "Development of a Methodology for Relating Machine Tool Errors to Workpiece Geometry," Masters Thesis, UNC Charlotte, Charlotte, NC, 1991.

Sartori, S., Cresto, P. C., Di Ciommo, M., Kancheva, T., Marques, D., "A method for the identification and correction of thermal deformations in a three coordinate measuring machine," VDI Berichte, 761, 1989.

Soons, J. A., Theuws, F. C., Schellekens, P. H., "Modeling the Errors of Multi-axis Machines: A General Methodology," Prec. Engrg. 14, 1, Jan. 1992.

Swyt, D., "Challenges to NIST in Dimensional Metrology: The Impact of Tightening Tolerances in the U.S. Discrete-Part Manufacturing Industry," NIST, NISTIR 4757, Jan. 1992.

Tipnis, V. A., Rpt. Comm. Chn., "Research Needs and Technological Opportunities in Mechanical Tolerancing - Results of an International Workshop," ASME CRTD-15, 1989.

Tlusty, J., "System and Methods of Testing Machine Tools," Microtechnic 13, 162, 1959.

Tlusty, J., "Testing of Accuracy of Machine Tools" (supplement to the Machine Tool Task Force reports; available from the same sources), UCRL-52960-Supp. 1, 1980b.

Weckenmann, A., Heinrichowski, H., and Mordhorst, H. J., "Design of Gauges and Multipoint Measuring Systems using Coordinate Measuring Machine Data and Computer Simulation," Prec. Engrg. 13, 1991.

Whitehouse, D. J., "Some theoretical aspects of error separation techniques in surface metrology," J. of Phys. E: Sci. Instr. 9, 1976.

Wilson, Jr., H. Ray, "Form Errors of Cylindrical Features for Specific Manufacturing Processes," Masters Thesis, UNC Charlotte, Charlotte, NC, 1992.

Wirtz, Adolph, "Coordinate Measurement Technique - A Unique Tool for Geometric Quality Control in Flexible Manufacturing Systems," Richard Hook, Editor. Neu Tecknicumbuchs, Buchs, Switzerland, 1986.

Woo, T., "Hammersley Sampling for Efficient Surface Coordinate Measurements," Proc. NSF Des. & Mfg. Sys. Conf. 2, Charlotte, NC, 1993.

Zhang, G., Veale, R., Charlton, T., Borchardt, B., Hocken, R., "Error Compensation of Coordinate Measuring Machines," CIRP Annals 34, 1, 1985.

Zhang, G., Ouyang, R., Lu, B., Hocken, R., Veale, R., and Donmez, A., "A Displacement Method for Machine Geometry Calibration," CIRP Annals 37, 1988.

CRTD-Vol. 27, International Forum
on Dimensional Tolerancing and Metrology
ASME 1993

THE IMPACT OF TOLERANCING ON GAGE REPEATABILITY AND REPRODUCIBILITY STUDIES

Ronald C. Schall
Powertrain Operations
Ford Motor Company
Livonia, Michigan

Dennis G. German
Powertrain Operations
Ford Motor Company
Dearborn, Michigan

Abstract
Tolerancing certainly has an immense impact on a Gage R&R study. This paper will explain what a gage R&R is, how and why it is performed. We will then illustrate the impact of tightening tolerances on performing and passing a gage R&R on the manufacturing floor.

This paper will illustrate what type of gage repeatability is required as the tolerance tightens. We will also give estimates of the cost impact associated with tighter tolerancing. Keep in mind this cost impact is from a gaging viewpoint only, there will also be increased cost from the machine tool aspect.

The important factor is that the engineer be aware of these issues when designing and tolerancing a part.

The trend towards tighter tolerances has brought new challenges to the manufacturing floor. One of them is gaging. It becomes increasingly more difficult to adequately measure the dimensional stability of our processes. We need a way to measure our measurement systems. The Gage R&R is one method of accomplishing this. Hopefully this paper will lend some insight to the design community of the impact ever tightening tolerancing has on manufacturing.

What is a Gage R&R ?
A Gage R&R is a Gage repeatability and reproducibility test. It is quite simply part of the measurement system analysis we apply to either accept or reject the measurement system on the manufacturing floor. A Gage R&R is to be run after the gage accuracy is established and the gage is properly mastered. Depending on the method employed the number of parts and the number of operators will vary.

REPEATABILITY
Gage repeatability is the variation in measurement obtained with one gage when used several times by one operator while measuring the identical characteristics on the same parts. This is also known as equipment variation.

REPRODUCIBILITY
Gage reproducibility is the variation in the average of the measurements made by different operators using the same gage when measuring the identical characteristics on the same parts. Reproducibility is also called the appraiser variation

Different Types of Gage R&Rs

Range method
This is also known as the short form of performing a Gage R&R. The range method is a very quick way to perform an R&R study. It will tell you whether or not the measurement system is performing. But it will not separate the EV (Equipment Variation) from the AV (Appraiser Variation).

Average and Range
Also called the long form of Gage R&R. The Average and Range method is more lengthy and a little more complex to run. However it will give you some indication of where you have problems within the measurement system. This differs from the short form because it allows independent analysis of repeatability and reproducibility.

ANOVA

ANOVA is the analysis of variance. This is the most comprehensive form of a Gage R&R currently in use within the industry. It can be extremely useful in targeting problems within the measurement system.

Attribute Gaging

This method is used for go, nogo type of gaging. We will not look at attribute gaging as it has no relevancy to the thrust of this paper.

The recommendation for a new gage delivered to the manufacturing floor is to use either the Average and Range method or the ANOVA method to accept or reject the measurement system. The Range method may be used to audit the system on a regular basis. If the Range method fails to accept the system at anytime, then the Average and Range method or the ANOVA method should once again be employed to find and correct the problem.

Acceptance Criteria

The criteria for accepting is based on the percent GR&R. The levels of acceptance are somewhat debatable. The AIAG recommends the following:

10% and under, accept the system

10%-30% may be acceptable

30% and over, the system is unacceptable

For the purposes of this paper lets study the range method. Our main concern is what impact the design tolerance has on a Gage R&R. We really don't need to separate the errors generated by the various components of the measurement system. The Range method will be much simpler to study. We will see the same type of impact on the other types of gage studies.

Employing the Range method we pick five parts and two operators. We have each operator measure the parts. For each part we take the value of the difference from each of the two readings. This is the range for each part. We then add the ranges up and divide by 5. This gives us our average range. We then utilize the following formulas to determine our %GR&R.

$GRR=4.33*\bar{R}$

Process Variation = Tolerance

%GR&R=100[GR&R/Process Variation]

There are two values here of interest to us. The tolerance and the average range of readings from the gage. The design engineer controls the tolerance. The Gage Engineer on the manufacturing floor has to live with the tolerance on the product part description, whether it is in the form of a B/P or a CAD model is not relevant. This tolerance is the legal contract under which the part must be manufactured.

The gage engineer tries to control the average range of readings coming from the gage. As the tolerance tightens this becomes increasingly more difficult.

Tol. **GR&R**	.127	.102	.076	.051	.025	.013
10%	.0029	.0024	.0018	.0012	.0006	.0003
20%	.0059	.0047	.0035	.0024	.0012	.0006
30%	.0088	.0071	.0053	.0035	.0017	.0009

Units are mm

Table 1

By using some simple algebra we solve for $\overline{R}$

$$\overline{R} = \frac{TOL*(\%GR\&R/100)}{4.33}$$

Using this simple formula we can start to substitute values for the tolerance to predict what sort of average range we need to get from the gage study to pass a Gage R&R. Table 1 illustrates the values we need to obtain for the average range as we substitute values for the tolerance. Some very interesting things jump out at us right away. Using a .127mm (.005in) tolerance we are already chasing microns for an average range. This is no simple task in the manufacturing environment. As we tighten the tolerance the average range continues to tighten. With a .076mm tolerance we need a range of under 2 microns. At .025mm (.001) we are at the sub micron level for average range. It is at this point some gage vendors will not give you a quote on gaging. They will quote on a time and materials basis only. The reason for this is that passing a Gage R&R at this level is highly uncertain. We can easily see how difficult it can be pass a Gage R&R study when the average range of the readings is required to be so small. It can prove to be extremely time consuming to make the system perform to these standards.

Cost

Somewhere around the .076mm tolerance level we make the jump to what is considered a high accuracy CMM. The exact point is unclear due to many factors. some of the factors to consider are machine configuration, probing systems, algorithms, environment, size of features, datum systems, and more. This can be an extremely expensive jump. A standard accuracy CMM starts at around $100,000. A high CMM accuracy starts at around $300,000. There are of course many variables in these prices but these are good figures to start with.

Dedicated gaging follows a similar pattern. We recently purchased a relatively simple gage to measure a pump bore. The cost of the gage was $2,500. The tolerance on the bore was .075mm. If the tolerance was tightened to .050mm the price would have been $4,000. Tighten the tolerance a little more to .025mm and the vendor would not quote. They would quote only on a time and materials basis.

Conclusion

The conclusion we would like to draw from this is to became aware of the ramifications of ever tightening tolerances on gaging. We certainly don't want to tell the design engineers how to tolerance their products. However we do want them to be fully informed of the impacts on gaging as well as machining.

References

American Society for Quality Control (ASQC)/Automotive Industries Action Group (AIAG), "Measurement Systems Analysis Reference Manual", October 1990

CRTD-Vol. 27, International Forum
on Dimensional Tolerancing and Metrology
ASME 1993

THE PRACTITIONER'S VIEW OF MEASUREMENT SYSTEM APPLICATION

Dean E. Beutel
Department of Manufacturing Engineering
Caterpillar, Incorporated
Peoria, Illinois

ABSTRACT

The measurement function is a process. It has variation caused by a variety of factors that detrimentally affect the reliability of the process. Historically the metrological engineer's basic function has been to design and/or apply measurement systems that have relatively small variation (or bias) when compared to blueprint tolerance. This paper will address important issues in ensuring that the measuring system is applicable from a global viewpoint.

The five critical stages of development in proper gage application are summarized in the following list:.

1. *Pre-Design Review*
2. *Blue Print Review*
3. *Manufacturing System Analysis*
4. *Process Control System Analysis*
5. *Miscellaneous Procedural Issues*

The involvement of the metrological engineer in the pre-design phase of product development will ensure that economical analysis is performed when assigning tolerancing to part features. Costs associated with excessively "tight" tolerancing can be astronomical (from both the gaging and the manufacturing process perspective). The metrological engineer will have much needed early input into decisions affecting cost, availability, reliability and capability of measuring systems applied during the production planning phase of the product development cycle.

Blueprint analysis consists of reviewing prints for all features that require measurement and control and creating a list of features and associated tolerances. Additionally, it is at this stage that precision ratios will be selected.

Analysis of the manufacturing system is another critical stage of the gage application process. Many issues such as the volume of parts being measured, system manning, facility thermal conditions, and process capability (CP) must be applied during stage three of measurement system design.

The process control strategy selected affects data collection methods, analysis techniques, and the need to transfer data. The requirement of communications interfaces for data transfer affects gage system design and cost. Unmanned applications require special considerations to prevent "over-control" problems. Special procedural concerns such as cleanliness, thermal equilibrium of gage system components, traceability of gage masters, and many other related issues must be addressed also.

Once the five stages have been completed a measuring system can be specified, purchased and installed with a high level of confidence in the success of the system. Elimination of any of the five steps brings unacceptable risk and/or cost to the process of gage selection.

REFERENCES

Fundamentals of Dimensional Metrology, Wilkie Brothers Foundation, Ted Busch, Copyright 1964

Temperature and Humidity Environment for Dimensional Measurement, ANSI B89.6.2, The American Society for Mechanical Engineers, Copyright 1974

Interest in Simultaneous Engineering or Concurrent Product & Process Design is rising as design and manufacturing engineers realize the benefits of early collaboration. Optimization of functional design needs, and manufacturing requirements early in the product introduction cycle is not only cost effective, it allows much quicker response to market needs. The key to competing successfully in today's competitive environment is to develop the best combination of functional performance at the lowest possible manufacturing cost. A key issue to consider in this process is the measurement function.

The measurement function is a process with inherent variation. The variation is caused by a variety of factors that detrimentally affect the reliability[1] of the process. Historically, the metrological engineer's basic function has been to design and apply measurement systems that have relatively small variation (or bias) when compared to blueprint tolerance. This was typically performed after the design was completed. This paper will address current issues of importance in ensuring that the measuring system is capable, economical, available and reliable.

To realize this goal there are five critical stages of development in proper gage application. They are summarized in the following list:

1. Pre-Design Review
2. Blue Print Review
3. Manufacturing System Analysis
4. Process Control System Analysis
5. Procedural Issues

PRE-DESIGN REVIEW

The involvement of the metrological engineer in the ***pre-design*** phase of product development will ensure that economical analysis is performed when assigning tolerancing to part features. Cost associated with excessively "tight" tolerancing can be astronomical (from both the gaging and the manufacturing process perspective). The metrological engineer should have early input into tolerancing decisions affecting cost, availability, reliability and capability of measuring systems applied during the production planning phase of the product development cycle.

The primary concern of the metrological engineer must always be the capability of the measuring system. To ensure good data various application rules must be followed regarding the part tolerance. These rules, normally called precision ratios, are dictated by feature tolerance and become extremely difficult to fulfill with unnecessarily tightly toleranced parts. Purchasing less expensive gages that do not meet required precision ratios will result in increased incidence of accepting bad parts and rejecting good parts.

As mentioned, other concerns such as availability and reliability of gages must also be considered. In situations where new product is being introduced and gage availability is a problem alternate sources for gages may be required. Typically there is a correlation between precision requirements and gage availability (the higher the precision requirement the lower the availability). Gage reliability may also be of critical importance in the selection process if gage downtime causes manufacturing machines to sit idle in FMS or transfer line applications.

Assuming that the decision is made to follow proper application rules, the next issue becomes an economic issue. Refer to Figure 1 for an example of two designs, equivalent in function that result in unequal gaging cost and cost to manufacture.

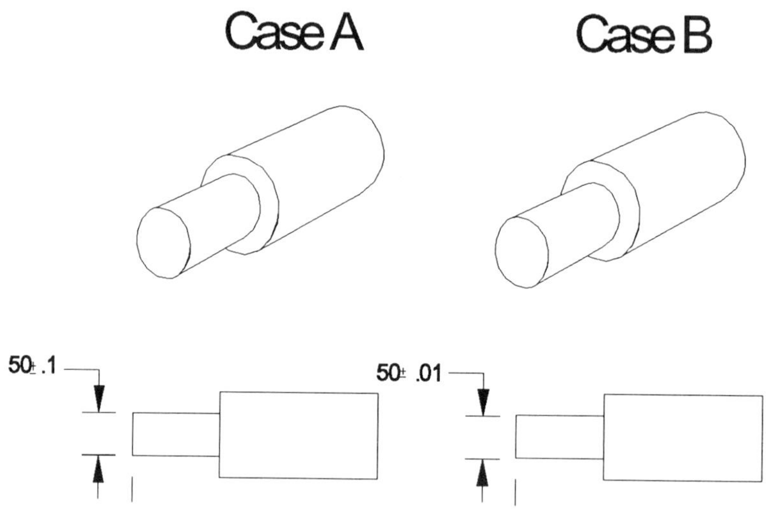

FIGURE 1

In Case A the design engineer, manufacturing engineer, and metrological engineer carefully examined functional requirements, manufacturing capability, and gage capability. They then applied tolerances to clearance dimensions which reflect the functional requirements. In Case B the design engineer neglected to consult with other team members and assigned tolerances that required a higher gaging class and additional machining operations. An examination of the cost to manufacture the shafts follows:

Case A

	Mfg.	Gage		
Saw to Length	1.00	.05		
Rough Turn	3.00	.07		
Finish Turn	1.80	.25		
TOTAL	5.80	.37	=	$6.17

Case B

	Mfg.	Gage		
Saw to Length	1.00	.05		
Rough Turn	3.00	.07		
Finish Turn	2.10	.50		
Grind OD	1.00	.35		
TOTAL	7.10	.97	=	$8.07

This very basic example shows how manufacturing to the tighter diametric tolerance in Case B adds cost to the turning operation, and adds a grind operation resulting in an overall increase in cost to manufacturing of 24%.

BLUEPRINT REVIEW

Blueprint Review consists of reviewing prints for all features that require measurement and control and creating a feature list with associated tolerances. At this stage final precision ratios will be selected, which as previously discussed have tremendous influence on gage design requirements and cost. Key factors that drive the selection of precision ratios are normally related to product performance. Safety, reliability, cost and other "product performance" issues must be considered and ratios will be assigned accordingly. It is uneconomical, and in many cases impractical, to always assign a 10:1 or 4:1 ratio. Note: If the decision has been made to manufacture with a process capability greater than one (CP > 1) this will also have a direct impact on gage selection but does not affect precision ratio selection (more detail later).

Application of precision ratios consists of comparing blueprint tolerances to several gage design characteristics such as gage resolution, repeatability, and accuracy. The following example illustrates how this works, using nominal ratios accepted for use in the industry today.

Assume: Blue print tolerance = .050 (bandwidth)

Requirement	Nominal Ratio	CP factor (if req'd)	Gage
Resolution	10:1	2	.0025
Repeatability	10:1	2	.0025
Accuracy	4:1	2	.00625

MANUFACTURING SYSTEM ANALYSIS

Analysis of the ***manufacturing system*** where the gage will be used is another critical stage of the gage application process. Many issues such as the volume of parts being measured, if the system is manual or unmanned, or if the facility is controlled thermally all have significant affect on proper gage selection. As mentioned earlier when the target for process capability (CP) exceeds one, this factor must be applied during stage three of measurement system design as well. If "CP" is not factored into the selection decision the gage will have insufficient resolution to correctly monitor the process.

In systems where a high volume of similar parts will be manufactured, speed and reproducibility of gaging is critical. It must be assumed that the gage will be used by personnel with varying levels of training and precision appreciation. Therefore gages applied in high volume applications should be very easy to use properly. Additionally, interpretation of gage response should be similarly easy. Obviously the gaging process must be very fast. (See Figure 2 for an example of a gage designed for use in a high volume application)

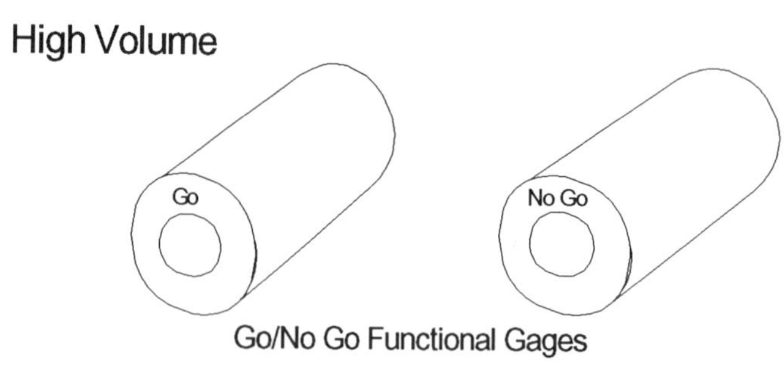

FIGURE 2

In low volume or job shop applications, speed of the gaging process is much less important. Critical attributes of gaging applied to low volume situations are flexibility, low linearity error, and high stability. Stability is important because calibration of these flexible instruments can be difficult and may be neglected for extended periods of time. A Coordinate Measuring Machine is a good example of this type of gaging instrument. (see Figure 3)

LOW VOLUME

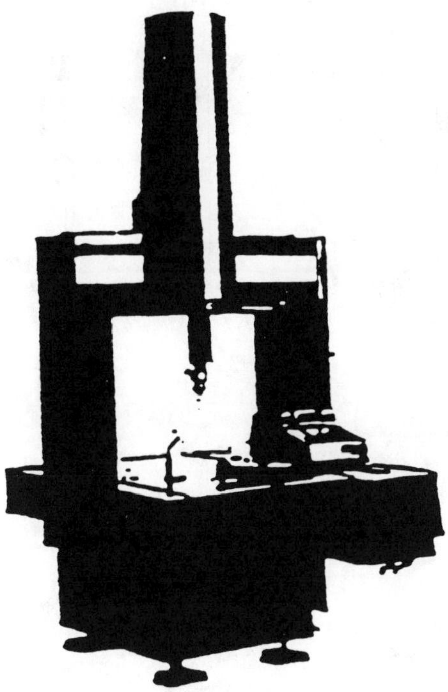

FIGURE 3

Another major concern in the analysis of the manufacturing system is thermal variation. The three primary components of the measurement system - the gage instrument (comparator), the gage master, and the part being measured must be in thermal equilibrium for optimal gaging accuracy. In situations when this is not possible, rigorous analysis of the "differential expansion"[2] must be performed to ensure that the loss in gaging accuracy is acceptable.

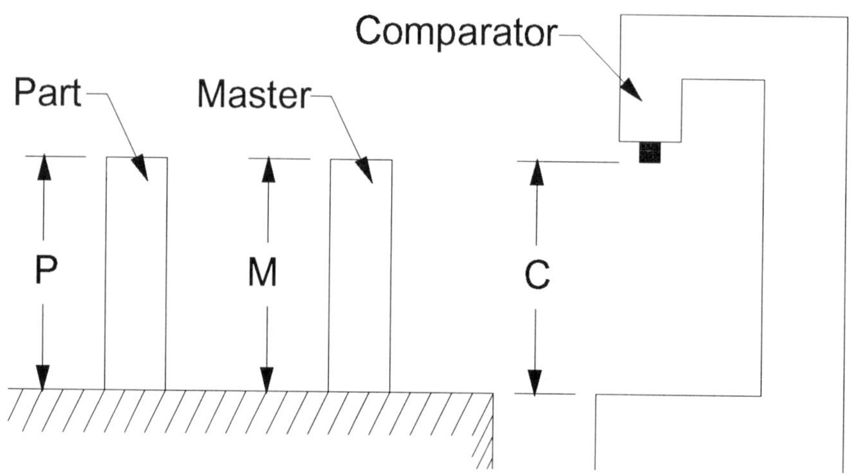

FIGURE 4

Calculation of the Thermal Error Index (TEI)[3] will determine if steps must be taken to correct for the differential expansion in the part and the master. The TEI can be particularly problematic when the Nominal Differential Expansion (NDE)[4] term is very high. This normally occurs because the (NE)[5] nominal expansion of the part being measured and the (NE) of the gage master being used consist of different materials (aluminum and steel), coupled with the fact that the measuring system is not in thermal equilibrium. See example below.

EXAMPLE

TEI = [(NDE + UNDE + TVE) / Working Tolerance] 100

$NDE = (NE)_{part} - (NE)_{master}$
UNDE = Uncertainty of Nominal Differential Expansion
TVE = Temperature Variation Error (drift of part or master comparator)

Given:

- Part Material = Aluminum (22 PPM coefficient of expansion)
- Master = Steel (11 PPM coefficient of expansion)
- Temp. Change = 10 degrees Celsius
- Feature Size = 300 mm with a +/- .100mm tolerance
- For ease of discussion assume the UNDE is approximately 15% of the Nominal Differential Expansion (difference in the NEpart and the NEmaster)
- Drift arrived at by empirical test (TVE)

$NEp = (10 degrees \times 22PPM \times .3M)_{part} = .066mm$
$NEm = (10 \times 11 \times .3)_{master} = .033mm$
NDE = .066 - .033 = .033mm

Assumed UNDE = (.066 + .033) .15 = .0099

TVE = result of empirical testing of comparator = .005

TEI = [(.033 + .0099 + .005) / .200] 100 = 24%

The conclusion of the example is that the measurement error (bias) can be assumed to be no smaller than 24% of the feature working tolerance due solely to thermal issues. For most manufacturing concerns this situation would be totally unacceptable.

There are several ways to alleviate the problem. The easiest is obvious - create thermal equilibrium in the system either mechanically or through compensation techniques. This immediately eliminates the NDE term. If utmost speed is a must there may be insufficient time to mechanically create a thermally equalized system, thus making compensation the technique of choice. The limitation to compensation is the UNDE and the TVE terms. The only way to completely eliminate the UNDE term is to have the entire system at exactly 20 degrees Celsius (since the UNDE term is caused by the inherent inability to accurately know the average coefficient of expansion in any material). Naturally, another benefit of maintaining a constant ambient environment is a reduction in the "drift" (TVE) term.

One additional issue needs to be considered at this time - form error. The presence or absence of straightness, flatness, roundness, cylindricity, and conicity is generally controlled by the manufacturing process. In many cases the amount of form error is negligible and doesn't need to be considered in the gage application process. However, in cases where form is a critical characteristic special attention must be given to how it is measured.

The major concern is that certain types of form error are difficult if not impossible to detect with incorrectly applied gages. A short list follows to give the reader several specific examples to be aware of, and increase awareness of the problem in general:

1. CMM - all types of form error are difficult to determine accurately due to: 1) lobing error in some types of probes, 2) discrete point repeatability error in most machines (as high as .010mm - .020mm), and 3) insufficient probing points taken to correctly determine form error since diameter of a bore can normally be determined adequately using four to ten probing points (analysis of form should use at least one point per ten radial millimeters - 32 points on a 100mm diameter bore);

2. Dual Point Gages - (dial bore, dial snap, micrometer, two point air jet, etc.) will not detect certain types of lobing error in part features;

3. Attribute Gages - (plug gages for diameters), will not differentiate between roundness error and diametrical error.

PROCESS CONTROL STRATEGY

Conducting the Blue Print Review and Manufacturing System Analysis phases in a comprehensive manner will ensure that reliable data is produced. When selecting a ***process control strategy*** one must determine what to do with the data. Gage type (attribute or variable), data collection technique, analysis methods, and the need to transfer data must all be considered.

If attribute gages will be used this will limit the types of control charting that can be applied and also the ability to perform trend analysis in the operation. Data collection techniques may be manual or automatic. This decision is normally driven by the volume of parts being manufactured and the cost of implementing the system. Analysis methods for the data that is collected vary widely. Additionally the requirement of communications interfaces for data transfer will affect gage system design and cost. If the system is an unmanned application, it will require a "knowledge" database that controls decisions on when to make adjustments to the machine's system as care must be taken to prevent "over-control" problems.

Two key issues which deserve special emphasis when identifying an appropriate process control strategy are the volume of parts being produced and the level of experience of the operators. Large production runs lend themselves well to many well known charting techniques, however in small lot size batches (under 30 parts) care must be given to select charting techniques that 1) really assist the operator in making good parts, and 2) provide statistically valid information. If the system will be used by inexperienced operators, or if there is high turnover of operators in a particular work station, the system must be user friendly and complicated charting techniques should not be used as they will only confuse the personnel. Historically precontrol charts have been used very successfully in these type applications.

PROCEDURAL CONCERNS

There are several ***procedural concerns*** that must be addressed to ensure an effective measuring system. Ensuring that the part is adequately clean before measurement may seem to be trivial, but cleanliness is certainly one of the most common causes of measurement error. Proper calibration techniques and adequate frequency of calibration of gages is of critical importance in systems with frequent and large thermal gradients. Ensuring that gaging masters or reference lengths are calibrated with appropriate techniques in a timely fashion, and with artifacts that are certified traceable to national standards laboratories is also a must.

Following rigorous procedures will ensure long-term reliable performance of gaging systems. Documented procedures are also becoming a requirement rather than "nice to have" for companies choosing to pursue ISO 9000 certification. Case histories of many companies who have become certified show that periodic cleaning and preventive maintenance of their measurement instruments pay good dividends.

Critical in the pursuit of ISO certification and good management in general is the availability and execution of proper training. Employees required to use and maintain precision measuring systems must be trained properly. Training in a variety of key areas is important, such as: proper application of the measuring system, calibration frequency and techniques, interim checking techniques (used to ensure gage system stability and overall performance is unchanged from the original calibrated condition), and periodic preventive maintenance.

SUMMARY

Once the five planning stages have been completed a measuring system can be specified, purchased and installed with a high level of confidence in the success of the system. Elimination of any one of the five steps brings ***unacceptable risk, potential delay in implementation of the system,*** and ***unnecessarily high cost*** to the process of gage application, and more importantly to the overall cost of the manufacturing process.

It is assumed by the author of this paper that there will be two basic types of readers: design engineers and metrological (or manufacturing) engineers. The critical point to the designers is that the process of measuring features properly is very difficult and potentially very *costly* - and that YOU have the ability to make it easier by working with the metrological engineer early in the product introduction cycle. The point to the metrological engineer is that you must not live in a vacuum, there are many critical issues to consider in addition to the blueprint tolerance - *look at everything*. To both engineers the importance of working together as a team is simply a must!

REFERENCES

1. Fundamentals of Dimensional Metrology, Wilkie Brothers Foundation, Ted Busch, Copyright 1964, "**Reliability** - the likelihood that actual performance will be the same as planned performance"

2. Temperature and Humidity Environment for Dimensional Measurement, ANSI B89.6.2, The American Society for Mechanical Engineers, Copyright 1974, "**Differential Expansion** - the difference between the expansion of the part and the expansion of the master from 68 degrees F to their time-mean temperatures at the time of the measurement"

3. Temperature and Humidity Environment for Dimensional Measurement, ANSI B89.6.2, The American Society for Mechanical Engineers, Copyright 1974, "**Thermal Error Index (TEI)** - The sum of all approximate thermal effects error components expressed as a percentage of the total permissible error"

4. Temperature and Humidity Environment for Dimensional Measurement, ANSI B89.6.2, The American Society for Mechanical Engineers, Copyright 1974, "**Nominal Differential Expansion (NDE)** - The difference between the nominal expansion of the part and the master"

5. Temperature and Humidity Environment for Dimensional Measurement, ANSI B89.6.2, The American Society for Mechanical Engineers, Copyright 1974, "**Nominal Expansion (NE)** - assuming that an object in a convective environment has a uniform temperature equal to the environment mean temperature in its immediate vicinity its Nominal Expansion is estimated by the equation: k(L)(t-68)". k = the estimate of the coefficient of expansion of a body, L = length of the body, t-68 = the temperature departure from the *standard* temperature of 68° F (20° C).

CRTD-Vol. 27, International Forum
on Dimensional Tolerancing and Metrology
ASME 1993

SESSION 4
QUESTIONS AND DISCUSSION

Moderators: **Frank Bakos**
Eastman Kodak Company
Theodore Hopp
National Institute of Standards & Technology

Reporters: **Steffen Parratt**
Cornell University
Dean Robinson
Cornell University

INTRODUCTION

This section provides a summary transcript of questions and discussion directed at the presentations in Sessions 1–3. The talks in these sessions were tightly scheduled, and only brief questions for clarification were admitted after each talk. Substantive questions were deferred to this session (#4), and were submitted in signed, written form. The session Moderators ordered the questions and managed the ensuing discussion.

The Reporters produced this record from written notes and tape recordings; it has been edited lightly by Herb Voelcker, acting for the *Proceedings* Editor, Vijay Srinivasan. The questions, responses, and remarks attributed to individuals have not been checked by those individuals because of the compressed publication schedule for the *Proceedings*. The Reporters and Editor apologize for any misinterpretations.

QUESTIONS AND DISCUSSION

Ted Hopp (to Neal Laurance): How does STEP relate to design and manufacturing as practiced today? How much culture change will be needed to make STEP useful?

Laurance: That's hard to answer. Some companies are beginning to use STEP. Industry needs a better appreciation of the importance of information modeling. A shift in thinking is beginning, but it will be an evolutionary change.

[unknown] (to Richard Walker): You mentioned three steps in mathematizing a tolerance specification: defining the tolerance zone, defining conformance to the zone, and defining "actual value". If the tolerance zone defines a limiting boundary, of what use is actual value to design intent?

Walker: It's useful in quality control, for one thing.

Ken Sheehan: Actual value is needed by B89.3.2.

Georg Henzold (to Alvin Neumann): Parts of Y14.5 contradict ISO standards. Other parts need to be discussed to determine if there is a conflict with ISO. Do you plan to issue the new standard before or after these issues are resolved?

Neumann: Y14.5 goes out for public review this year. If all goes well, it will be issued in January of 1994.

Henzold: So you will issue it before ISO completes its review?

Neumann: In the past, parts of Y14.5 have been adopted later by ISO. Some of the concepts in Y14.5 are still evolving in ISO. We need to move ahead now with some form of mathematization.

John Hertrich (to Al Neumann): With the new definition of composite position as I understand it, the upper control frame defines location and the lower control frame defines orientation requirements. How will a designer describe requirements on location of pattern and location of individual features within the pattern?

Neumann: The composite position tolerance specification has been expanded. The new specification explicitly allows the use of a secondary datum in the lower control frame, to control pattern orientation.

Frank Bakos: What is a 'statistical' dimension, according to the new Y14.5 standard?

Neumann: There is much interest in statistical tolerancing, but Y14.5 had no provisions for this. We added a symbol, but the underlying statistical formulas are up to you. The rule for interpreting a statistical dimension would be stated someplace on the drawing.

Neil Bernstein (to Neal Laurance and Richard Walker): To what extent have Y14.5, Y14.5.1, and STEP part 47 [covering tolerances] been harmonized?

Laurance: Regarding STEP part 47, the question is premature. It is the intention that STEP will be harmonized, but part 47 is in working draft form at best. Part 47 will reflect Y14.5.1.

Bud Krayer (to Fred Parsons): Should the measurement plan be integrated with the design intent definition? All on the same drawing? Is it possible to have one design intent document related to multiple measuring plans?

Parsons: B89.3.2 does not now intend to require this information on the drawing. If you put this kind of information on the drawing, you might be forced to change it as better measurement methods come along. A better alternative is to put this information in a separate document.

Ken Sheehan (to Paul McKim): What, if any, incentives to industry exist to encourage participation in the standards process? (Examples: tax credit/write-off for labor; matching funds.) What efforts are underway to improve this?

McKim: There are no direct and immediate financial incentives that I'm aware of. There is the pride and satisfaction of making technical contributions to published standards. Standards work makes it easier for us to do business. I would also make a plea for more involvement in the ISO standards-making process. There are two informational videotapes from ANSI that are intended to educate management on the benefits of industrial participation in standards work.

Jim Hurt (to Paul McKim): Please expand on your statement about the availability of standards on CD-ROM and over computer networks. What standards are available, e.g. Y14.5, ISO-1101? How, and from whom, does one order? Are licenses sold to permit redistribution, e.g. as part of a CAD product? Is there a catalog listing available standards?

McKim: There are many standards available today, including ISO, ANSI, ASTM, SAE. There is a company that provides these on CD-ROM. In the future, some standards will be licensed to companies over data networks.

Curt Larson (to Robert Hocken): What is aliasing? What is your recommendation to ensure undersampling does not occur in a production environment?

Hocken: If you have a periodic signal of some frequency and sample it at a lower frequency, you can produce an apparent lower-frequency signal that doesn't exist. That phenomenon is called aliasing. The general rule is that you must sample at twice the frequency of the largest frequency component you want to pick up.

This rule is usually viewed in the time domain, but it can be viewed in the spatial domain, as those in surface finish measurement have known for years. To measure a particular wavelength of surface roughness, we apply filters to ensure that a higher frequency isn't artificially recorded as a lower frequency signal.

On avoiding undersampling: you have to look at the particular geometry you're measuring. Speaking only of CMM's, and the specific geometries we've checked.... If you're dealing with circles with suspected lobing, I would suggest up to 19 points. In our survey of lobing in different production processes, we found few parts with more than 8 or 9 lobes. One needs about twice this number of samples, which is analogous to the Nyquist rule in the frequency domain.

We are experimenting with other geometries. I worry about planes, which have very low order form error dominated by boundary conditions. Circles and spheres seem to have symmetry properties that allow us to develop strategies that work in most cases. We have yet to find similar strategies for planes.

There are other problems in stability of algorithms. For example, consider a barrel-shaped cylinder sampled on cross-sections normal to its axis. The minimum circumscribed cylinder depends only on the circular section measured at the widest point. The computed mating cylinder is unstable, and very sensitive to measurement errors. Using the computed cylinder as a datum would cause problems. Similar problems occur with "rocking" datums for planar surfaces. The fitting algorithms in these cases also have poor convergence.

Walker: There's been some movement in ISO to define datum surfaces on the basis of least-square fits. From the research you've done, would you be comfortable with that?

Hocken: I think you'd get more stable results if you did that. Now whether they would be stably *wrong* is a different matter. I wouldn't want to do it where I had to fit a cylinder in a hole, for instance. But if it were just to establish datum surfaces, I think least-square fits would be stable.

Jim Bryan: You and I have talked for years about using an optical technique, a hybrid CMM, to survey Moire fringes using a diffraction grating, to look for discontinuities (on the part) and then intelligently select those areas to inspect with the CMM.

Hocken: I have always liked that scheme conceptually and I think it's coming with new scanning machines that can look at large areas quickly. You would scan the part for bad areas, and then measure there with slower-speed higher-accuracy probes. If the part is generally good, then you need to inspect far less area that way. There are Moire fringe systems on the market, and there are triangulation scanners, which, while not very accurate, can sample large areas very quickly.

Ted Hopp (to Ari Requicha): You asked, "Should measurement be explicitly acknowledged in the theory?" What do you mean by measurement – current measurement technology or some aspect of measurement theory? Also, what kind of acknowledgement do you have in mind?

Requicha: In current standards and practices, you say something like, "Here is a tolerance zone and your entire surface must lie in this zone". It's pretty obvious that there is no way we can ever check that with a CMM. So we have a tolerance specification, and we have measurement techniques such as CMMs, and the tol-spec is such that from the beginning we know we can't check it. That's one approach, but it looks slightly peculiar.

My question: Is that right? Or is there some way we could provide a semantics for tolerancing which acknowledges that things have to be measurable? I don't know how to do that, but I am curious as to whether that might be a good thing to do. There are theories, such as quantum mechanics, in which measurement is central.

George Allen (to Richard Walker or Fred Parsons): How are CAD/CAM software vendors participating in standards committees like ANSI Y14, B89, ISO, etc.? How should they participate?

Walker: We have committee members from SDRC, Schlumberger, Brown and Sharpe, etc. In ISO, there's a participant from Autodesk. Now that we have a complete draft of Y14.5.1, vendors are showing interest in developing algorithms consistent with it. There still hasn't been as much involvement as we would like. We encourage CAD vendors to take this document and use it.

Walker (to Jim Bryan): Do you sense that the point of your talk may have been missed? It seems to me your examples were for high precision parts, but the principles can economically be applied elsewhere (e.g. to radiators, which GM and Ford have lots of). I thought it was interesting that American automakers may be missing the point, whereas you have ten Japanese joining your deterministic group.

Bryan: That's a good comment. The reason we talk about high precision is to show that the method works at that level. If so, it must work at lower levels. We are continuing to work at all levels of precision. In an undergraduate project, a 50-year-old lathe with plane bearings was modified, with an investment of $100, to repeat within tenths of microinches. We can exploit these methods at all levels of economic significance.

Ray Wheeler (to Vijay Srinivasan): How does the theory you discussed apply in the actual world where parts are made and *people* deal with actual parts and equipment?

Srinivasan: I'll try to answer that in three parts. First, much of the theory that Ari and I presented covers what you currently have in Y14.5M-1982. It unifies the concepts of offset tolerance zones, sweeps, etc., under operations like Minkowski sums. The concepts are already there and people are actually using them.

Second, with issues such as size, the mathematization requires a deviation from shop floor practices like micrometer measurement. But the new definition doesn't tell you to *not* to use micrometers. It's okay to use micrometers, but if we also want to use more sophisticated instruments such as CMMs, optical scanners etc., the mathematical definition gives us powerful means of developing assessment algorithms.

Third, ideas such as extending current syntax, using Minkowski sums for tolerance accumulation, etc., are still in the research stage. We won't know how good and useful this work will be until the research matures.

Wheeler: On the second part of your response, 90% of the people here probably didn't understand the theory. That's fine, the theory's good, but we can't use it in the workplace. What feedback can you give to the people in quality and manufacturing, who may be using a micrometer? Should they now take 24 readings instead of 2? How does the theory relate to actual practice in the shop?

Srinivasan: At the level of tolerance specification, there is no hint as to how you should measure it. The way you measure must come from the metrological community. You have to turn to people like Bob Hocken, to find out how many points you need to assess a form tolerance on a particular feature within a certain precision.

Walker: Y14.5.1 is not written for the shop floor. We thought about adding a statement to the Foreword saying something like, "if you're reading this document and you don't understand it, there's a 95% chance you don't need it". Y14.5.1 is meant to be a bridge between CAD/metrology and tolerancing. The original GD&T standard was based on assembly and hard gaging practices. Much of Y14.5.1 is just the addition of formulas to describe these practices.

Parsons: I disagree when you say that the average man on the shop floor won't use Y14.5.1. He will use it without knowing it, because we will build it into measuring equipment. What you've heard today are geometric principles that are putting on very sound footing the kind of metrology equipment that is going to be needed for the future.

Dean Beutel: I think it's inadequate to say users don't have to know anything about Y14.5.1. You must know the general form of the part, and thus what type of algorithms to use. You have a choice of using least squares or maximum inscribed cylinder, etc. You have to know something about what's going on to make the right choice. You don't have to be an expert in the theory, but you do have to know something about your manufacturing process, or your results will be garbage.

Walker: A few people have spoken of measurement being "in line" in the process. I think it's a sin to take a part out of a process and into a separate room for measurement. You ought to know what your process is doing. We should be characterizing our manufacturing processes, as in the new B5 machining standard, and not worrying about go/no-go systems for whether the part's good or bad. CMMs should be used for process control.

Paul Jackson (to Rich Walker): Considering the differences between the mathematical definition of actual value and that of measured value, where actual value is *void* of measurement uncertainty, what is the purpose of defining actual value?

Requicha (to all): How relevant are "values" of geometrical tolerances to SPC (Statisitcal Process Control)? What should we measure for effective SPC?

Hopp: NIST's Guide to Measuring Uncertainty lists a number of things you have to know to define the measured quantity. The first is the definition of what you're trying to measure. The second is the measurement process. Everything in B89.3.2 is concerned with the measurement process, and relating that to some formal definition of measuring that's independent of the process. If you define what you're measuring in terms of how you're going to measure it, it's a bootstrapping problem.

Walker: With flatness, you have a tolerance zone of thickness t. The actual value is the minimum separation of parallel planes about the feature. The actual value tells the person doing the measurement what value he's looking for. It doesn't imply that the measurement can determine this value. Every measurement is an estimate. ... The actual value is a benchmark. That's probably the best way to look at it.

[unknown]: You can't do anything with it. You can't even talk about a value for it unless you associate uncertainty with it, because you can't ever have the infinite set of points.

Tom Charlton: The typical person doing measurement wants a number: what is the size of that cylinder? The philosophy that separated Y14.5.1 from B89.3.2 is as follows: Y14.5.1 was to say, suppose we had the infinite set of points on that surface, what answer should we get for the size of that cylinder? Totally impractical from a measurement point of view, but it was considered to be the "right" answer.

On the measurement side, you sample the surface somehow and you get a number. The issue for the measurement side is – how close is that to the "true" number as specified by Y14.5.1? For many applications, you just want the true number and how close you are to that number. You can then make statements about the uncertainty of the measurement method. You can make statements about the accuracy of the measurement method without infinite sampling; you can view it as experiments with a computer.

Requicha: My question is, why would you want to know the size? You want to know that something fits somewhere. You want to know that your process is stable. You can answer those questions without producing that number. I think that's my claim.

Charlton: I agree. However we have a manufacturing culture that depends on concepts like size, and we can't change it all overnight. ... I don't disagree philosophically with what you said.

Sheehan: I do. For purposes of, for example, projecting tool wear, you need something to track besides go/no-go.

[unknown]: Aren't those the measured values?

Sheehan: My measured value has to have the same philosophical basis as his measured value, if we're both suppliers. So we create Y14.5.1. Then we know that we're all thinking the same way.

CRTD-Vol. 27, International Forum
on Dimensional Tolerancing and Metrology
ASME 1993

UNIT MANUFACTURING PROCESSES

W. L. Otto, Jr.
Industrial Programs
Concurrent Technologies Corporation
Johnstown, Pennsylvania

Iain Finnie
Department of Mechanical Engineering
University of California
Berkeley, California

ABSTRACT

A critical factor in the manufacture of products with superior quality at competitive prices is an understanding of the manufacturing system by which the products are produced. The manufacturing system can be broken down into a series of unit processes that impart both physical shape and structure to the product. The unit processes are intimately linked to one another by the fact that the output of one process becomes the input for the next process. The quality of the final product is dependent not only upon the capability of each unit process, but also upon the unit processes working together. Continuous improvement of the manufacturing system involves creation of a physical understanding of each process by itself as well as the influence of each unit process upon subsequent unit processes.

A current study by the Manufacturing Studies Board (MSB) of the National Research Council (NRC) is developing the future process science research needs for unit processes. The information developed by the study is the basis for this presentation, which discusses the concept of a unit manufacturing process and presents a taxonomy of five unit process families that categorizes unit processes by their prime change mechanism: mass change, structure change, phase change, consolidation, and deformation. The research needs of each unit process family are reviewed in terms of several areas of research which are crucial to the improved understanding of unit processes. These areas, know as enabling technologies, offer challenges and opportunities for future advances in processing science. Specific emphasis is given to the connecting role of process precision and metrology among the enabling technologies.

INTRODUCTION

Manufacturing is a significant portion of the U.S. economy and has experienced a loss in its global competitive position. One of the key factors contributing to the loss of manufacturing competitiveness and productivity is the corresponding reduction in research and development investment related to manufacturing. This extends to the processing equipment industry and has been shown to also hamper competitiveness of the manufacturing industry. Finally, the areas of advanced materials and manufacturing have been identified as technologies critical to the future economy and security of the U.S.

In addition, several other critical technologies are related to manufacturing, either as areas of application of manufacturing technology, or as technologies that support or enable future development of manufacturing.

Future technology driven competitiveness will depend more on new processes and less on new product technologies, because advances in reverse engineering techniques have made cloning of successful products possible. Coupled to this process technology base, an educated, skilled workforce will be the dominant competitive weapon.

Unit process science and the development of the understanding of unit processes are critical elements in the future growth of these critical technology areas. Improved understanding of the enabling science and technology issues will result in improved advanced manufacturing technology products and a more competitive U.S. manufacturing industry. These issues form the basis of a major study conducted by the MSB of the NRC focused on the future process science research needs of unit processes. The observations developed in the study are used as the foundation of the present discussion, which provides a high level summary and specific opportunities for process precision and metrology. The reader is referred to the final study report, available in early 1994, for detailed findings and recommendations of the committee.

UNIT MANUFACTURING PROCESSES

Manufacturing involves the conversion of raw materials, usually supplied in simple or shapeless forms, into finished products which have been designed to fulfill specific requirements. This conversion of materials into finished products is accomplished using a great variety of processes. These processes involve the application of energy to produce controlled changes in the configuration or properties of materials, or both. The energy applied during processing may be mechanical, thermal, electrical, or chemical in nature. The resulting properties and configuration changes are based on functional requirements defined during the product design stage. In the past, design, materials, and manufacturing were often treated as three independent engineering specialties. However, for cost effective and timely manufacturing, design for production, maintenance, and eventually recycling, may be as important as design for the efficient and reliable functioning of a product.

In many cases, product manufacture includes a series of sequential unit processes, referred to as a process stream. Each individual unit process of the stream has the output of the preceding unit process as its input material and is influenced by the characteristics of this material. The intermediate products may then be considered as the carrier of the process history of the preceding unit processes. The final product properties, including microstructure, are the summation of the individual unit process experiences, both positive and negative, and define the final part quality and performance in application. For example, the first stage may be "casting" in which liquid metal is poured into a mold to produce a desired shape. Next, the solidified metal may be machined with cutting tools to produce surfaces of specified form. Finally, some type of surface treatment may be employed to improve the durability of the part. These three individual operations -- casting, machining, and surface treatment -- are described as unit manufacturing processes. For brevity, we will refer to them as "unit processes". They are the individual steps along the route for production of finished goods and represent the value added to the original material as it proceeds through the transformation from raw material to a finished product.

The information and material flow associated with a typical unit process are shown in Figure 1. Raw material or parts from a previous unit process are the input. The outputs consist of parts, which are one step closer to their final form, and an influence on the environment, such as particulate or noise pollution. The information inputs to the unit process include product data, process information, process control methodology, and the resource requirements of the unit process such as equipment, energy, and people.

A unit process is optimized when the value added in terms of the required configuration and property changes are delivered to the workpiece in a cost effective manner. This involves minimization of several factors: energy use, scrap generation, labor costs, and capital equipment requirements. In addition, a rapid response to the needs of customers and a safe working environment are essential. Sequential unit processes, known as process strings, include cost factors which may result from previous unit processes, such as repair operations required by quality lapses of intermediate process steps. Thus, many factors enter into cost effectiveness which may be defined as "minimization of input and resource costs per unit of output product value."

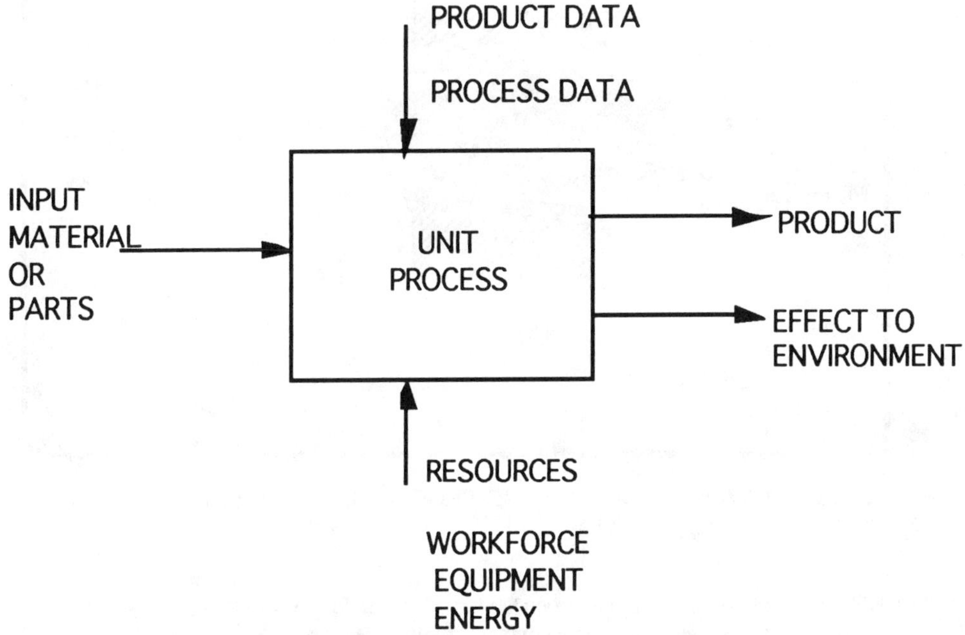

FIGURE 1. UNIT PROCESS INFORMATION AND MATERIAL FLOW

The following discussion considers the future research needs of unit processes. There is a vast number and a great variety of individual unit processes, both classical and novel. Compilations have been prepared which identify several hundred individual processes. In order to identify technical issues, common to unit processes, a schematic model which identifies five components common to all unit processes may be considered. Discussion of these components leads to a taxonomy which divides unit processes into five major families. The unit processes in the families may be applied to metals, polymers, ceramics, and composites of these three classes of materials. This three-dimensional framework of process components, process families, and materials provides a concise description of the broad topic of unit processes. From this framework, one can see that the understanding of essentially all unit processes depends on a number of critical technologies which we refer to as enabling technologies. These technologies focus the direction of future research in unit manufacturing processes.

A schematic model of a unit process is shown in Figure 2. Energy is delivered to the workpiece material by means of the process equipment and its tooling, and is transferred through an interface region between the tooling and the workpiece, often involving a medium such as a coolant or lubricant. The specific changes in the material workpiece configuration and structure often occur in a localized area of the workpiece, designated as the workzone. For example, a group of metal removal processes, loosely known as machining, includes the operations of turning, milling, drilling, boring, etc. Each of these processes is distinguished by the tooling design, the interface (represented by the cutting fluid), and the equipment design and characteristics, such as degrees of motional freedom, rate of workpiece or tool feed, and machine rigidity. The workzones of these processes are localized on the workpiece surface and involve shear deformation, which imparts a change in shape to the workpiece.

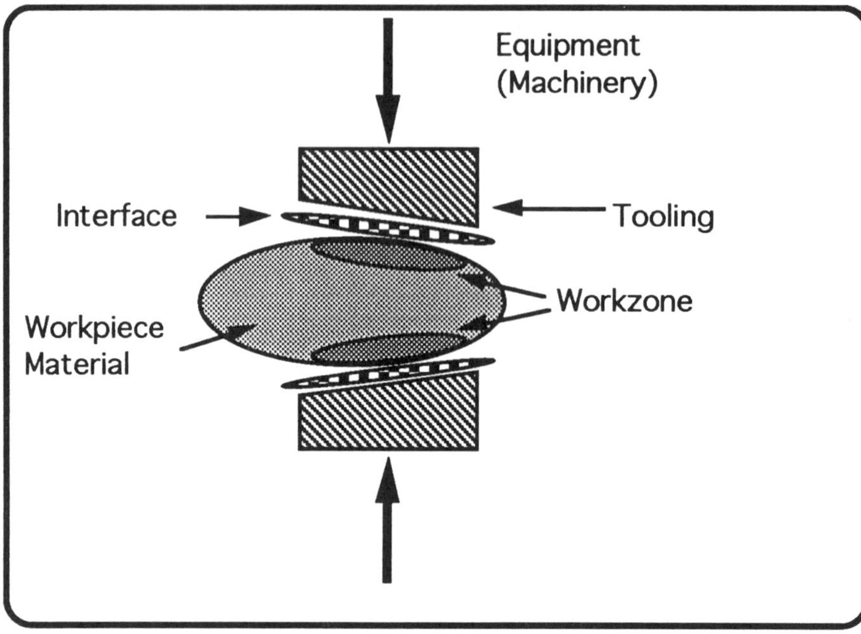

FIGURE 2 UNIT MANUFACTURING PROCESS MODEL SHOWING THE FIVE COMPONENTS OF A PROCESS

The .wide diversity of unit processes (such as machining, forging, casting, and injection molding) incorporates equally diverse groups of equipment, tooling designs, interface materials, and workzone mechanisms. The process equipment may belong to the groups of mechanical, thermal, chemical, photonic, and electrical equipment. Tooling elements include cutting tools, grinding media, dies, molds, forms, patterns, electrodes, and laser beams. Similarly, the array of interface materials typical of unit processes include lubricants, coolants, insulators, electrolytes, hydraulic fluids, and gases. The operative mechanisms found in the workzones of unit processes include deformation, solidification, fracture, conduction, convection, radiation, diffusion, erosion, vaporization, melting, microstructure change, phase transformations, chemical reactions, and many others. Examples of the five unit process components for several specific unit manufacturing processes are shown in Table 1.

Each of the five process components -- equipment, workpiece, tool, interface, and workzone -- are influenced by the other process components. For example, the interface conditions may govern the rate of energy transfer from the equipment to the workzone and may control the extent of the workzone localization and the uniformity of the changes in the workpiece shape and structure. In the machining process, variation in the thermal behavior or effectiveness of the cutting fluid may impose thermal distortions in the workpiece or equipment and result in a loss of process precision, manifested in products of poor quality.

Most processes involve several competing workzone mechanisms, with one mechanism overriding the others, at any given instant in the process. The design and selection of the process components and operating conditions are usually predicated on the assumption that a prime mechanism will remain dominant during process operation. In some instances, the process conditions may change such that an alternative mechanism becomes dominant, impacting the process operation and the resulting product quality. For example, in hot forging of jet engine disks from elevated temperature alloys extended contact time between the heated workpiece material and colder forging dies leads to increased heat flow to the dies. As a result, the workpiece material near the die-workpiece interface cools and does not deform as readily as the hotter zones of the workpiece. This non-uniformity in deformation adversely affects the microstructure and the properties of the forged disk. The solution to this problem is to control the speed

TABLE 1 EXAMPLES OF UNIT PROCESS COMPONENTS

Unit Process	Equipment	Workpiece	Tool	Interface	Workzone Mechanism
Machining	Lathe	Bar Stock	Single Point	Cutting Fluid	Deformation, Fracture
Surface Modification	Furnace	Part	Atmosphere	-- -- -- --	Diffusion Phase Change
Casting	Furnace	Melt	Mold	Release Agent	Solidification
Forging	Press	Preform	Die	Lubricant	Deformation
Powder Compaction	Press	Powder	Die	Lubricant	Deformation Consolidation

of the forging operation, thus minimizing the contact time of the workpiece and dies and the temperature loss at the workpiece surface. The end result is uniform deformation, and uniform microstructure and properties in the forged part.

The identification of process components provides a useful overview. However, several hundred individual unit manufacturing processes are commercially used in manufacturing operations. In order to discuss these processes in more detail, it is necessary to classify them using some common features. Many such classifications or taxonomies have been presented in the literature. The formation of a taxonomy is important in that it plays a vital role in subsequent discussions of unit processes. The taxonomy chosen for this study emphasizes the physical process by which the configuration or structure of a material is changed. This results in the selection of the following five families of processes.

Mass change processes involve the removal of material by mechanical, electrical, or chemical means. Included in this family are conventional processes of machining and grinding, as well as nonconventional removal processes such as electro-discharge and electro-chemical machining.

Phase change processes produce a solid part from material originally in the liquid or vapor phase. Casting of metals, infiltration of composites, vapor deposition, and injection molding of polymers represent this family. Phase changes in the solid state are included in the following family.

Structure change processes alter the microstructure of a workpiece, either throughout its bulk or in a localized area, such as its surface. Heat treatment and surface hardening are commercial processes representative of this family.

Deformation processes alter the shape of a solid workpiece without changing its mass or composition. Classical bulkforming metalworking processes of rolling and forging are in this category, as are the sheet forming processes of deep drawing and ironing.

Consolidation processes combine materials such as particles, filaments, or solid sections to form a part or component. Powder metallurgy, ceramic molding, and polymer matrix composite pressing are examples of consolidation processes. Joining processes, such as welding and brazing, also belong to this process family.

The processes in each family of the unit manufacturing process taxonomy can be applied to any material, metal, ceramic, polymer, or to the many composite formulations of these materials with polymers, metals, or ceramics as the matrix material. For example, processes of the consolidation family are used in the production of metals (powder compaction), ceramics (hot pressing), and polymer composites (autoclaving). In addition, each combination of process and material requires consideration of the five process components for successful production.

Figure 3 illustrates this interaction of process families, materials, and process components. This three-dimensional array was used as the framework to examine the current research needs and recommendations for developing criteria for selecting critical research topics for unit processes. In doing so, it became apparent that a thorough understanding of any unit process is dependent on six critical or key technologies that enable the correct design and operation of any process. These enabling technologies are simulation and modeling, material characterization, equipment design, sensors, process control methodology, and process precision and metrology.

Simulation and modeling involves the analytical and numerical representation of the five components of a unit process, as shown in Figure 2. If successful, this approach can eliminate time consuming and expensive "trial and error" process development and lead to rapid development of processes for new materials and new types of parts. In the past, this enabling technology was limited by the extensive amount of computation required. Now, with the rapid advances in high speed computers, the limitations lie in our physical understanding of both the process and material behavior.

Material characterization involves an understanding of the material properties and microstructure at the start of the process and as the material is modified during the process. A large database for properties under extreme conditions of temperature, strain-rate, etc., may be required. The evolution of microstructure, conditions under which fracture occurs, and an understanding of interface conditions such as friction are among the factors which have to be understood. There is a strong linkage between this enabling technology and the previous one dealing with simulation and modeling.

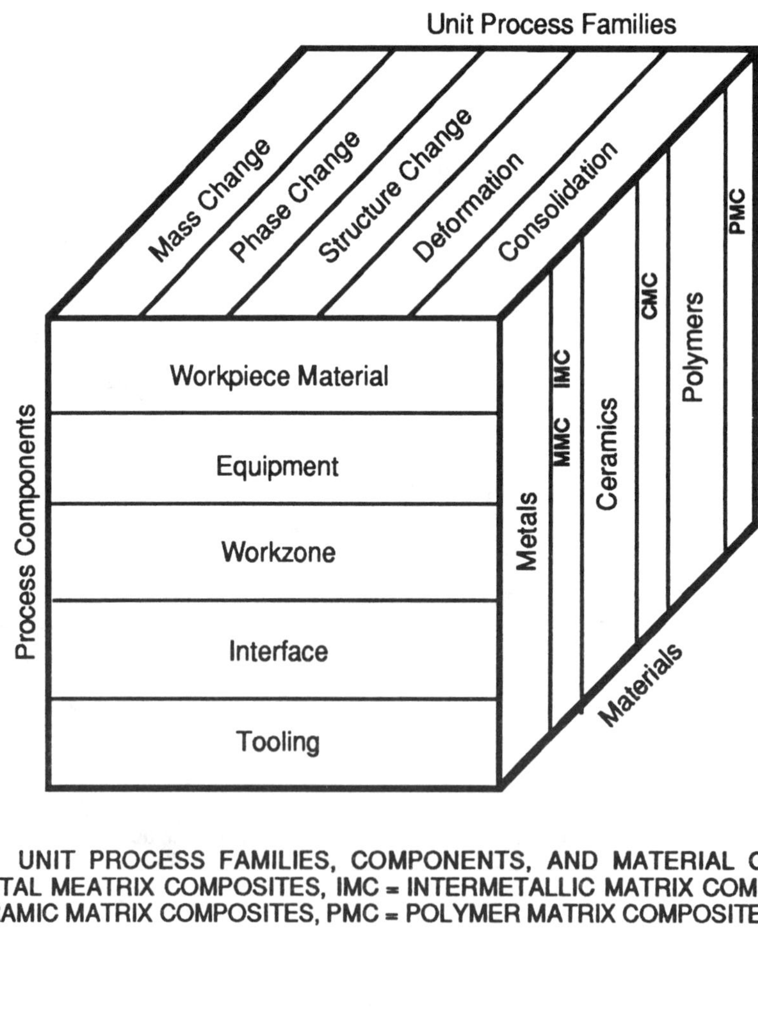

FIGURE 3 UNIT PROCESS FAMILIES, COMPONENTS, AND MATERIAL CLASSES. (MMC = METAL MEATRIX COMPOSITES, IMC = INTERMETALLIC MATRIX COMPOSITES, CMC = CERAMIC MATRIX COMPOSITES, PMC = POLYMER MATRIX COMPOSITES)

Equipment design is the central feature of any unit process. The equipment and associated tooling have to be designed to fulfill a specific function in a production environment. In addition, factors such as cost and time of fabrication, maintenance, flexibility of use, production rates, and quality of the parts produced are important aspects. Equipment may involve mechanical, electrical, chemical, and optical devices. Innovative design concepts have led to dramatic improvements in unit processes.

Sensors are independent devices for measurement of process conditions and the response of the material and interface. They may be remotely located, part of the equipment, contained in the workpiece, or in the interface between the workpiece and the tooling. Sensors are based on a great variety of physical phenomena and the development of innovative new sensing devices could greatly enhance the operation of unit processes. By providing real-time measurement of processing conditions, sensors enable control of processes so that they may be optimized during operation.

Process control involves the development of improved software and hardware for the optimum operation of process equipment. For example, recent advances in "intelligent" process control methods introduce spontaneous, mid-cycle process changes to ensure high quality parts. Past control methodology often used the "black box" approach and described the process with a simplified transfer function. An improved understanding of the unit process, together with improved sensors, should lead to improved control procedures for processes.

Process precision and metrology is listed as a separate enabling technology because of the pressing need to develop a better understanding of the technical factors which control the accuracy or precision of manufacturing. In addition to the need to build precision into a process, subsequent measurement is required to ensure that satisfactory parts are being produced. The specialized, but important, topic of metrology or precision measurement certainly requires greater attention from researchers.

The six enabling technologies are also dependent upon each other. For example, advanced equipment design efforts often utilize simulation of the equipment operation, including accurate modeling of the workpiece and interface materials behavior during processing. In addition, the design of process equipment is highly dependent on an understanding of the process precision and metrology which are, in turn, integral with the sensor selection and process control methodology. These relationships between and among the six enabling technologies and the five major process components are depicted schematically in Figure 4. The concentric (hierarchical) structure emphasizes that knowledge flows from the six enabling technologies to the process components, which, in turn, form the unit process.

For example, development of process equipment and tooling requires knowledge in the areas of equipment design, sensor formulation and selection, identification of process precision limits and metrology, and appropriate control methodology for production of high quality products. Understanding of both the interface and workzone components of unit processes and their influence on the workpiece requires detailed simulation of the workzone and interface physics, coupled with concise characterization of the behavior of both the workzone and the interface materials. In addition, the initial and boundary conditions for a process simulation are based on an understanding of the process equipment design, control, and precision.

RESEARCH NEEDS

Research and development on unit processes enhance the knowledge level of the unit process and allow the production of better, cost effective and competitive products. Over the long run, understanding of manufacturing processes improves our standard of living and provides jobs for our citizens in a competitive global environment. Research and development activities are located in the enabling technology groups and are typically focused on a particular process and material. The understanding developed by such a focused activity often may be applied to other materials and processes.

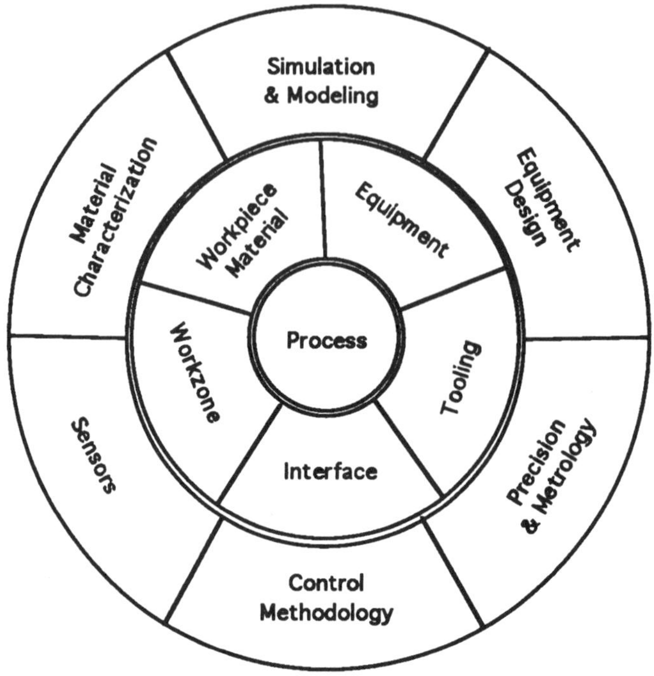

FIGURE 4 UNIT PROCESS COMPONENTS AND ENABLING TECHNOLOGIES

Unit process research and development may involve the processing of traditional or emerging materials by either conventional or novel processes. That is, significant benefits may result in optimization and improvement of conventional processing of high-volume conventional materials. Improved quality and costs, although incremental per unit of material, are significant when applied to large production quantities. Application of conventional processes to advanced or emerging materials is often cost effective because existing equipment and facilities are utilized. The extension of current process practices to these materials, however, requires developing a new understanding of the process requirements specific to the new material. Since these materials typically display high-performance characteristics, their use in manufacture often requires high-performance processes. Novel processes may be applied to either conventional or emerging materials. An example of the former is the development and application of net shape processes to structural materials, such as the near-net-shape casting of steel structural shapes. In certain situations, advanced materials require novel processes for production of quality products. Many of the metal matrix and ceramic matrix composite materials are in this group and utilize novel processes such as plasma spraying, atomization, vacuum hot pressing, hot isostatic pressing, and isothermal deformation processing.

The enabling technology of process precision and metrology has several areas of research that will advance unit process understanding. These include extension of process modeling and simulation to include prediction of part geometric variations and definition of tolerancing standards in mathematical terms. These issues must be addressed by technical representations from government, industry and academics. Specific actions are listed below.

1) A major effort is needed to devise models and metrics for characterizing the precision of unit processes in ways useful to process planners and designers. A two-pronged effort is likely to be needed, in which models and metrics are first sought in the natural parameters of each process, and then means are sought to translate or interpret these in terms of design tolerances.

2) A long term effort is needed to assess quantitatively, for the most important members of the various process families:
   a) their scalability;
   b) their intrinsic or ultimate precision, as set by the underlying physics or chemistry of the process; and
   c) the precision currently available in typical industrial implementations.

The results of such studies should be incorporated in the models and metrics developed under Action 1.

3) Academics should be encouraged to participate in all of this work, including that of the standards committees, and to lead some of it. One of the most important contributions academics can make is codifying systematically the currently scattered knowledge in these important fields.

4) Academic should be encouraged to teach the essential principles of tolerancing, metrology, and process modeling in engineering design and manufacturing courses. At present only process modeling receives more than cursory attention.

ACKNOWLEDGMENT

The information contained in this manuscript was supported by funding of the National Science Foundation and the Department of Defense through NSF Grant DDM-9022041. The valuable contributions of the MSB study committee members are also acknowledged.

CRTD-Vol. 27, International Forum
on Dimensional Tolerancing and Metrology
ASME 1993

COMPARISON OF VECTORIAL TOLERANCING AND CONVENTIONAL TOLERANCING

Georg Henzold
Siemens AG
Mülheim an der Ruhr, Germany

ABSTRACT

Coordinate measuring machines probe points of the workpiece surface. They store the coordinates of these points and evaluate the parameters of the workpiece geometry by calculation using appropriate computer programs.

This method enables to achieve a new approach to assess the workpiece geometry called vectorial dimensioning and tolerancing which is dealt with in ISO TR 10 360-1 (under preparation).

This new method is described. The differences between this method and the usual conventional method according to ISO 286, ISO 1947, ISO 5459, ISO 1101 etc. are shown. The advantages and disadvantages of each method are explained.

The vectorial method defines the components of size and geometry to be influenced in the manufacturing process in a clear and distinct way. It enables proper manufacturing control.The vectorial method separates form deviations from orientation deviations. Thereby it provides proper means to specify particular functional requirements, e.g. of kinematic linkages.

GENERAL

The technique of coordinate measuring machines enables to achieve a new approach to assess the workpiece geometry, i.e. to assess geometrical properties which in the past could hardly be assessed.

Coordinate measuring machines probe points of the workpiece surface. They store the coordinates of these points and evaluate the parameters of the workpiece geometry by calculation using appropriate computer programs. Provided the availability of appropriate computer programs various properties of the workpiece geometry according to various geometrical definitions can be assessed. So the deviations of workpieces according to the conventional method of dimensioning and tolerancing as well as to the vectorial method of dimensioning and tolerancing can be assessed.

CONVENTIONAL DIMENSIONING AND TOLERANCING

Conventional dimensioning and tolerancing is based on the outward or inward of the material directed extreme points of the surface.

The relevant International Standards are:

ISO 129	Dimensioning
ISO 286	Linear dimensional tolerances
ISO 1947	Angular dimensional tolerances
ISO 1101	Geometrical tolerancing
ISO 5459	Datums and datum systems
ISO 1660	Tolerancing of profiles
ISO 3040	Tolerancing of cones
ISO 5458	Positional tolerancing
ISO 2692	Maximum and least material requirements

The corresponding American National Standard is ANSI Y14.5M.

VECTORIAL DIMENSIONING AND TOLERANCING

Introduction

Vectorial dimensioning and tolerancing is based on substitute elements. Substitute elements are geometrically ideal features (e.g. straight line, circle, plane, cylinder, sphere, cone, torus). On a real workpiece the actual substitute element is calculated from the assessed points of the workpiece surface (e.g. by using a coordinate measuring machine) using the Gaussian algorithm. So the vectorial dimensioning and tolerancing is based on all points of the surface, not only on the extreme points as in case of conventional dimensioning and tolerancing.

The locations and orientations of the substitute elements are expressed by location vectors and orientation vectors in a workpiece coordinate system (substitute datum system). Form deviations are considered to be superimposed on the substitute element.

Actual substitute elements with their actual location and actual orientation vectors, their actual substitute sizes (diameters of cylinders, spheres, tori, apexes of cones) and form deviations provide a description of the actual workpiece geometry as assessed by measuring.

Nominal substitute elements with their nominal location and nominal orientation vectors and their tolerances and the nominal substitute sizes and their tolerances and the form tolerances may be specified by the designer.

Description of the Principle

Points in space are defined by their coordinates Xo, Yo, Zo (location vector P) and stored for further data processing. Orientations in space are defined by the components Ex, Ey, Ez of the unit vector (orientation vector N). It therefore applies that $N = \sqrt{(Ex^2 + Ey^2 + Ez^2)} = 1$. The orientation vector is always directed out of the material, Fig. 2.

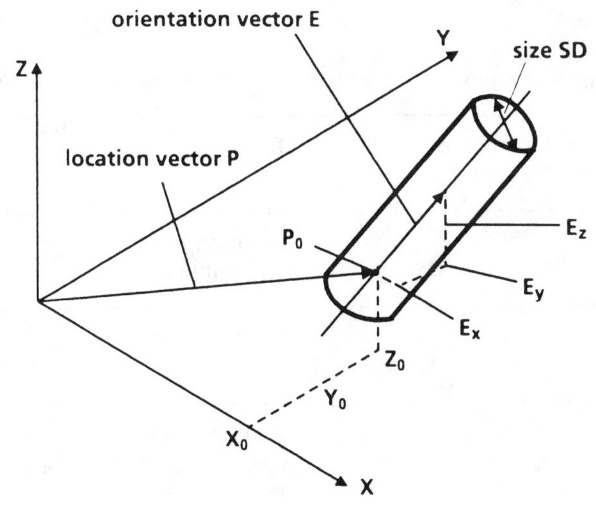

Fig.1: Definition of a substitute cylinder (location vector, orientation vector, substitute size)

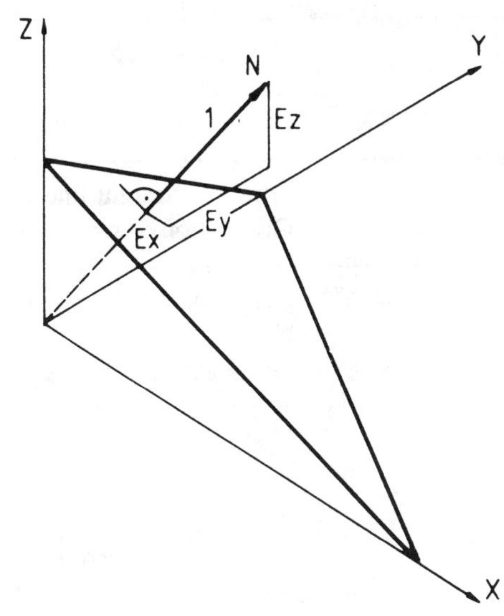

Fig. 2: Orientation vector of a plane

Table 1 shows which vectors are needed to define the substitute element. From the stored coordinates of the sensed workpiece points the actual substitute element is calculated, Fig. 3.

Fig. 4 shows the principle of specifying the nominal geometry and sizes of a part and their tolerances. The features of the part are enumerated. To each feature the nominal vectors and their tolerances are specified. The workpiece coordinate system is specified by substitute elements.

Table 1: Substitute elements

Element	Location Xo Yo Zo	Orientation Ex Ey Ez	Size Size	Remarks
Plane	X			Point on plane
		X		Normal to plane
Sphere	X			Centre point
			X	Diameter
Cylinder	X			Point on axis
		X		Orientation of axis
			X	Diameter
Cone	X			Point on axis
		X		Orientation of axis
			X	Cone angle
Torus	X			Centre point
		X		Orientation of axis
			XX	Diameters of pipe and ring
Straight line	X			Point on straight line
		X		Orientation of straight line
Circle	X			Centre point
		X		Normal to plane
			X	Diameter

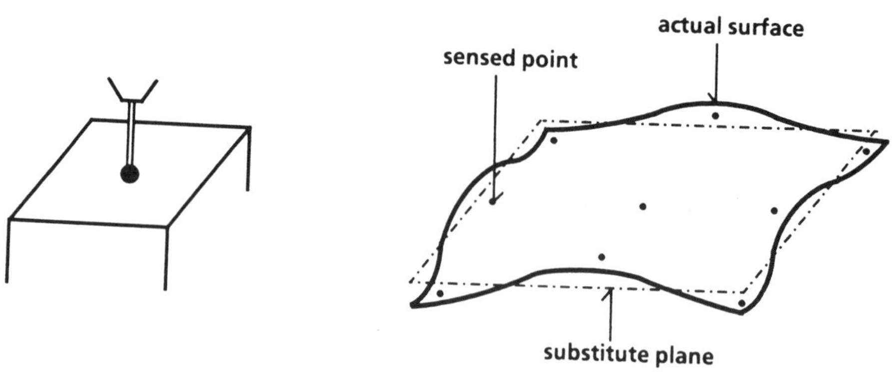

Fig. 3: Sensed points and actual substitute element of a plane workpiece surface

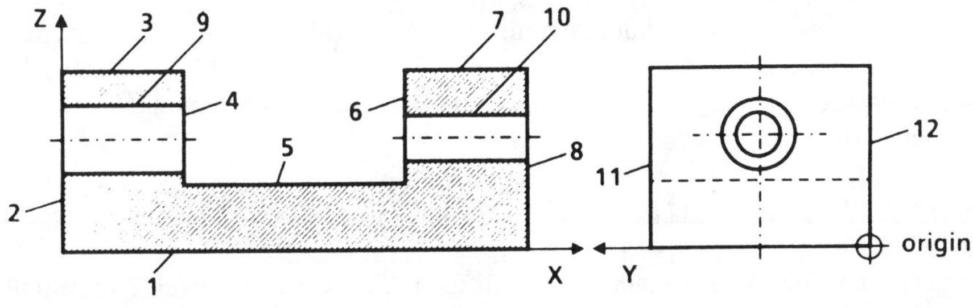

No	Location v.		Orientat. v.		Size 1		Size 2		Form Tol.
	Nom.	Tol.	Nom.	Tol.	Nom.	Tol.	Nom.	Tol.	
	XoYoZo		ExEyEz		M 1		M 2		
1									
2									
3									

Fig. 4: Principle of vectorial dimensioning and tolerancing

Definitions

Substitute Datum System. The substitute datum system is the coordinate system of the wokpiece derived from designated substitute elements. It is defined by the primary, secondary and tertiary substitute datums as follows:

Primary substitute datum (xy plane) defined by:

- specified substitute plane or
- plane containing the substitute axis of the primary datum feature and the specified substitute intersection point (apart from the substitute axis, intersection of another substitute axis with a substitute plane)

Secondary substitute datum (xz plane) defined by:

a plane perpendicular to the primary datum and containing

- the intersection line of the substitute plane of the secondary datum feature with the primary datum or
- the substitute axis of the primary datum feature (which becomes thereby also the secondary feature, see Fig. 9) or
- the line between two substitute intersection points specified as secondary datums,

this line being not normal to the primary datum

Tertiary substitute datum (yz plane) defined by:

a plane perpendicular to the primary and to the secondary datums and containing

- the point of intersection of the substitute plane of the tertiary datum feature with the primary and secondary datum or
- the substitute intersection point specified as tertiary datum.

Location and orientation of the substitute elements are related to this coordinate system (substitute datum system).

Linear Substitute Size. Linear substitute sizes are the diameters of substitute spheres, cylinders, cones, tori.

The linear substitute size deviation is the difference between the actual substitute size and the nominal substitute size.

The linear substitute size tolerance is the difference between the upper and lower limit of the linear substitute size deviations.

Angular Substitute Size. Angular substitute sizes are the cone angles (apexes) of substitute cones.

The angular substitute size deviation is the difference between the actual substitute cone angle and the nominal substitute cone angle.

The angular substitute size tolerance is the difference between the upper and lower limit of the angular substitute size deviations.

Substitute Location. The substitute location is defined by the substitute location vector. It indicates the location of a specified point of the substitute element in the substitute datum system. The specified point can be:

- substitute intersection point
- point of the substitute element specified to ly

  in case of a plane on a straight line defined by the two coordinates and perpendicular to the plane defined by the two pertinent coordinate axes, see Fig. 6

  in case of an axis on a plane defined by one coordinate and parallel to the plane defined by the two other coordinate axes, see Fig. 7
- centre of a substitute sphere
- centre of a substitute torus

The substitute location deviation is the difference between the actual substitute location vector and the nominal substitute location vector. It can be expressed by its three components in the directions of the axes of the substitute datum system.

The substitute location tolerance is the difference between the maximum permissible substitute location deviations expressed in the three directions of the axes of the substitute datum system. In case of a specified point of a substitute plane two coordinates are theoretical exact and only one is to be toleranced, Fig. 6. In case of a specified point of a substitute axis one coordinate is theoretical exact and two are to be toleranced, Fig. 7.

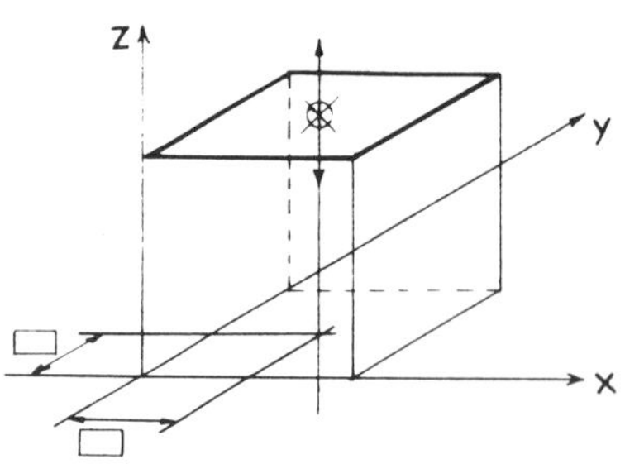

Fig. 6: Directions of substitute location tolerance of a plane

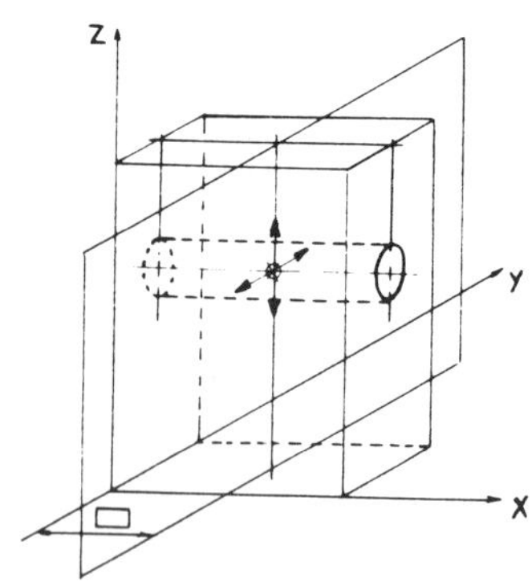

Fig. 7: Directions of substitute location tolerance of a cylinder

Substitute Orientation. The substitute orientation is defined by the substitute orientation vector. It indicates the orientation of the substitute plane or the substitute axis within the substitute datum system. The substitute orientation vector is normal to the substitute plane or parallel to the substitute axis (cylinder, cone, torus) and points

- in case of a plane away from the material
- in case of a cone in the direction of the greater size.

The substitute orientation vector N has the value 1 and is defined by its components Ex, Ey, Ez which are in the directions of the axes of the substitute coordinate system (substitute datum system).

The substitute orientation deviation is the difference between the actual substitute orientation vector and the nominal substitute orientation vector. It can be expressed by its three components in the directions of the axes of the substitute datum system.

The substitute orientation tolerance is the difference between the maximum permissible substitute orientation deviations expressed in the directions of the axes of the substitute datum system. In order to avoid overspecifying only two of the three components of the substitute orientation vectors can be toleranced (the third is already determined by the difference to 1).

Form. The nominal form is defined by the drawing or by indication in a table. In case of a bounded form (sphere, cylinder, cone, torus) and indication in a table it must be stated whether it is an inner form (hole) H or an outer form (shaft) S.

The local form deviation is the distance of the workpiece feature from the substitute element at the considered location. The total form deviation is the range of the local form deviations, see Fig. 8. The form tolerance is the value of the permitted total form deviation.

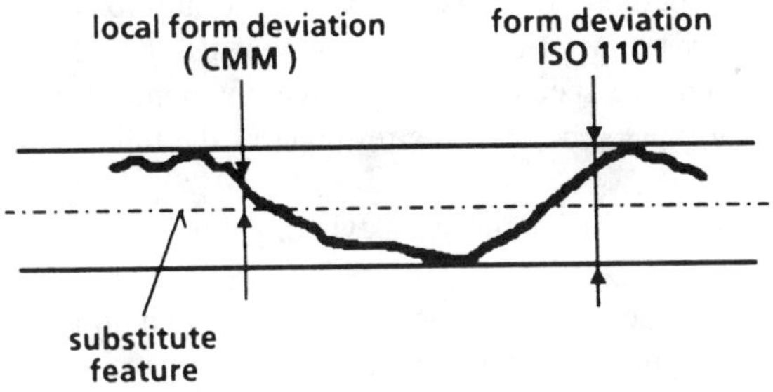

Fig. 8: Local form deviation and total form deviation

Example of Vectorial Dimensioning and Tolerancing. Fig.9 shows an example of vectorial dimensioning and tolerancing. In this example the following symbolization is used:

substitute element number (No 2)	e.g. O-2
specified points of substitute elements (No 5 and 6)	e.g. ⊠ 5,6
substitute intersection point of a substitute axis and a substitute plane	e.g. ⊠ 2+3
substitute datum established by an actual substitute element	e.g. SA
substitute datum established by a combination of an actual substitute axis and an actual substitute intersection point or by two actual substitute intersection points	e.g. SA-SB

In the example, Fig. 9, the workpiece coordinate system is determined by:

- axis of substitute feature No.1, intersection of axis of substitute feature No. 2 and substitute feature No. 3 (SA-SB) to establish the xy-plane
- axis of substitute feature No. 1 (SA), perpendicular to xy-plane to establish the xz- plane
- intersection of substitute feature No. 2 and substitute feature No. 3, which is the specified point 2+3 (SB), perpendicular to xy-plane and xz-plane to establish the yz-plane.

From these three planes (xy, xz, yz) derives that the specified point 1 is the origin of the co-ordinate system.

With the enumeration of specified points the sign "+", e.g. "2+3", indicates that it is the point of intersection of two substitute elements, the sign ",", e.g. "5, 6" indicates that there are two different specified points of the same nominal location.

The location tolerances apply to the specified points of the substitute elements, e.g. for the substitute element 2 determined by the specified point 2+3 in the y-direction (because in the z-direction it coincides by definition of the workpiece coordinate system system with the coordinate z = 0) and for the substitute element 5, determined by the specified point 5 in the y- and in the z-direction (in the x-direction the point is specified by the nominal value 45 which determines at which point of the substitute element axis the location deviation shall be assessed).

The orientation tolerances apply to the substitute elements themselfes. The axis of substitute element No. 1 has no tolerance because it establishes the x-direction of the workpiece coordinate system.

Regarding orientation the permissible deviation are only specified for 2 components of the orientation vector, because the permissible deviation of the third component is already determined by the condition that the orientation vector (resultant of the 3 components) has the value 1.

When the nominal orientation vector is not parallel to any axis of the coordinate system permissible deviations may be specified for two components only or for all three components. When permissible deviation are specified for all three components the extreme deviation can not occur on the same workpiece because the resultant of the three actual components has the value 1 (by definition).

With form the indication "Zyl H" indicates a cylindrical hole, "Zyl S" indicates a cylindrical shaft.

The vectorial tolerancing specifications can be amended by additional specifications on envelope requirement, maximum material requirement, surface roughness, surface treatment condition etc.. In Fig. 9, e.g., the maximum material requirement is specified by the maximum material virtual size (boundary) ø 19,94 to the elements No 1, 2, 4 and 5 (No 1 relative, i.e. coaxial, to No 4 and No 2 relative, i.e. coaxial, to No 5).

COMPARISON OF VECTORIAL TOLERANCING AND CONVENTIONAL TOLERANCING

With vectorial dimensioning and tolerancing sizes, distances and deviations of orientation and location relate to the substitute elements and not to the actual features exhibiting form deviations. Therefore these sizes, distances and deviations are to be distinguished from those according to ISO 8015, ISO 286, ISO 1947 and ISO 1101.

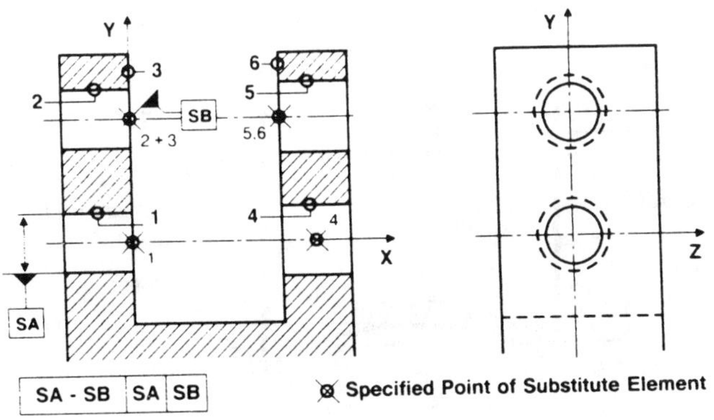

Primary Datum	1 - 2 + 3	(Substitute El. Intersection Point)
Secondary Datum	1	(Substitute Element)
Tertiary Datum	3	(Substitute Element)

Substitute Element No			1	2	3	4	5	6
Location	Nom.	Xo Yo Zo	0 0 0	0 35 0	0 35 0	55 0 0	45 35 0	45 35 0
	Perm. Dev. ±	x y z	– – –	– 0,1 –	– – –	– 0,1 0,1	– 0,1 0,1	0,1 – –
Orientation	Nom.	Ex Ey Ez	−1 0 0	−1 0 0	+ 1 0 0	+ 1 0 0	+ 1 0 0	−1 0 0
	Perm. Dev. ± x 10^{-3}	ex ey ez	– – –	– 0,2 0,2	– 0,2 0,2	– 0,2 0,2	– 0,2 0,2	– 0,2 0,2
Size	Nom.		15	15		20	20	
	Perm. Dev. ±		0,05	0,05		0,05	0,05	
Form	Nom.		Zyl. H	Zyl. H	Plane	Zyl. H	Zyl. H	Plane
	Tol.		0,005	0,005	0,01	0,005	0,005	0,001
Waviness	W_t µm		–	–	–	–	–	–
Roughness	R_z µm		6	6	16	6	6	16
Further Requirements			X	X		X	X	

No	**Further Requirements**
1	MMVS DIA 14.94 rel. to No 4 MMVS DIA 19.94
2	DIA 14.94 rel. to No 5 MMVS DIA 19.94
4	See No 1
5	See No 2

Fig. 9: Example of vectorial dimensioning and tolerancing

Sizes

Figure 10 shows a cylinder with form deviations. The actual local sizes according to ISO 8015 and ISO 268 vary between the minimum and the maximum actual local sizes. The actual substitute size in this case is the diameter of the substitute cylinder.

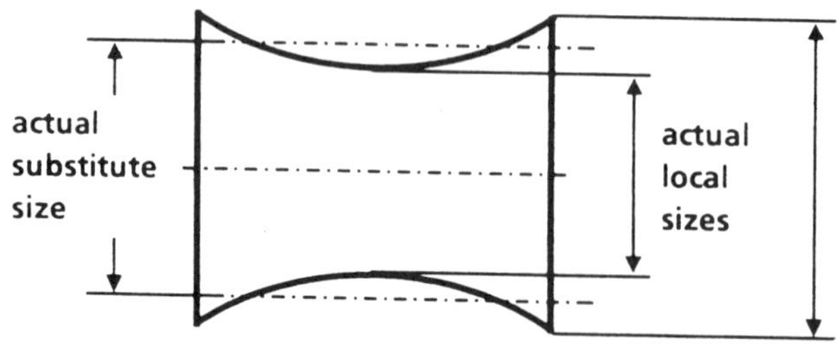

Fig. 10: Cylinder with form deviations, actual local sizes and actual substitute size

Figure 11 shows a cone with form deviations. The actual local cone angles according to ISO 8015 and ISO 1947 vary between the maximum and the minimum actual local cone angles. The actual substitute cone angle is the cone angle of the substitute cone (which deviates from the nominal cone angle).

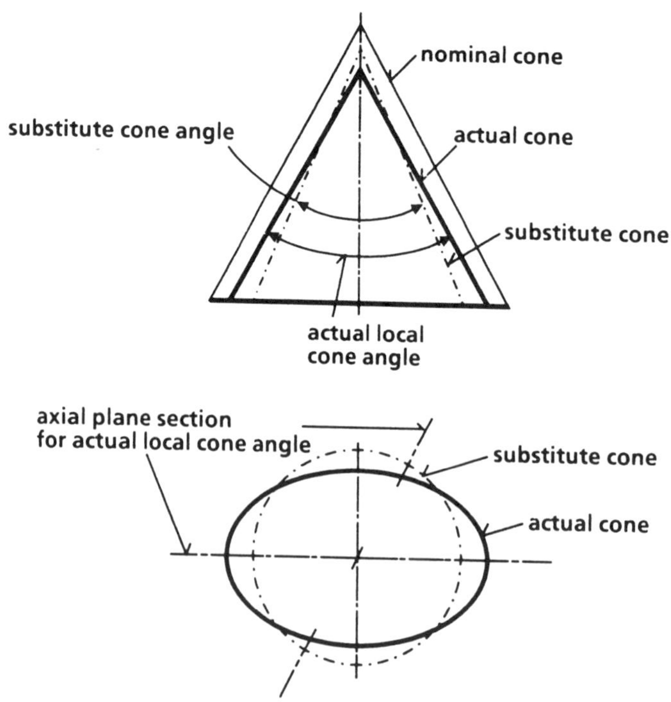

Fig. 11: Cone with form deviations, actual local cone angles and actual substitute cone angle

Figure 12 shows a bar with holes. The bar has deviations from straightness. The actual local distances between the edge and the hole centers according to ISO 8015 are approximately the same at all locations. The distances of the hole centers from the substitute plane, however, differ throughout.

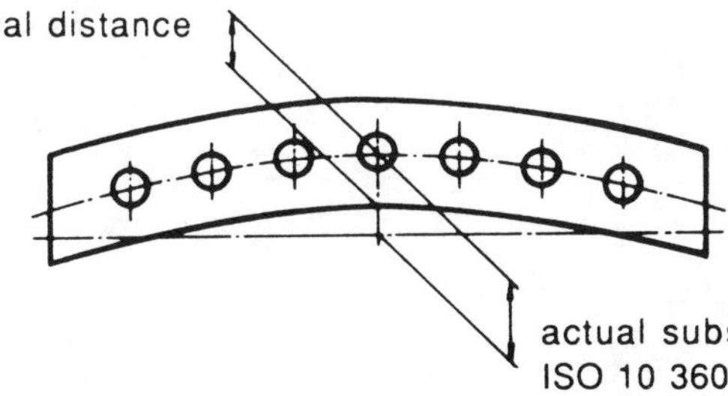

Fig. 12: Bar with holes and deviations from straightness, actual local distances and distances from the substitute plane

Depending on the data processing program incorporated the coordinate measuring machine can assess both types, actual local sizes (length sizes and angular sizes) according to ISO 8015 and actual substitute sizes and distances.

Deviations of Orientation and Location

With vectorial tolerancing deviations of orientation are obtained as an inclination of the substitute element Fig. 13(difference of orientation vectors). In this procedure the form deviation is eliminated. The sign of the orientation deviation indicates the direction of the orientation deviation. These are the main differences to the deviation of orientation according to ISO 1101.

Figure 14 shows a workpiece with deviations of coaxiality. The deviation of coaxiality according to ISO 1101 also includes the form deviation of the axis of the cylinder to be measured and is therefore greater than the coaxiality deviation of the substitute element.

Figure 15 shows a workpiece with a positional deviation . The positional deviation according to ISO 1101 is greater than the positional deviation of the substitute elements.

The positional deviation according to ISO 1101 includes the form deviation of the toleranced feature and is related to the datum which contacts the datum feature. The positional deviations of the substitute elements differ from those according to ISO 1101 because the form deviations of both features, the toleranced feature and the datum feature, are eliminated when the positional deviation of the substitute element is assessed and the substitute element intersects the actual feature.

Coordinate measuring machines assess substitute orientation and location deviations. When appropriate data processing programs are available they can also calculate deviations of orientation and location according to ISO 1101.

Therefore, however, it is necessary to specify the definition of the actual axis of a cylindrical or conical feature. According to ISO 1101 it is assumed that the actual axis follows the form of the feature and therefore deviates from a straight line.

Datum systems

With vectorial dimensioning and tolerancing the datum system is based on substitute elements and their intersections. Workpiece datum systems according to ISO 1101 and ISO 5459, however, are based on contacting elements contacting the highest points of the workpiece surfaces and directed according to the minimum rock requirement. Fig. 16 shows the differences.

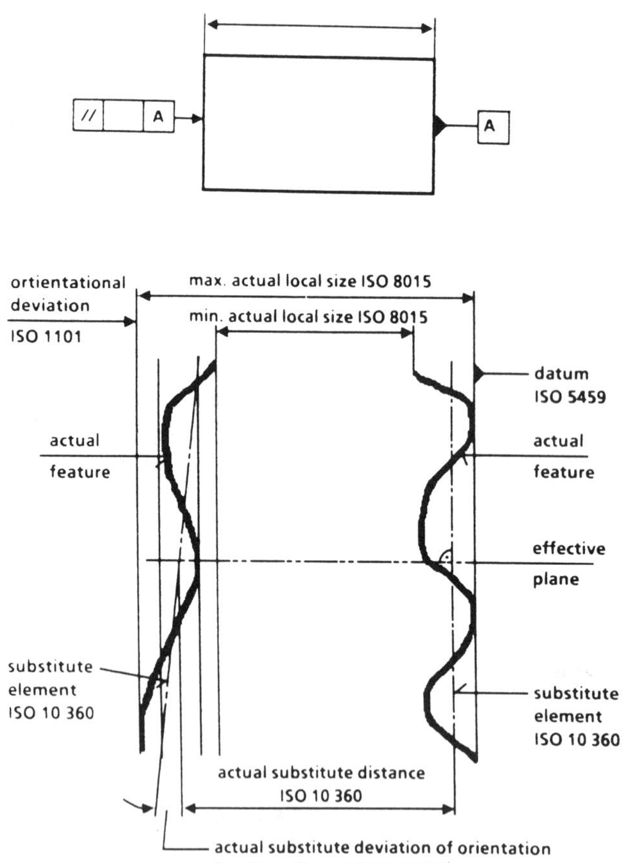

Fig. 13: Deviation of orientation (parallelism) according to ISO 1101 and orientational deviation of the substitute element

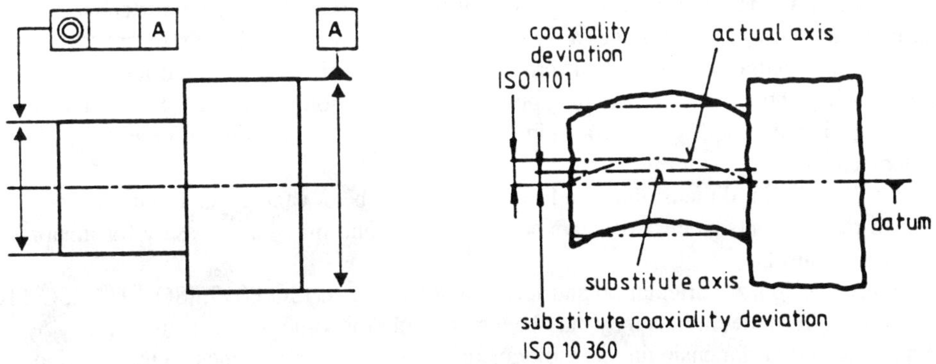

Figure 14: Coaxiality deviation according to ISO 1101 and coaxiality deviation of the substitute element

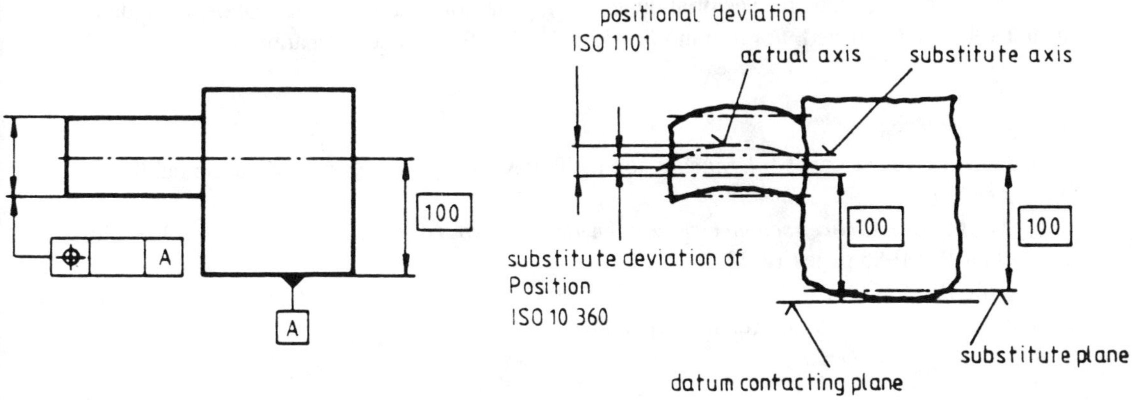

Fig. 15: Positional deviation according to ISO 1101, positional deviation of the substitute element

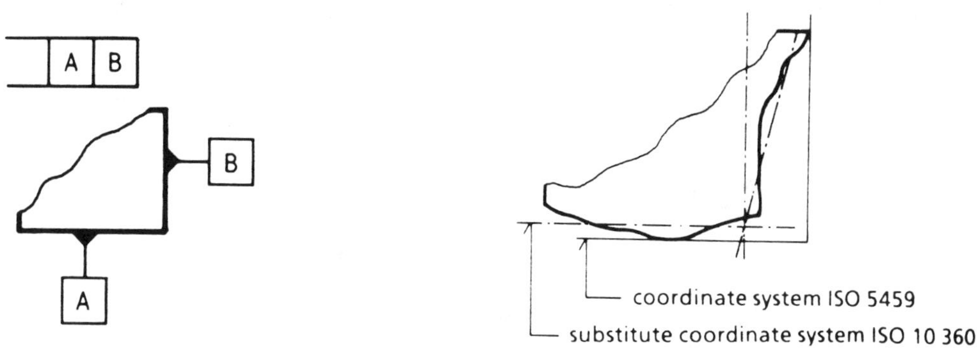

Fig. 16: Datum systems according to vectorial dimensioning and tolerancing and according to conventional dimensioning and tolerancing

APPLICATION OF VECTORIAL DIMENSIONING AND TOLERANCING

As form deviations are eliminated by calculation from the substitute elements vectorial dimensioning and tolerancing provides a clear distinction between substitute size, substitute location, substitute orientation and form. It provides further the direction of the orientation and location deviation and gives therefore the appropriate information for the direction of the correction of the production equipement (machine tool). This contrasts to conventional dimensioning and tolerancing.

Therefore vectorial dimensioning and tolerancing may be advantageous for specifying some functional requirements as e.g. for gears and gear boxes and may be necessary for manufacturing process control.

As deviations in size, orientation and location according to ISO 8015, ISO 1947, ISO 1101 include always form deviations they may be less suitable for some of these purposes.

Which method of dimensioning and tolerancing should be used depends on the functional requirements and on the requirements of the manufacturing process control.

As size, orientation and location according to both methods may have quite different measuring results at the same workpiece there must be a clear distinction between both methods for the drawing indications as well as for the communication in the field of inspection.

For drawing indications for vectorial dimensioning and tolerancing an International Standard has not yet been published. For terminology ISO TR 10 360 is under preparation.

REFERENCES

G. Henzold, *Geometrical Tolerancing,* John Wiley & Sons, Inc., New York, to be published December 1993

A. Wirtz, *Die theoretischen Grundlagen der Dreikoordinatenmeßtechnik*, Technische Rundschau 38 / 1985 pages 12-23

ISO TR 10 360-1, *Co-ordinate metrology - Part 1: Definitions and applications of the fundamental geometric principles*, (under preparation)

CRTD-Vol. 27, International Forum
on Dimensional Tolerancing and Metrology
ASME 1993

DIMENSIONAL CONTROL THROUGH THE TOLERANCING OF THE SOLID MODEL

Benjamin R. Taylor
Renishaw plc
Wotton-under-Edge, Gloucestershire, United Kingdom

ABSTRACT

The use of conventional tolerancing methods such as the ANSI Y14.5 or BS308 requires the use of detailed drawings, but does provide a very comprehensive means of defining the nominal geometry of the part and the allowable variations of position, size and form.

If the part manufactured is being accomplished within a single system which encompasses the concept part configuration and tolerancing, and also the manufacturing process, it is possible to utilise other methods to provide the same function.

Renishaw has developed a process in which the solid model is developed within the Computer Aided Design system that has been toleranced utilising a specially developed tolerancing scheme, whereas individual gauge points are located on the surface with an allowable variation specified.

As a result, no conventional detailed drawings are developed or produced and the process control method is then part of the specification of the part itself. This vastly reduces the design time and allows a comprehensive on machine process control method to be used.

RENISHAW'S NEED

Renishaw manufactures a wide range of products for the metrology industry and has been implementing a program to significantly reduce product development lead times. This program addressed all aspects of product development, including part conception, the generation of drawings, the development of the required NC programing for making the components and the methods of measurement used for process control and part acceptance. In the past, most of these tasks were usually carried out twice, once for the pre-production batches and then again for the production runs because different machines were usually used for the small batches than were used for the larger production quantities.

The initial effort was focused on reducing machining time by developing an in-house automatic load and unload system for milling machines, which incorporated tool handling and artefact comparison measurement, to significantly improve unmanned part component machining accuracy, as well as drastically reducing the set-up time. This resulted in a machining system that can effectively be used not only for "one offs" but also for our largest production runs.

During this effort, and in conjunction with the purchase of a new Unigraphics CAD/CAM system, we also addressed another factor which was the time taken to generate engineering drawings and the associated interpretations of these drawings in regard to the inspection for process control and part acceptance.

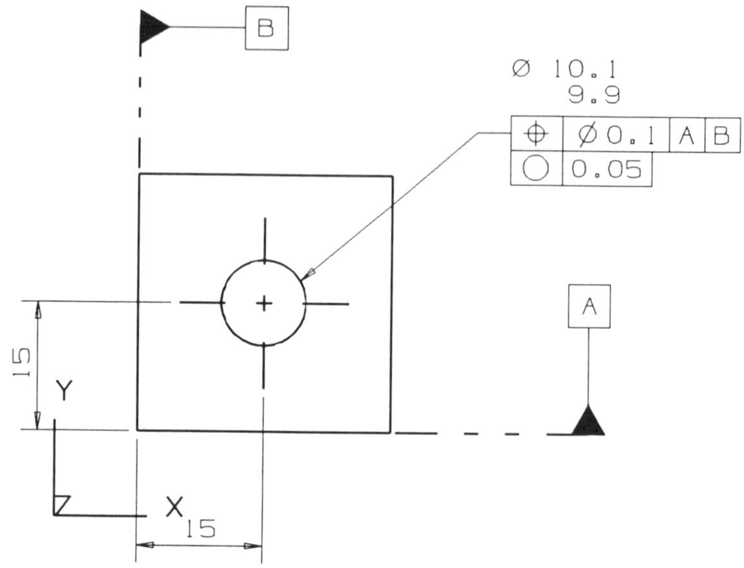

FIGURE 1

The traditional process of designing a product, building samples, inspecting them using conventional tools and then testing these small numbers might prove the product concept but in no way proves out the manufacturing methods that will be used for production. These methods used to make the pre-production samples are normally different to the methods used for production manufacturing. Consequently we felt that the standard production methods must be utilised for the production of model batches and these methods must then lend themselves to much larger batch sizes so that the initial batch can be large enough to determine the production process variations, as well as prove the concept. As an example, instead of a small batch of 5 using other than standard production methods, ie the model shop, 100 samples could be made using the standard production methods. This required collapsing the time taken so that the difference between the two is minimal.

In conjunction with this, we looked at redefining the methods used to specify and tolerance manufactured components. The process was to establish a closer relationship between the conception of the components, the specifications of these concepts and the manufacturing and inspection methods used to produce these components.

COMPONENT SPECIFICATION

First of all, standard engineering drawings are presently used to specify the part in its nominal configuration, as well as the allowable deviations of features from those nominals. These drawings are then used to determine the measurement processes that are used for both process control and part acceptance. However, prior to manufacturing the part the manufacturing processes need to be established. It is the choice of processes that determines the accuracy to which the parts are manufactured.

During our review it became apparent that if a method of manufacturing is analysed carefully enough, the specification and tolerancing of the components need not necessarily be on the engineering drawing and the whole concept of tolerancing features, in relationship to other features, can be modified to be more closely related to the process of manufacturing.

A standard NC machining center is commanded to move through a series of points and, with the cutting tool, will generate the surface of a feature. All these methods are referenced to a theoretical X=0, Y=0, Z=0 position within the machine working envelope so, during the process of producing surfaces, the machine is always attempting to pass through the commanded positions to give an acceptable resultant surface. In essence, the measurement of the resulting surface against its accepted value is the actual error that the machine generates and is the one that must be controlled to be able to make parts successfully.

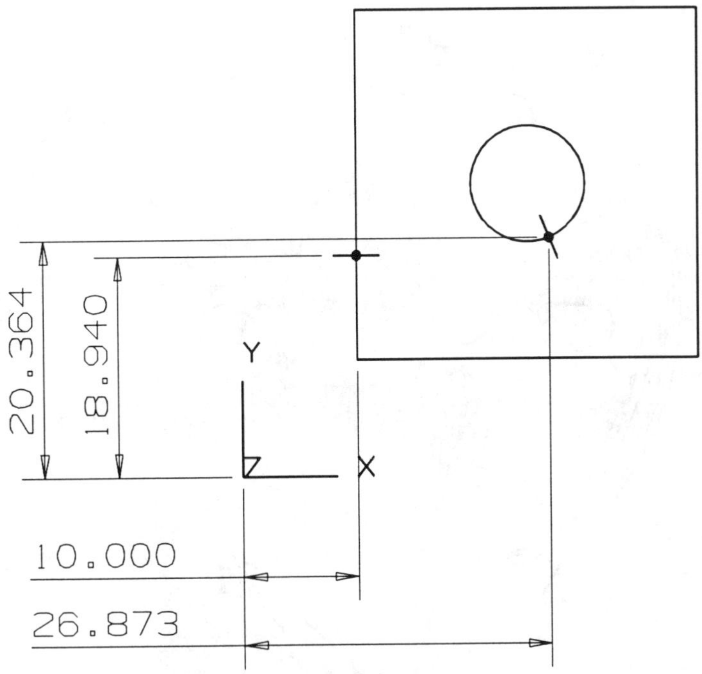

FIGURE 2

Normally, the location, size and form of a hole might be specified as in Figure 1, but in actuality the generation of this feature is done by using a method that is referenced from the 0, 0, 0 of the machine tool and results with surface points that are also referenced to 0, 0, 0 (Figure 2). Currently the process that we are implementing is one in which we define the capability of each of the allowable processes that are associated with an individual machine tool. This will include milling, drilling, boring, circular interpolation of holes, etc, and, after our study, a process capability was determined for each of these. This process capability is very specific and relates to the actual tool used and the feeds and speeds of the machine for each material.

As an example, if milling a circle by circular interpolation is called for, then the maximum excursion of the surface from the nominal prescribed value for the machine was determined. This then becomes the allowable tolerance for a surface that is generated using this machining method, so in concept if the component definition calls for a nominal hole in a specified position and the process to generate this hole is also specified, and since the process specifies the process capability for surface point generation accuracy, the entire feature specification, its location and tolerance is now realised.

What we are now doing is to utilise the conceptual modeller of the CAD/CAM system to generate the nominal values for features on a component. Along with the process of concepting the part for function, the designer will also choose the process for how that feature will be generated and that process will contain in it the tolerance for the feature generator and hence the feature itself. To be assured that the process maintains control, the standard machining practices have been augmented by applying metrology principles to the machine tool.

As part of the process to control the machining of features, which allows us to meet the process capability specified in our process manuals, the machine tools are regularly measured for some of their fundamental errors. Calibrated artefacts are automatically loaded to the machines and are used to determine the linearity and squareness of the axes. Because these artefacts are resident on the machine during the machining processes, they can be referred to on a very regular basis to regularly update these fundamental errors.

Also, to significantly enhance the measurement capability when the component itself is being measured, the measurement process consists of indexing the artefact, using the index table on the machine, to the same position in which the part will be measured and then measuring the feature of the artefact that mimics the feature to be measured in size and location. It is then indexed out of the way and the component itself is brought into the same position and then quickly measured.

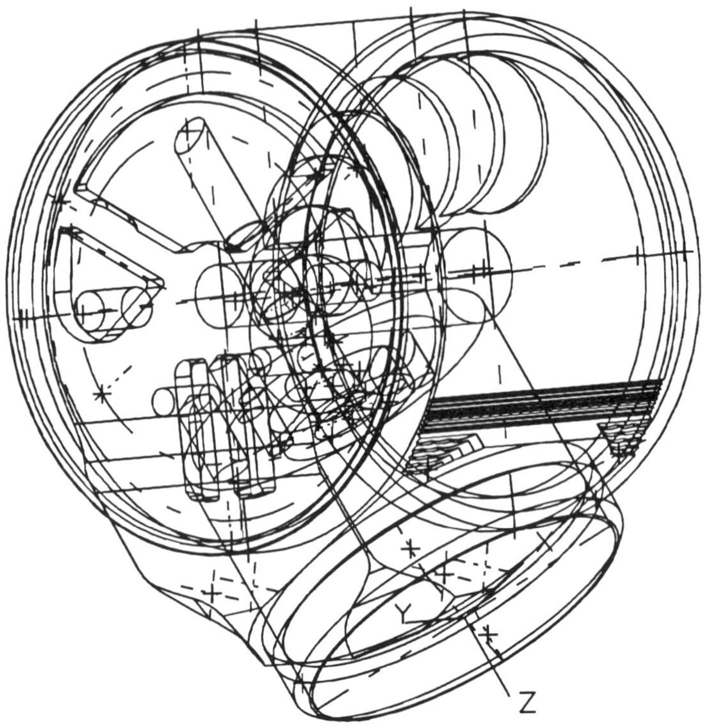

FIGURE 3

By comparing the measured value of the artefact with its calibrated value against the measured value on the component, a measurement with much lower uncertainty can be derived and many of the machine tool errors are removed. This then allows the machine to establish the offsets needed for the part program to perform to the part process control specifications, which is the tolerance for the feature being machined.

TOLERANCING THE SOLID MODEL

In addition, the gage points are established on the solid model for the component. The gage points are assigned a tolerance and these gage points are instrumental in providing the needed control to further ensure the process capability. For example, when the Renishaw Manual Indexable Probe (MIP) was being developed, all the machined components were produced utilising this new system. The center swivel, which houses the probe, has assigned to it, on the solid model, many gage points that the machine will probe during initial setting up, to determine the offsets required for that particular batch.

Figure 3 shows, in a very complicated form, a model of this center swivel and also indicates a range of gage point locations which are specified. This would be done on the first part to be machined in a batch and, utilising these measurements and the fundamental error updates that were determined from the artefact, tool offsets are generated for machining the remaining parts on the fixture.

Figure 4 shows a simpler view which, by the use of small cylinders which protrude from the part, indicates the location and, in greatly scaled magnitude, the value for the gage points used for part inspection. There are normally far less points used to ensure that the process stays in control than are needed for set up. These protruding cylinders only show the value for the gage point allowance outside the material. These cylinders extend the same amount into the part as well. To assist in identifying this, the colour of the cylinder indicates its allowable value. Figure 3 has had its cylinders replaced by crossed intersecting lines to minimise the confusion on an already confusing picture. It is the designer's responsibility to choose the location of these gage points and their allowable tolerance.

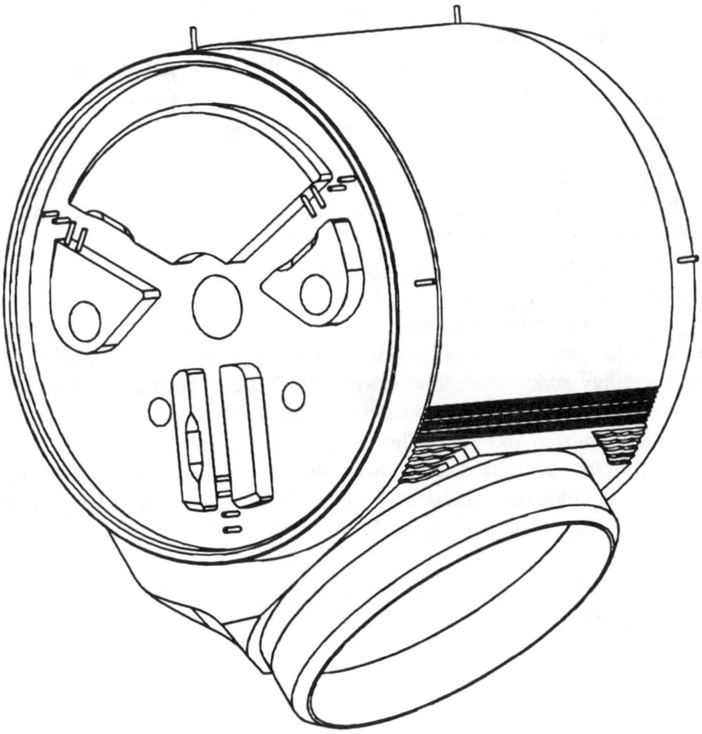

FIGURE 4

Again, because the machine is commanded to pass through co-ordinated points to generate the feature, the system is then asked to measure the surface to see if it is in the prescribed location and by adding additional gage points to a single feature, such as a hole, you can better control the excursion of the surface than would normally be done with less gage points and is the interaction of the process control tolerancing of the choice in location and size of the gage point allowances that control the feature.

It was very important to Renishaw that the person responsible for the conceptual work of the part, who has the best understanding of its function, be the same person who determines not only the allowable excursion of the feature surfaces, but also the method of manufacture and the method of inspection, and the schedule for measurement that results in acceptable parts.

Since each point on the surface is an X, Y, Z co-ordinate based on the machine's 0, 0, 0 position, the ability to measure the part, once removed from the machine, is now very difficult since datums cannot be established off of features. However, while still on the machine, the requirement to locate a datum, with the uncertainty associated with doing this, is also removed from the process, so the measurement of a feature location has an uncertainty that is fundamentally less than the measurement uncertainty when referenced against datums if they need to be established as they would with conventional practice.

When features are produced through circular interpolation, especially at high speeds, their out of roundness can have a very repeatable but non circular shape and this is determined by the orientation of the axes of the machine with the component. If a fudged program is necessary to drive the machine, so that the resultant feature is within tolerance, then gage points should be chosen at locations on the feature which is expected to be deviating the most from nominal are measured. This then allows the most beneficial offsets to be used when updating the process. This means that if the designer, who is also responsible for the fixturing of the part on the machine, is aware of its orientation, in respect of the axes, the interaction of tolerancing and choice of locations for measurement, which are part of the gage point strategy, result in the highest possibility of producing good parts.

FUTURE EFFORTS

The initial success of the development of the MIP were very positive. There is much more to be done to prove this out further. Presently it is being used on four Mazak AJV-18s machining centers. We plan to transfer this same concept over to our mill turning machines so that again we can apply this more rapid development method. The time we save in not producing the traditional engineering drawings and the benefit we gain from having a close relationship with our component specifications, inspection methods and process control methods, have resulted in a much reduced product to market time. From concept to first deliveries was six months.

Because the general concept of this means that the tolerance effectively is located in the process documentation, ie the part program, as well as being affected by the gage point tolerances, there is work being done to allow for a quick determination of the exact total tolerance specified for a component. What we are doing is limiting the choice that the designer has to only one proven option of speeds and feeds with any single tool and no NC program is allowed to vary from this limited set of options.

Also, since after machining features on a component there are many times that the component must be removed from the work holding fixture and flipped to allow machining on the other faces, there is an additional process uncertainty added when this is required. This also becomes part of the uncertainty that is included in the process capability so that the total part uncertainty which represents the tolerance on most engineering drawings is immersed and part of the complete component specification.

We are very pleased with the results to date and two complete products have been launched utilising this method and the number of problems associated with this method have been very minimal and we are looking forward to further implementation in the future.

CRTD-Vol. 27, International Forum
on Dimensional Tolerancing and Metrology
ASME 1993

GEOMETRIC VARIATION EFFECTS ON ROLLING ELEMENT BEARING LIFE

Michael R. Hoeprich
The Timken Company
Canton, Ohio

ABSTRACT

Variations from the nominal geometry in bearing components, combined with those of housings and spindles, provide a variety of interactions affecting the fatigue life of bearings. While tapered rolling element bearings do not appear to be complex devices, determining bearing fatigue lives can be very involved. The large number of components in a bearing and the wide variety of application conditions result in numerous possibilities of internal load distribution. To avoid unfavorable stresses, minute modifications to the basic conical shapes of tapered rolling element bearing components must be made to accommodate the elastic deflections occurring in bearings and machinery components during operation. These modifications to the rolling surfaces are designed for the expected ranges of load and other application conditions.

The calculation of contact stresses between the rolling elements and races is the basic starting point for determining bearing fatigue life. Since bearing fatigue life is related to contact stresses raised to the 6.66 power, changes in contact stresses from desired levels and distributions can have a significant effect on life. Geometric variations in bearings, housings, and spindles can alter loads on bearing rolling elements and change the distribution of contact stresses along the lengths of the rolling elements. The degree to which these geometric variations influence fatigue life is affected by the elastic deflections in bearings, housings, and spindles when the machinery is being operated. Some geometric variations become less important and others become more important as load increases. Since machinery generally operates at a variety of load levels, the determination of tolerances may require the evaluation of their net effect over the entire load cycle.

INTRODUCTION

Figure 1 shows the basic components of a single row tapered rolling element bearing. The inner race is usually mounted on a rotating shaft and the outer race is usually mounted in a stationary machinery housing. Rolling elements transfer the load while rolling between the inner and outer races. A rolling element retainer (cage) keeps the rolling elements together as a set with one of the races (usually the inner race). Figure 2 shows the basic design concept for tapered rolling element bearings. The rolling surfaces are portions of cones whose apices meet at a common point. This design allows the bearing to rotate freely. Because the rolling elements are tapered, a small resultant force develops which would push the rolling elements out of the bearing if it were not for the thrust rib (Figure 1) which is generally part of the inner race.

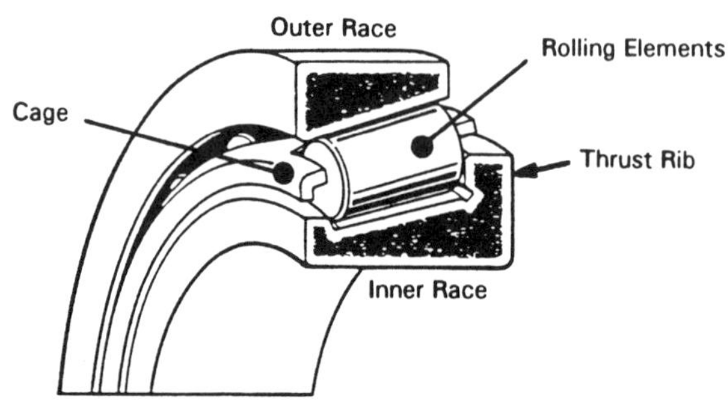

FIGURE 1 Tapered rolling element bearing components.

Elastic deflections in machinery components or variations in the accuracy of manufactured machinery components can result in a misaligning of bearing inner and outer races and a reduction in life (Zantopulos, 1972 and Harris, 1969). The centerlines of the raceways are no longer parallel. To accommodate this misaligning of bearing races and to keep the edges of the rolling surfaces from contacting (resulting in high local stresses), rolling surfaces are curved slightly along their length. If necessary, bearing component surface angles can be designed to compensate for larger amounts of misalignment.

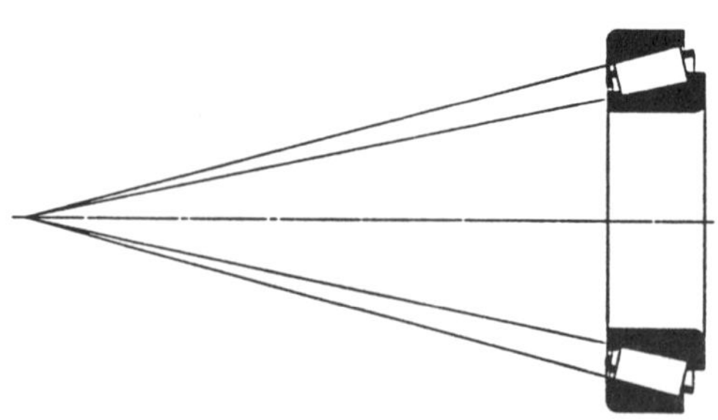

FIGURE 2 The basic principal of the tapered rolling element bearing. Apices of the tapered surfaces meet at a common point on the axis of the bearing.

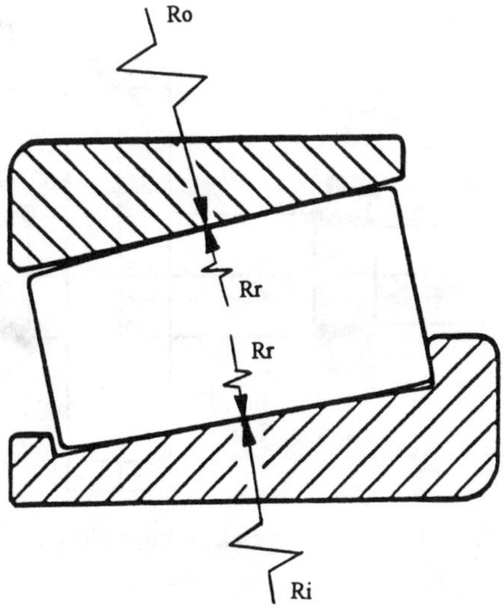

FIGURE 3 Rolling contact surfaces are curved to accommodate misalignment and avoid high stresses at the raceway edges.

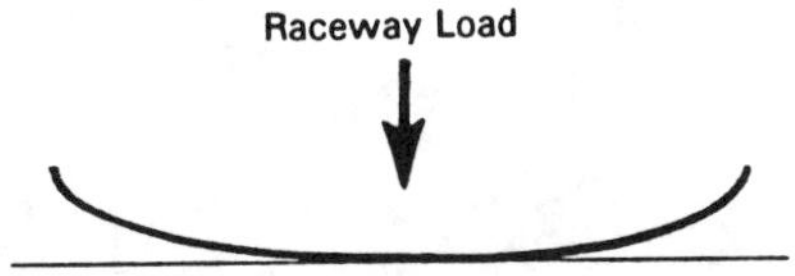

FIGURE 4 Complex contact geometry with continuously changing curvature.

The curvatures (crowning) applied to rolling elements and raceways are illustrated in Figures 3 and 4. Curvature applied to these rolling surfaces can be simply circular arcs shown in Figure 3 (on the order of tens of meters in radius) or the more effective complex shape shown in Figure 4 (Lundberg, 1939; Hoeprich, 1985 and 1989). The more complex geometry in Figure 4 can be approximated with a crown radius over the middle portion with a sequence of smaller radii increasing the curvature at the ends of contact. Contact geometry profiles are designed according to the rolling element loads and bearing misalignments experienced either in general classes of applications or for specific applications which may be more demanding. It is the total initial separation between the contacting bodies before elastic deformation that is important. Therefore, trade-offs can be made between the curvatures applied to contacting bodies. Variation from the desired contact curvature will be used in this paper to illustrate the relationship between acceptable tolerance bands and required performance.

Figure 5 shows a representative application, a geared shaft supported by two bearings. In general, each bearing would be subjected to different loads and the magnitudes of these loads fluctuate as machinery is being operated. The load on a bearing is composed of an axial force directed along the

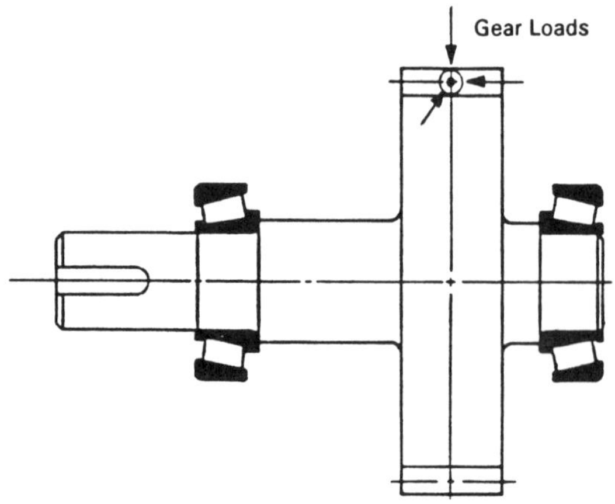

FIGURE 5 Representative bearing-spindle application

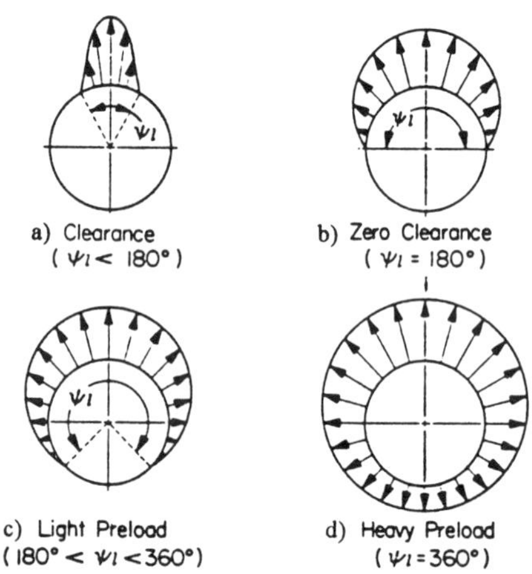

FIGURE 6 Rolling element load distributions.

axis of the shaft and a radial force perpendicular to the axis of the shaft. Depending on the relative magnitude of radial and axial forces, the number of rolling elements carrying the load can range from few to all of the rolling elements (Harris, 1991). This is depicted in Figure 6 where the ratio of thrust load to radial load is increasing from a) to d). The angle of the arc on a bearing race over which the rolling elements are loaded is called the load zone (ψ_l). Loads on the rolling elements vary over the load zone with the heaviest rolling element load occurring at the center position. Even when machinery loads such as gear loads do not change, the size of the load zone and rolling element loads can vary between "identical" machines due to differences in bearing preload or clearance settings during machine assembly. These load zone size changes affect bearing fatigue life primarily through changes in rolling element loads.

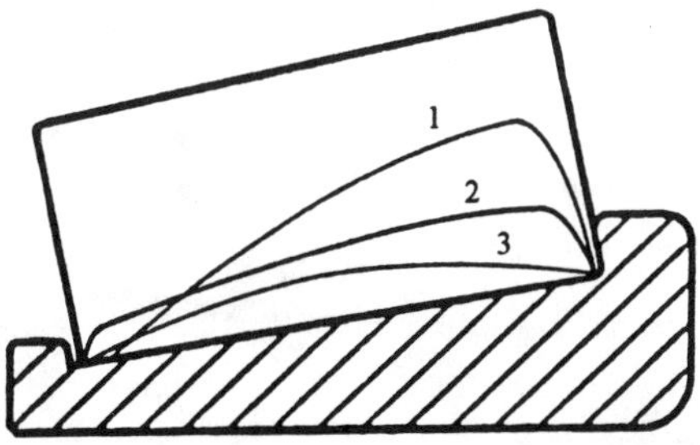

FIGURE 7 Inner race contact stress distributions for three different loads and raceway alignments.

BEARING APPLICATION LOAD CYCLES

As stated above, machinery is subjected to a variety of loads during use. Thus, the loads on rolling elements, alignment of races, and bearing component elastic deflections will also fluctuate. If these loads are monitored during machinery operation, a representative load cycle can be determined and used for bearing design. Figure 7 depicts three different stress distributions which might develop on an inner race during bearing operation. Stress distribution 1 would be for the heaviest load and stress distribution 3 the lightest. Heavier loads cause more elastic deflection and misalignment of races; therefore, the pressure distribution for load 1 is shown significantly shifted, less so for load 2 and a negligible amount for load 3. For each load, there may be an optimum raceway design which distributes the load along the raceway in a uniform manner. However machinery and bearings must be designed to best accommodate the entire load cycle. Bearing load, inner and outer race alignment and the percent time at each load are the load cycle variables used in the examples below.

CONTACT CURVATURE VARIATIONS VS. LIFE

A 45 mm bore tapered rolling element bearing will be used to illustrate the effect of contact geometry curvature change on bearing cycle life. The curvature applied to the rolling surfaces of the outer race and rolling elements are single radius arcs. The radius is different for the outer race than for the rolling elements. The inner race has the more complex geometry shown in Figure 4. In the following examples, contact radii of curvature for inner (Ri), outer (Ro), and rolling elements (Rr) will all be changed in the same proportion . Only the middle crown radius will be changed for the inner race. The more complex curvature at the ends of contact for the inner race will remain the same for simplicity. Changing the curvature on all components at the same time represents a somewhat worse case. The curvature for all components would not normally be expected to change in the same direction and percentage at the same time. This simplified situation is being used because the possible combinations of taper, curvature, placement, size and other bearing dimension variations are very extensive and beyond the scope of this paper.

Three load cycles, given in the following table, were used to calculate fatigue lives. Load cycles I and II use the same loads and misalignments but the heavier loads are applied for a greater percentage of time in load cycle II. Load cycle III is the same as II except the misalignment is reversed to more heavily stress the end of the raceway away from the rib where the internal bearing geometry is different. One half of the rolling elements are assumed to be carrying the load (180 degree load zone).

BEARING LOAD CYCLES

Load Cycle	Bearing Load (percent radial rating)	Bearing Misalignment (radians)	Percent Time at Load
I	(1) 200	.0025	1
	(2) 125	.0010	10
	(3) 50	.0005	89
II	(1) 200	.0025	10
	(2) 125	.0010	25
	(3) 50	.0005	65
III	(1) 200	-.0025	10
	(2) 125	-.0010	25
	(3) 50	-.0005	65

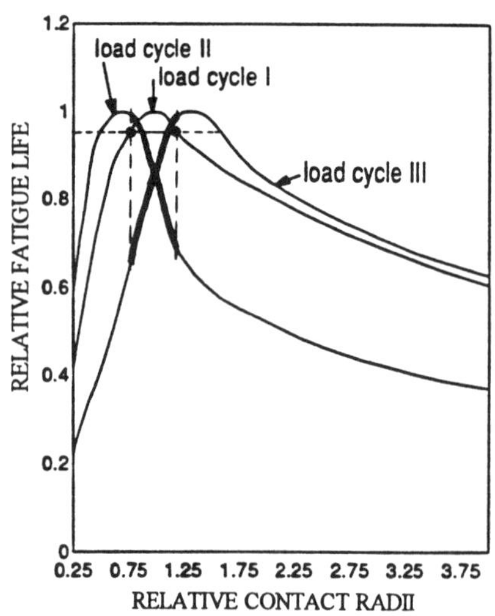

FIGURE 8 Relative load cycle life vs. change in contact radii for load cycles I, II and III.

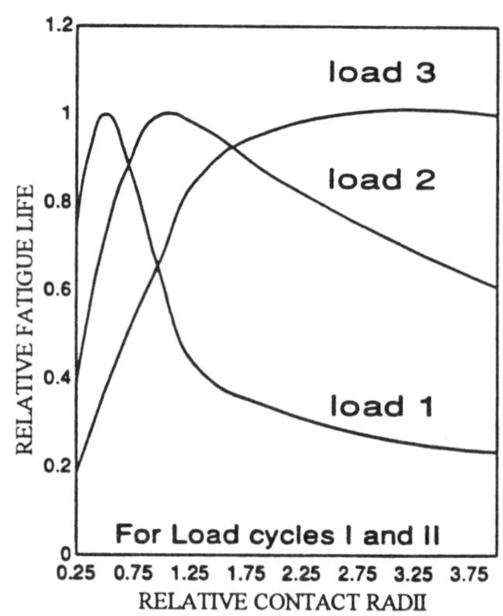

FIGURE 9 Relative life vs. contact geometry for loads 1, 2 and 3 for load cycles I and II.

Figure 8 shows the calculated change in life for each load cycle as the rolling contact crown radii change relative to the nominal design value (Hoeprich, 1989). Maximum lives for each load cycle are different. To allow an easier comparison of the percent change in fatigue life, the life plotted for each load cycle is normalized relative to the maximum life in each load cycle. The response to changing curvature is not symmetrical and the drop off rate is not the same. For any selected life reduction limit, the tolerance band width of acceptable change in curvature for each load cycle would be different. The

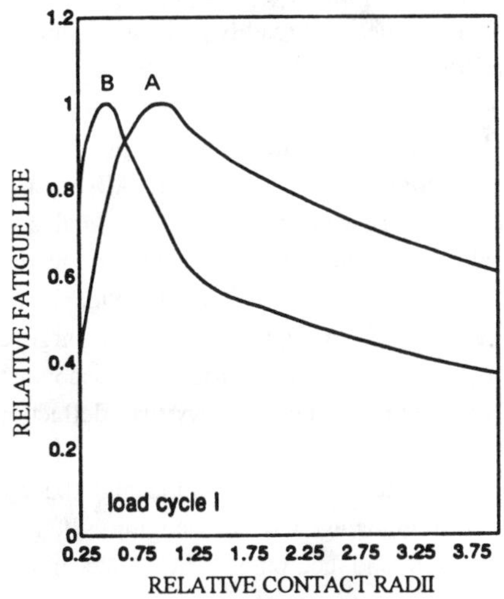

FIGURE 10 Relative life vs. contact geometry for inner race radius curvature and complex curvature.

more perplexing part of the problem would involve selecting a single range of contact radii when the curvature for obtaining maximum life can change significantly even with only the percent time at each load changing. To obtain a range of acceptable radii when multiple load cycles are possible, the probability of a piece of equipment being subjected to one of the load cycles could be used in conjunction with the life prediction tools. It is also seen that if a tolerance band is selected to minimize life variation for one set of conditions, the life may vary widely under another set of conditions. For example, a five percent limit on life loss for load cycle I gives a curvature tolerance band bounded by the vertical dashed lines located at the intersections of the horizontal dashed line (located at .95 life) and the load cycle I line. The possible life variation for load cycles II and III for this same range of curvature values is significantly wider (approximately 40 % of their respective median curvature lives). Life variation for load cycles II and III is shown in Figure 8 by the bold line segments of their respective life curves.

The effect of different machinery operating conditions on bearing life can be much greater than shown above since a particular bearing may be utilized in a number of applications. Figure 9 shows the relative life versus contact geometry for the three loads and misalignments used in load cycles I and II. If these three loads were each the highest or most influential load in three different load cycles, it is obvious that the acceptable tolerance band width around the design geometry could be radically different. The lighter the load the greater the acceptable crown radius variation. Loads 2 and 3 would be common in bearing applications. Load 1 is more typical of accelerated test conditions. Designing and tolerancing for accelerated testing can be in conflict with specifications for actual operating conditions.

The acceptable range of curvature variation is also affected by the type of contact geometry used. If a simple crown radius is used on the inner race in place of the more efficient geometry shown in Figure 4, bearing life is more sensitive to curvature changes. The data in Figure 10 shows this as follows. Curve A in Figure 10 is the same "load cycle I" plot shown in Figure 8 where the inner race has the more complex raceway geometry shown in Figure 4. Curve B is derived from exactly the same conditions except the center radius of curvature on the inner race is extended to the ends of contact (simple crown radius). High edge stresses can now develop readily and significantly affect fatigue life as the radius of curvature becomes larger and the clearance between contacting bodies is reduced. Consequently the acceptable tolerance band width is reduced for the simple crown radius style of contact geometry.

These examples were fairly simple since the crown radii of all components changed in the same proportion and in the same direction at the same time. All other geometric dimensions were held constant. If other geometric variables were included in the analysis, very nonlinear and confounded results would be had. As already noted, affects of geometric variations depend on the application conditions. Variations in raceway angles may offset or add to machinery misalignment. Heavy loads on bearings which are not designed for heavy loads may cause large elastic deflections at the rolling element contacts which bring edges of rolling surfaces into contact, resulting in high undesirable stresses. However, unlike at lighter loads, larger heavy load deflections may also make some raceway surface irregularities relatively insignificant.

With the advent of inexpensive high speed computing and the complexity of geometric variation effects on life, the direction of tolerancing will be changing. It may become reasonable to measure bearing components in production and compute a life factor as a tolerancing procedure on life. However, with all the interacting geometry variation effects on life, it could be difficult for a machinist to determine exactly which geometric feature to adjust. Conversely, it may give the machinist more options to bring the part into compliance. Tolerancing rolling contact geometry involves more than just the value of the primary radius of curvature used in the examples above. The essence or important characteristics of the desired geometry must be captured by the tolerancing procedure. It may be more important to have the contact geometry fall within a given geometric envelope and have a smooth continuous shape rather than try to deduce from the finished product the probable values of dimensions originally used to define the desired shape. Various levels of inspection sophistication can be simultaneously applied to aid in product inspection.

SUMMARY

The basic conical geometry of tapered rolling element bearings is modified slightly (crown radii or more complex geometry) to minimize high stresses at the edges of raceway surfaces and to accommodate machinery manufacture or load induced bearing misalignment. Tolerances on these profile modifications can be established as a function of the operating conditions. The widths of the tolerance bands as well as the actual tolerance values change with changing application conditions. A tolerance band established to minimize cycle life variation for one load cycle cannot be expected to hold the life to the same degree of variation for other operating conditions. The acceptable tolerance range of a geometric feature changes when other geometric features which also affect the distribution of stresses change. Modern analytical models permit the specification of an acceptable range of product performance as a function of several interacting geometric features. These models provide an additional level of inspection which allows an accountability for certain complex shape deviations that are difficult to control with geometric tolerances.

ACKNOWLEDGMENT

The author wishes to thank The Timken Company for support of this work and permission to publish.

REFERENCES

Harris, T. D., Rolling Bearing Analysis, John Wiley & Sons, Inc., New York, 3rd ed., 1991.

Harris, T. A., "The Effect of Misalignment on the Fatigue Life of Cylindrical Roller bearings Having Crowned Rolling Members," ASME J. Lubrication Tech. 294-300, April 1969.

Hoeprich, M. R., "Numerical Procedure For Designing Rolling Element Contact Geometry as a Function of Load Cycle," S.A.E. Paper 850764, presented at Earth Moving Conference, Peoria, Ill., April, 1985.

Hoeprich, M. R,. "Process for Deriving the Contact Geometry for Raceways and Rollers of a Rolling Element Bearing" Patent No. 4,877,340, Oct. 31, 1989.

Lundberg, G., "Elastische Beruhrung Zweier Halbraume," Forsch Gebiete Ingenieurw, 10(5), 1939, pp. 201-211.

Zantopulos, H., "The Effect of Misalignment on the Fatigue Life of Tapered Roller Bearings," ASME Journal of Lubrication Tech., April 1972, pp. 181-187.

CRTD-Vol. 27, International Forum
on Dimensional Tolerancing and Metrology
ASME 1993

TAGUCHI METHODS AND TOLERANCE ASSIGNMENT: THE WAVELENGTH ASPECTS OF SURFACE METROLOGY

Samir Bhargava and Marion B. Grant
Corporate Metrology
Cummins Engine Company Incorporated
Columbus, Indiana

I. Abstract

A study of a crankshaft bearing failure was conducted using surface metrology techniques in conjunction with a Taguchi designed engine test. Preliminary results indicated that the failure mode was not a function of surface roughness as was initially expected. Software was developed on a DOS based PC to acquire unfiltered total profile data from a surface profiling stylus instrument. The software also implements phase-correct Gaussian filtering, according to ISO TC57 draft standards[1,2,3], in a band pass configuration to isolate waviness from both roughness and form. With these tools, the effects of all three components of the surface profile: roughness, waviness and straightness (form), could be isolated and evaluated separately. The Taguchi ANOVA results indicated that the failure mode was strongly correlated to waviness (W_t - evaluated using ISO band-pass Gaussian filtering), but weakly or not correlated to any aspect of roughness or straightness. The specification for the crank pin and main bearing surface profile was changed to include tolerances for W_t and W_c. The failure mode has not been observed since these changes were implemented. This work shows the wavelength spectral structure of the dimensional aspects generated by the honing process, and the dimensional wavelength dependence of the component performance. The indication is that for a form call-out to be effective (i.e. correlate to surface performance or be useful in controlling a manufacturing process), it must have associated with it an explicit or understood surface wavelength regime to which it applies.

II. Observed Problem Symptoms:

In 1990, Cummins Engine Company had a problem on what it calls the 91N14, its largest on-highway truck Diesel engine. Engine rating is 460HP (345KW) at 1800 RPM with around 1400 ft-lbs (1900 Nm) of torque. The engine had received a 5% upgrade in power arid torque and had significantly improved fuel economy from the 88NT version. These results were obtained in part by increasing combustion pressure.

The earliest failure problem on the engine was that on around every 5th engine in early testing, the upper half-shell bearing on one of the connecting rods failed in a short duration engine test. Upon disassembly of the engine, the product development engineers found this half-shell to have the inner bearing coating removed down to the copper substrate coating. The crankshaft journal that this bearing was running on was suspected to be the cause, but the inspectior the size, form and finish of the journal showed the crankshaft to be within specification fo existing tolerances.

A study was conducted over a one month interval by Atlas crankshaft, a subsidiary of Cumn that supplies this crankshaft, and the product development engineers involved, in which f crankshafts with progressively higher Ra values were studied relative to surface metrology (Table 1). The two crankshafts with the highest roughness values were outside of specification. R_a, R_z, R_p, R_{max}, R_{sk}, Tp 90% depth and W_t were analyzed on the cranksha and they were run in a 50 hour engine test. The two crankshafts with the lowest and n highest R_a failed, while those at the highest and next to lowest R_a had minimal damage. T was contrary to the expectation of the product engineers running the test. As a result, no cl cause could be resolved by the engineers involved.

III. Problem Definition Research:

Cummins Corporate Metrology was called upon to help, because this problem would hin the product introduction if not solved. The inspection of the destroyed half-shells by Corpor Metrology staff showed the coating to be removed in bands that ran circumferentially on eit side of the oil hole anywhere from 3 to 6 cm in circumferential length and 1.5 to 4.5 mn axial width. Visual inspection of the crankshaft journals showed that even though the R_a val of axial traces on some of the journals looked to be within specification, all the journals lool fairly rough ground and depicted a visible circumferential banding, a corduroy appearance width corresponding to the wear bands on the half-shells. This banding which did not app in the roughness traces. The only specification for surface metrology on the print for crankshaft was for R_a and straightness. The total profiles for the traces were then analyzed in the unfiltered profiles there was clear evidence of hollow in the straightness component (Figure 1 for example). In addition, there was clear presence of waviness, whose predomin wavelength apparently corresponded to some level, to the banding widths of the bear coating removal on the failed bearing shells.

When the data from the four crankshaft study was reviewed, it was clear that although the R_z, R_{sk}, R_p, and 90% material ratio depth did not correlate to the failure mode, the W_t resu did correlate (see Table 1). The crankshafts with values of W_t below one micron did not ca bearing failure, the crankshafts with W_t above one micron did fail.

Atlas Crankshaft was visited. Upon talking there to product, manufacturing and qua engineers responsible for this component, and their gage lab supervisor, they said their curr lapper had been in use well beyond its expected life, and they were in the process of replacin with a new lapper. It could be that the clearance between the backing shoes for the tape a the journal had increased with time that the paper had the tendency to crimp with t characteristic wavelength to create this waviness on the journal in the lapping process. T hollow in the center of the journals (the journals had an hourglass shape) was caused by lapping oil filling the oil hole during the process. As the abrasive paper tape moved over the hole, it would pull fresh cutting fluid into the circumferential band in the axial center of ırnal. In this area, the lapping paper cuts more efficiently as a result, creating the hollow. las tries to control this effect to under 5 microns by limiting the lapping time and the amount cutting oil supplied while lapping. Yet evidently bearing shell failure has little to do with this m error unless it becomes excessive. They discussed how they measured waviness in the esence of this hollow for this four crankshaft study. By manually constructing a band that velopes the peak-to-valley waviness regardless of the local slope on which it resides, they are le to isolate the waviness from the form graphically. The problem with this approach is the)jectivity in deciding what peak-to-valleys are waviness and what is part of the overall aightness. It was clear that this caused some variability in their value they assigned to W_t.

. **Problem Solution:**

or to this effort, Cummins Corporate Metrology had developed the ability to link DOS-sed PC's to surface finish machines to be able to extract as unfiltered data as was obtainable m the instruments for later, usually off-line data analysis. This capability was developed marily to perform special processing on cylinder liner plateau finishes. New phase correct, ussian digital filters developed from metrology insight gained from ISO TC57 (the ISO mmittee on Surface Metrology) participation had been included as part of this analysis)ability.[1,3] Another routine had also been developed to provide the ability to use this new ussian filtering technique any number of times on surface profile data and at any filter quency. It had also been learned from ISO TC57, via work that the German delegation had ne, that a passive Gaussian digital filter can remove straightness error from a trace. This ftware was used to develop a dedicated analysis routine for this problem. An Perthen S6P s borrowed so as to have the same instrument as was at Atlas, an interface was developed to 86 based PC, a crankshaft sectioned and data sets were developed from unfiltered traces.

alysis of a representative set of profiles (see for example Figures 1 through 3) showed that ng a Gaussian filter with 0.8 micron cutoff to eliminate the roughness profile and a Gaussian er at 8.0 micron to eliminate the straightness profile worked very well to isolate the viness profile in order to determine waviness parameters such as W_t and W_c (the average ak to the average valley for all waviness peaks and valleys that lie between zero crossings). few profiles were observed (Figure 3) in which a 6.0 micron straightness filter was better, at ninating what was clearly straightness, but the 8.0 micron filter corresponded to the ISO e under consideration, that the form filter should be default at ten times the roughness filter ue used for the surface.[2] Also for those traces where a 6 micron filter would have been eferred, the use of an 8 micron filter simply causes the W_t value to increase, which errs in the nservative or safe direction.

th this development, another more extensive study was done using 8 crankshafts and oking at only W_t, R_a, R_{sk}, and lay (the angle between the lapping marks and the cumferential direction). The experiment was designed to give 3 levels of waviness, two els of roughness, two levels of skew and three levels of lay, distributed over the 8 cranks cording to an L8 experimental design. Eight engine tests were run and the upper bearing-ells evaluated on a scale of 0 to 10 for coating damage, by several product engineers cording to a predefined damage scale. None of these shells went to complete failure as in the previous tests. The damage results were the average of the ratings of all of the engineers afte removing those ratings that were statistically outside the distribution for each shell. The result are shown in Figure 4 where the average value of the evaluated parameters are as indicated i the figure.

It is clear that the parameter W_t (Al, A2, A3) correlates strongly to the bearing shell damag R_a (Bl, B2) shows some small correlation, but the difference maybe primarily within the nois of the rating process. Similarly for R_{sk} and lay.

V. Resulting Engineering Documentation Changes:

Tolerances on the crankshaft were set at values that would keep the process operating at thes lower damage values for R_a, R_{sk} and lay, but the waviness W_t value was set to a maximu value of 2.5 microns by this new method since further measurements have shown the ne lapping process acquired by Atlas has the capability for W_t to be 1 micron with 3 sigma at 0. to 0.7 micron. W_c was also added to the print and an acceptance criteria documented. N crankshafts produced after this new specification has been installed have exhibited this failur mode.

VI. Conclusion:

Here the functional correlation of the parameter W_t to the failure mode of the component ha been demonstrated to exist strongly, if bandpass filtering is used to isolate the wavelength considered in the evaluation of the parameter. The Perthen S6P used in the study has selection for W_t on it but it has no long wavelength filter associated with the selection, so tha the W_t that it reports includes the straightness error. As a result, the Perthen S6P W_t woul not have correlated to the failure mode. This suggests that print surface metrology (surfac texture and form) call-outs that correlate to performance must have an implied or explic surface wavelength limitation associated with them. In addition, correlation across instrumen makes will require that call-outs have very specific parameter definitions with very specifi wavelength limitations.

ISO TC57 and ANSI Y14.5.1 standards committees have resolved that form call-outs for suc things as cylindricity, flatness, roundness, and straightness will include waviness or will no include roughness.[4,5] This is a first step toward reaching the ability to reach correlation acros instrument makes and toward reaching functional correlation. Yet most consideration of for errors does not include waviness as one of its components, but rather includes only the longes wavelengths (hollow, hourglass, bellmouth, taper to name a few typically considered for errors). In addition, component performance and manufacturing process performance that i known to be effected by or to cause form errors are not those effected by or causing wavines down to the maximum roughness wavelengths. Limiting form errors to the longer wavelength can greatly reduce measurement and analysis time for digital instruments, to the point o making such assessment practically possible, with correlation and traceability and eve precision. Shorter wavelengths can be specified on the drawing in the form of roughnes and/or waviness in linear traces, or as short wavelength roundness (chatter) in circumferentia ices. Cummins will be supporting these directions for standards in order to reach nation-de and world-wide correlation in the assessment of such print specifications.

II. References

ISO CD 11562 :	Metrological Characterization of Phase Correct Filters and Transmission Bands for Use in Contact (Stylus) Instruments.
ISO CD 10479 :	Surface Waviness - Terms and Definitions.
ISO TC 57/SC4/N178 :	Rules and procedures for the measurement of surface waviness using stylus instruments.
ISO TC 3-10-57/JHG/N5 :	Resolutions of the 1st Meeting of the Joint Harmonization Group of ISO TC3, TC10 and TC57, Coventry, UK.
ANSI Y14.5.1 :	Report of the Committee on the Mathematization of Geometric Dimensioning and Tolerancing.

TABLE 1

Measurement results of first bearing shell test using manua graphical waviness evaluation methods.

Crank	Surface	Rslt*	Wt	Ra	Rz	Rsk	Rp	Rmax	Tp90%	Dpt
Shotts #1	Pin	Fail	2.00	0.27	2.64	-0.71	1.15	3.63	-1.02	
Shotts #2	Pin	Fail	1.07	0.19	1.96	-0.61	1.17	3.51	-0.71	
Test	Reg.Prcs	Good	0.77	0.30	2.64	-0.88	1.07	3.20	-1.11	
Test	SpclPrcs	Good	0.48	0.23	2.29	-0.48	1.61	2.97	-0.86	

* A test result of "fail" meant the bearing shell coating wa removed down to copper. A test result of "good" meant insignificant wear.

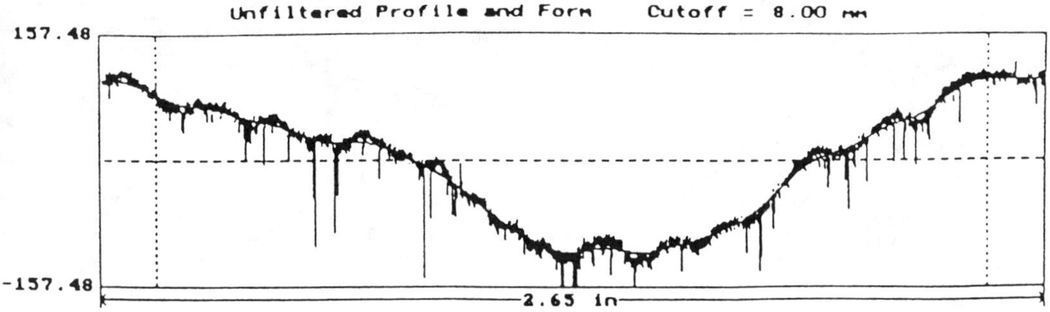

Figure 1A. Total profile: straightness + waviness + roughness

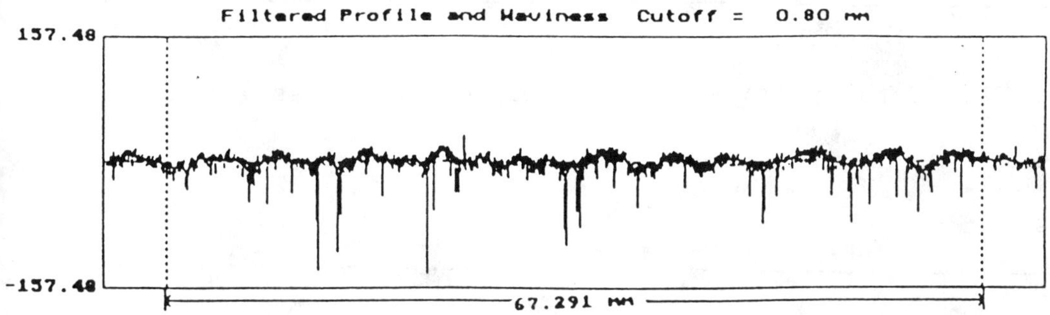

Figure 1B. Waviness + roughness profile

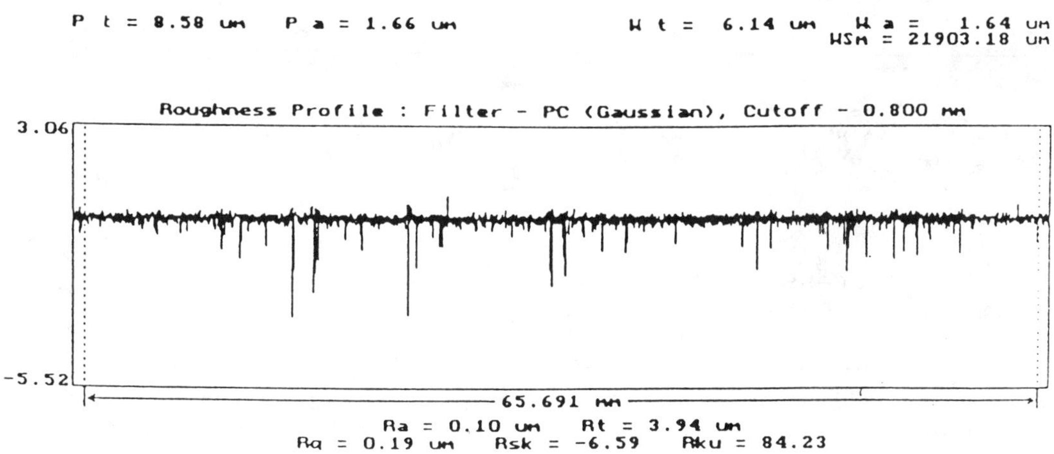

Figure 1C. Roughness profile

Figure 1. Separation of straightness ("hourglass" or "hollow") from waviness and waviness from roughness, using 8.0mm and 0.8mm cutoff Gaussian phase-correct filters respectively. Shows filter operation on a "V" shaped profile.

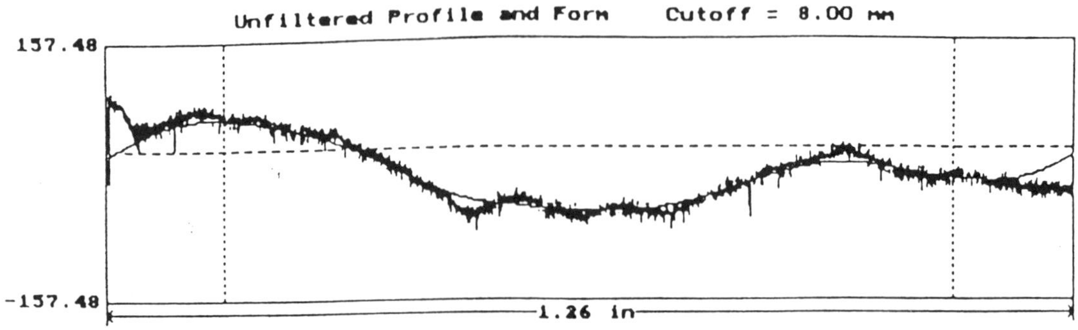

Figure 2A. Total profile: straightness + waviness + roughness

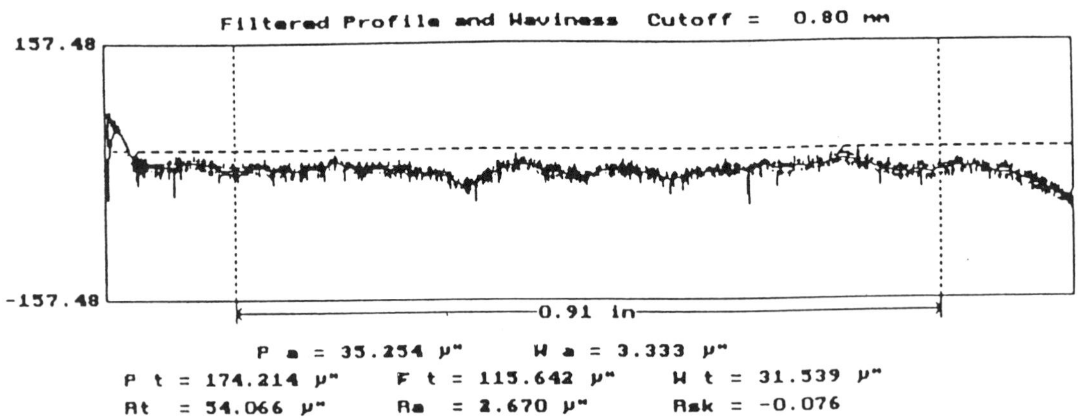

Figure 2B. Waviness + roughness

Figure 2. Smaller peak to valley straightness. Shows filter operation on "M" shape. Note that scale and parameters are in microinches.

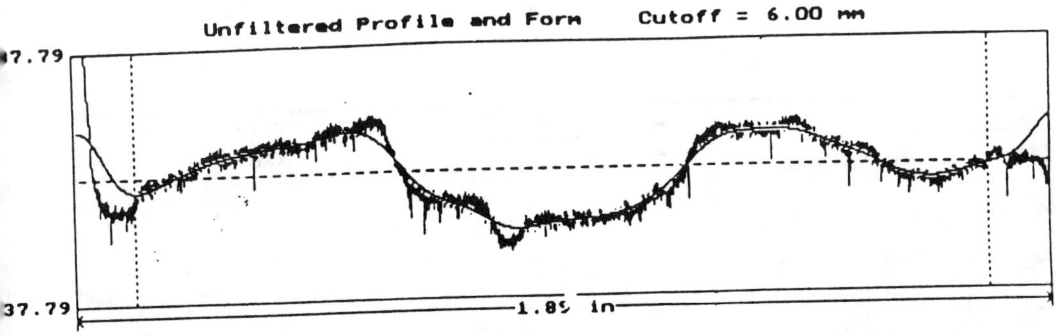

Figure 3A. Total profile: straightness + waviness + roughness

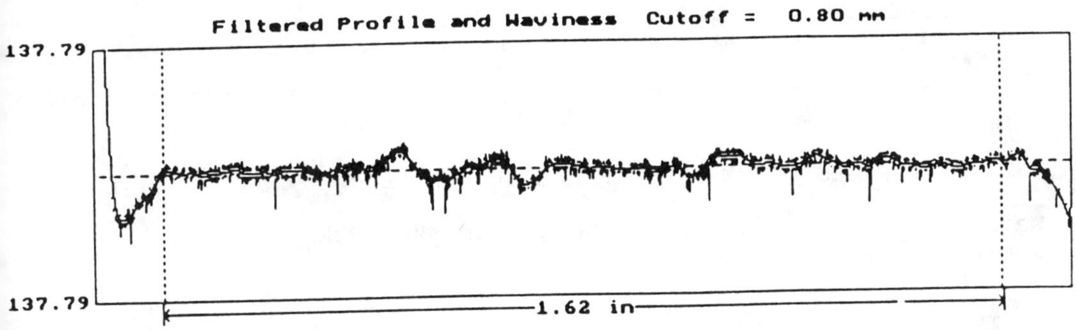

P a = 32.363 μ" W a = 3.485 μ"

P t = 723.590 μ" F t = 106.030 μ" W t = 38.147 μ"

Rt = 51.664 μ" Ra = 2.882 μ" Rsk = 444.818

Figure 3B. Waviness and roughness

Figure 3. Sharper peak to valley straightness. Shows filter operation on a sharper "M" shape. Note that paramater values and scales are in microinches. Also note that straightness to waviness filter is set to 6.0mm rather, than 8.0mm.

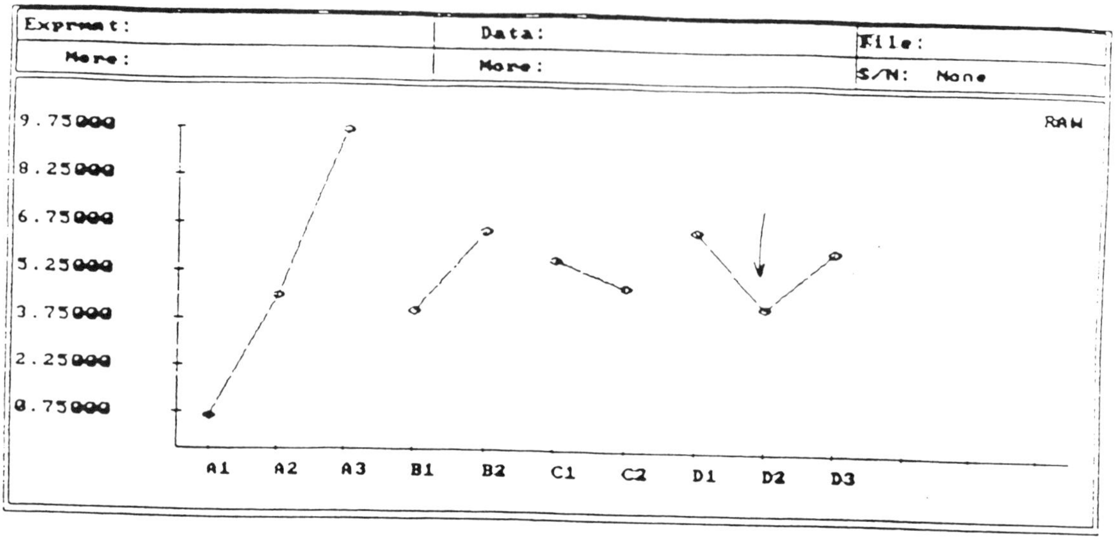

AVERAGES OF MEASURED PARAMETER VALUES:

Waviness, Wt (microns)		Roughness, Ra (microns)		Roughness Skew, Rsk		Lay (Degrees)	
A1:	0.6	B1:	0.09	C1:	-0.43	D1:	0
A2:	1.3	B2:	0.41	C2:	-1.27	D2:	8
A3:	2.8					D3:	15

RESULTING SPEC.CHANGE: Wt = less than 2.5 microns,

Ra = less than 0.3 microns,

Rsk = -0.8 or less

Lay = 8 degrees + or - 4 degrees.

Figure 4. ANOVA results

CRTD-Vol. 27, International Forum
on Dimensional Tolerancing and Metrology
ASME 1993

ADVANCED TOLERANCE ANALYSIS AND SYNTHESIS FOR GEOMETRIC TOLERANCES

James Guilford and Joshua Turner
Design and Manufacturing Institute
Rensselaer Polytechnic Institute
Troy, New York

ABSTRACT

With the increasing trend towards replacing engineering drawings with solid modeling systems, a means is needed for integrating tolerances within the solid modeling framework. Before this can be done, however, a clear understanding is needed of what a tolerance is and how it can be analyzed.

This paper compares parametric and geometric tolerances. It then describes the theory behind the GEOS tolerance analysis system, which is designed explicitly to handle geometric tolerances.

The GEOS system integrates a tolerance analysis capability within a solid modeler. The solid modeler actually defines a complete variational model; the tolerances become constraints on the geometric variations. This allows GEOS to derive relationships between the variational model and the tolerances, and between the variational model and the design function being analyzed. Using linear programming techniques, the numerical relationship between the design function and the tolerances can be derived.

Tolerance analysis is the process of taking known tolerances and computing their effect on a particular design function. Tolerance synthesis is the process of taking some desired behavior of the design function along with suitable models relating tolerances to manufacturing cost and computing a set of tolerances which minimize the cost. Instead of working from fixed limits on the design function, the GEOS system considers variations of the design function from nominal as a cost associated with a "quality loss" function from Taguchi. The tolerance synthesis algorithm then attempts to minimize the total cost, which is the sum of the manufacturing cost and the cost due to the quality loss.

INTRODUCTION

Traditionally, designs have been represented as engineering drawings, in which tolerance specifications are expressed as annotations. Tolerance analysis during design is often overlooked. When an analysis is performed, it often uses simplifying assumptions and little integrated support. Such "back of the envelope" calculations are error-prone and inaccurate (Guilford et al., 1992).

Because of this, designers tend to use unnecessary tolerances and to make tolerance values tighter than they have to be. These practices increase manufacturing costs. Designers also often fail to take into account some critical interaction of toleranced faces[1]. These critical interactions can go undetected until late in the product development cycle, when engineering changes are very costly and can result in major product delays.

With the increasing trend towards representing designs as solid models, there is the possibility of improving this picture dramatically. Since the design geometry is already completely represented, it should be possible to merely add the tolerance information, and then to have an automated system perform an accurate tolerance analysis. This requires a set of algorithms which can deal with the tolerances currently being used.

In current practice, designers do not want to deal with tolerances. It has traditionally not been part of their duties, and they are reluctant to accept new responsibilities. If they are to actually use a tolerancing system, it must not require much additional work. This implies that the solid model which currently exists must be exploited. The user should not be forced to manually create an equivalent model for the tolerance analysis. This also implies that the tolerances used by the analysis system must be equivalent (or nearly equivalent) to those tolerances which currently appear on the drawings and which the designer is used to. These tolerances are typically based on the ANSI standard Y14.5 geometric tolerances (or the ISO equivalents), which include geometric interpretations for the traditional "plus-minus" tolerances.

If the tolerancing system uses a different conceptual base for its tolerance representation, then the tolerances must be converted to this new form. This requires additional manual work, with its associated chance for errors, and may require special training or creative insight. In some cases, an equivalent form may not exist, requiring manual analysis and/or approximations to be used. It is clearly preferable to have the tolerancing system use the same conceptual basis as the tolerancing standard being used by the designer.

Given a particular tolerance representation, efficient and accurate algorithms are needed to actually perform the tolerance analysis. Unfortunately, as the geometric tolerances are complex, so too are algorithms using these tolerances.

However, in virtually all commercial tolerancing systems, the tolerance representation and the analysis algorithms are chosen for the convenience of the implementors rather than the user. The tolerance representation is typically based on *parametric tolerances* and is not directly compatible with the standard geometric tolerances.

Once a tolerance analysis capability is realized, this capability can form the kernel of a tolerance synthesis system. Whereas analysis attempts to predict the effect that the tolerances have on one or more *design functions*, tolerance synthesis attempts to compute a set of tolerance values which satisfy constraints on the design functions. Although the design constraints can be trivially satisfied by making the tolerances arbitrarily tight, this would result in an expensive or infeasible design. Since looser tolerances tend to result in lower manufacturing cost, the

[1] Actually, tolerances can be applied to other *topological features*, such as center lines and center planes; but to avoid confusion with "feature-based design" principles, we will use the term "face" to mean all topological features to which a tolerance can be applied.

tolerance synthesis problem can be stated as finding the lowest cost design that meets the functional requirements.

This paper proceeds in three parts: first the parametric and geometric tolerances are compared. Then the theory behind the GEOS system for performing tolerance analysis is described. Finally there is a description of what tolerance synthesis is, and how it can be implemented.

PARAMETRIC VERSUS GEOMETRIC TOLERANCES

There are two primary definitions for mechanical tolerances: *parametric* and *geometric*.

Parametric Tolerance: a limit or a pair of limits applied to some scalar parameter of an engineering model.

Geometric Tolerance: a constraint on the geometry of a face. The tolerance defines a tolerance zone in which the face must lie.

Parametric tolerances (Gossard et al., 1988; Hillyard and Braid, 1978; Light and Gossard, 1982; Lin et al., 1981; Martino, 1988; Turner and Anderson, 1988; Wang and Ozsoy, 1991) view each object as being defined by a set of scalar parameters. In a CSG-based or a feature-based system these are the parameters used in the interactive specification of the object. In a dimension-driven variational geometry system, the parameters are the part dimensions. Tolerances are viewed as limits on the parameter values. From a mathematical point of view, the tolerance limits define a hypercube in parameter space.

Geometric tolerances (ASME, 1982; Jayaraman and Srinivasan, 1989; Srinivasan and Jayaraman, 1989; Requicha, 1983a; Requicha, 1983b; Requicha and Chan, 1986; Rossignac and Requicha, 1985; Srinivasan and Jayaraman, 1985) define constraints on the geometry of a face. These are typically zones within which the faces must lie. Different geometric tolerancing schemes define the zones in different ways, but the significant effect is that the control is directly on the geometry and is not related to a particular model parameterization.

The advantage of the geometric tolerances is that they more accurately reflect "real-world" conditions, particularly mating-part relationships. They are more relevant to actual, physical parts, which have geometry but no intrinsic parameters. In view of these advantages, geometric tolerances form the basis of the major tolerancing standards, both in ANSI and in ISO.

The disadvantage of geometric tolerances is that they tend to be complex and difficult to analyze. The advantage of parametric tolerances is that the resulting analysis algorithms tend to be relatively simple. As a result, these tolerances form the basis for virtually all commercial tolerancing software (Turner and Gangoiti, 1991a; Turner and Gangoiti, 1991b).

The disadvantages of parametric tolerances are that they do not enforce functional requirements (such as ensuring fit of mating parts) as well as the geometric tolerances, they require a parameterized model, and that the results are heavily dependent on the parameterization chosen. This means that using a parametric tolerancing system often requires the user to redefine the product model in a parametric way. This step, in addition to being more work for the designer, requires skill and an understanding of the underlying principles behind the tolerances. If this step is done incorrectly, accurate results are impossible.

An additional problem with parametric tolerances is that they arise from a different conceptual basis than the geometric tolerances. Since the tolerances currently used by designers and currently found on drawings tend to be geometric in nature, a conversion process is needed to derive equivalent parametric tolerances. Unfortunately, this process is at best difficult, requiring skill and understanding. At the worst, it is impossible without a variety of approximations.

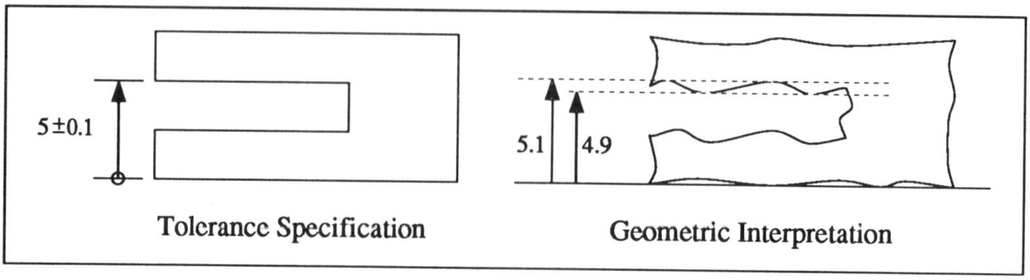

Figure 1: Geometric Tolerance

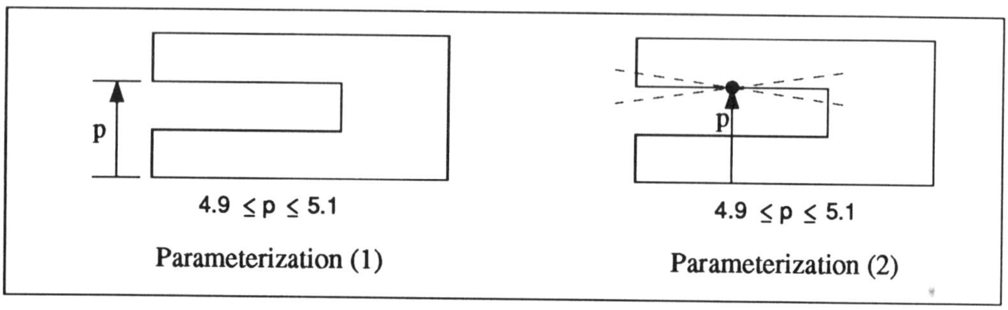

Figure 2: Parametric Tolerance

It is difficult to use parametric tolerances when two tolerances apply to the same face. An example of this is applying orientation and size tolerances to the same two faces (this example is further described in the following section). A size tolerance implicitly controls the orientation of two faces. If a tighter control is desired, then an orientation tolerance may be used in addition to the size tolerance. Note that this does not enlarge the tolerance zone; the second tolerance is more restrictive, not more inclusive. When two (or more) such tolerances interact, it becomes very difficult to convert these to limits on parameters.

Some of these points will be illustrated with simple examples. Figure 1 shows a simple asymmetric geometric locating tolerance. The interpretation is that a datum plane is constructed against the bottom face, and a zone is defined parallel to the datum, and extended away from it for a distance of between 4.9 and 5.1. The part is in-tolerance if the toleranced face lies within this zone.

The toleranced face can vary in location, orientation, and form. All of these variations are controlled by the tolerance.

Two different parametric interpretations are shown in Figure 2. In the first, the two faces are constrained to remain parallel as the parameter p varies. Effects due to orientation and form variations are not considered.

Alternately, the parameter can define the distance from the midpoint of the top face to the bottom face. In this parameterization, orientation effects can be parameterized as well, with an independent parameter, but the orientation of the top face would not be constrained by the tolerance. A separate tolerance would need to be defined for this second parameter. This is not easy to specify, as the limits on the orientation parameter would depend upon the value of the parameter p (in order to get the same semantics as the geometric tolerance) .

In summary, the two types of tolerances, geometric and parametric, are based on different conceptual models and result in different variational coverage. The parametric tolerances are based on a construction procedure that specifies how to construct a given model instance for given parameter values. If the designer's solid model is not in this form, a parametric model needs to be created. In contrast, the geometric tolerances are declarative constraints on the toleranced faces. No construction procedure is implied. This makes it easier for the existing solid model to be used.

The variational class for a parametric tolerancing system is defined by the parameters chosen. Any particular size, orientation, location, or form variation of a given face is treated as perfectly constrained unless a parameter is introduced for it. In contrast, the variational class for a geometric tolerancing system is independent of the tolerances used. Any particular size, orientation, location, or form variation of a given face is treated as totally unconstrained, unless a tolerance is introduced that constrains it. If insufficient tolerances are present, the design function is unbounded, and the designer can be instructed to add appropriate tolerances.

FEASIBILITY SPACE APPROACH

We have developed an approach to solving tolerancing problems based on variational modeling and geometric tolerances, called the feasibility space approach (Turner, 1987b; Turner, 1987a; Turner and Wozny, 1988; Turner, 1988; Turner and Wozny, 1990; Turner, 1993). Our approach is the basis of the GEOS tolerancing system.

The approach involves the following steps:

1. The declarative (boundary) model defined by the user's CAD system describes an ideal part. This model is converted to a variational model, which describes a range of parts that differ in some ways from the ideal. Each individual variation is governed by a *model variable*. Possible variations include variations in location, orientation, size, and form (Gupta and Turner, 1991; Gupta, 1991; Gupta and Turner, 1993).

   The entire collection of model variables define an n-dimensional Cartesian space. Any individual point (n-tuple) in that space corresponds to a particular set of values for the model variables and hence to a particular varianced part instance.
2. Each n-tuple can be classified against a given tolerance as to whether that part instance satisfies the tolerance or not. Effectively, each tolerance partitions the space into an acceptance region and a rejection region.
3. Collectively, all of the tolerances define a *feasible region* of the space. Any n-tuple in this region defines a part instance which satisfies all of the tolerances.
4. At this point, a variety of analysis algorithms may be applied. For example, optimization techniques can be used to generate a worst-case and linear statistical analysis.

   These analyses can also be used as the basis of other algorithms, such as tolerance synthesis applications.

The feasibility space approach will be illustrated with an example. For simplicity, this example will be described in two dimensions. Actual problems are defined using three dimensional solid models. Figure 3 shows a block with two tolerances applied to it: an asymmetric size tolerance, and a parallelism tolerance. These define the two zones shown in Figure 4. The

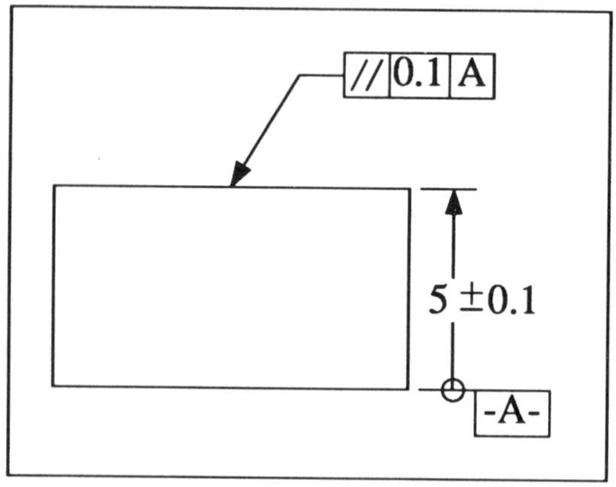

Figure 3: Example Drawing

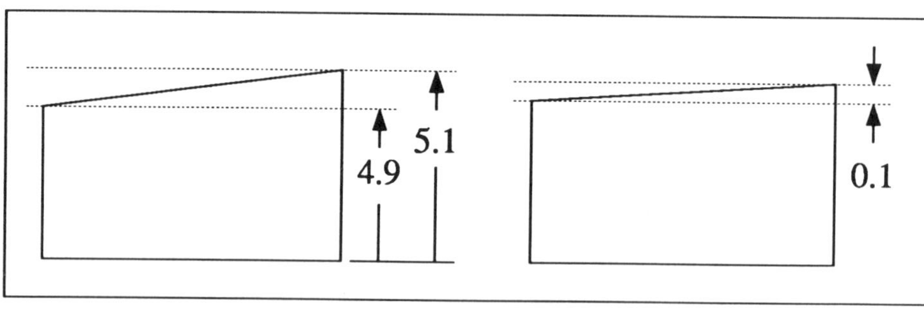

Figure 4: Example's Tolerance Zones

larger, fixed zone is defined by the size tolerance. The smaller zone is defined by the orientation tolerance. The smaller zone's location is not defined, so it is free to "float" up or down; but the zone is constrained to remain parallel to datum A. The actual face is within tolerance only if it lies within both of these zones.

The 2-D solid model for this part consists of four edges, joined at four vertices. The first step is to convert this declarative model to a variational model by applying model variables to it. In this case we are interested in location and orientation variations, and we will ignore form variations. We would normally apply model variables to every edge of the part, but for simplicity, we will only consider the model variables applied to the top edge. The model variables are illustrated in Figure 5.

There are two model variables: M_1 which controls the location of the top edge, and M_2 which controls its orientation. Note that M_2 is defined as the difference in height of the two endpoints, so that the units of both variables are distance. Both variables describe variations from nominal, so the nominal part instance is represented by both variables being equal to zero.

We note that the edge will be in the tolerance zone if both endpoints are within the tolerance zone. We therefore define some auxiliary variables (tolerance variables) which can more directly

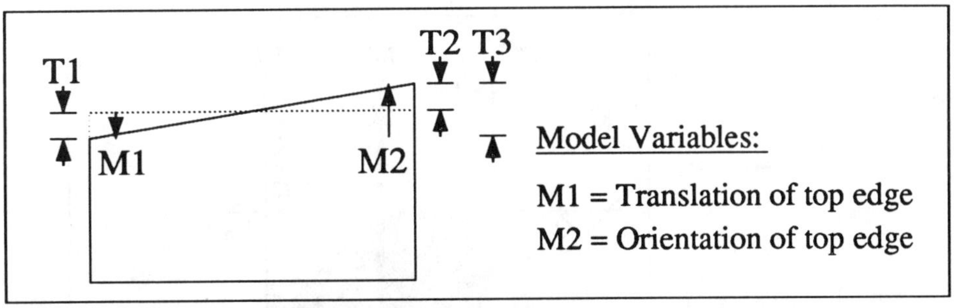

Figure 5: Example's Model Variables

be related to the tolerance limits. T_1 and T_2 are the displacements from nominal of each endpoint, and T_3 is the distance between two planes, both parallel to datum A, which contain the edge. These are related to the model variables by the following equations:

$$\begin{aligned} T_1 &= M_1 \\ T_2 &= M_1 + M_2 \\ T_3 &= M_2 \end{aligned}$$

In terms of these tolerance variables, the tolerance constraints can be expressed as follows:

$$\begin{aligned} -0.1 \leq\ & T_1 \leq 0.1 \\ -0.1 \leq\ & T_2 \leq 0.1 \\ -0.1 \leq\ & T_3 \leq 0.1 \end{aligned}$$

where the first two inequalities represent the size tolerance, and the last inequality represents the orientation tolerance.

Each of these constraints defines an acceptance region within the feasibility space defined by the model variables. The part instances represented by points within the acceptance region satisfy the tolerance constraints. The intersection of all of these regions yields a feasible region within the space (Figure 6).

A worst-case solution can be obtained by finding the maximum and minimum of any design function of interest over this region.

In summary, the feasibility space approach has the following characteristics:

Tolerancing problems can be solved without extensive interaction with the designer, since most of the required information is automatically derived from the designer's CAD model. It is unnecessary for the designer to create special-purpose models, to enter dimension loops, or to provide equations relating the toleranced properties.

Our methods can detect and analyze areas of excessive tolerance buildup that were unanticipated by the designer. The methods are suitable for the solution of problems in which tolerances can interact in all three dimensions. Thus the most general problems can be solved. Our methods support both worst-case and statistical solutions to the tolerance analysis and synthesis problems. Our methods support the geometric concept of tolerance, formulated in the ANSI tolerancing standard, which incorporates geometrical formulations of the traditional plus-minus tolerances as well as the newer geometric tolerances.

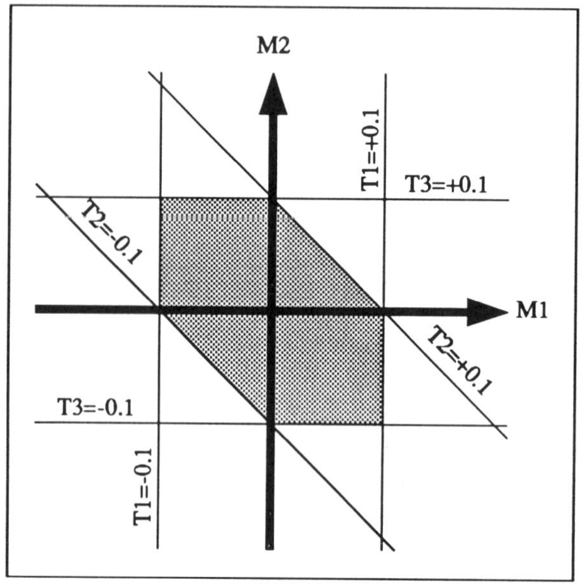

Figure 6: Feasible Region

The feasibility space approach has a number of advantages over the algorithms used by commercial software packages (Turner and Gangoiti, 1991a; Turner and Gangoiti, 1991b). The designer can use as many or as few tolerances as necessary to control critical design properties. The omission of a functionally-significant tolerance leads to an unconstrained variation, which can be detected and reported to the user. Extra tolerances simply add additional constraints to the feasibility space. We believe that all of the Y14.5 tolerances (both plus-minus and geometric) can be handled in this manner in a way which preserves the meaning and intent of the standard.

TOLERANCE SYNTHESIS

Tolerance Synthesis is the process of deriving tolerance limits which will result in the lowest-cost design which satisfies the design constraints. This requires both a mechanism for expressing the design constraints and a means of estimating the manufacturing cost associated with each tolerance.

The traditional way of expressing design constraints is by placing fixed constraints on the design functions. This turns the tolerances synthesis problem into a constrained optimization problem.

However, Taguchi and others (Phadke, 1989) have pointed out that as parts deviate from nominal, although they may still function, they often function less well (for example, they may have a lower reliability). A part that is just within specification is rarely perceived to be very different from a part that is just outside. Thus he proposes a "quality loss" function that is a continuous function of the deviation from nominal. This quality loss can be expressed as a cost, either to the user who must deal with a lesser product or to the manufacturer who may have to reject parts.

Using such a quality-loss model, the tolerance synthesis problem can be restated as an unconstrained optimization problem: finding the tolerances which yield the lowest total cost, where the total cost is the sum of the manufacturing cost and the quality loss.

A cost model needs to be associated with each tolerance, which relates the cost of holding that tolerance to the value of the tolerance limit. A variety of cost models may be used. A simple one is the *inverse power* model (Bennett and Gupta, 1969; Wu et al., 1988):

$$C = Kt^{-n}$$

where K and n are parameters set by the user, and t is the size of the tolerance zone. For the plus-minus tolerances, t is given by the difference in the tolerance limits.

A reasonable estimate of quality loss is the quadratic loss function (Phadke, 1989):

$$Q = \frac{A}{\Delta^2}(D - D_{nominal})^2$$

where Δ is the functional limit (the variation of the design function at which the product will fail approximately half of the time), A is the cost of repair or replacement, and D is the worst-case value of the design function.

Once the cost models are defined, and the design functions are related to the tolerance limits (through the tolerance analysis algorithm), a non-linear optimization approach, such as Newton-Raphson, can be used to find the minimum cost configuration.

In the GEOS implementation, the logic flow would look like Figure 7. In the initialization phase, the geometric model is converted to a mathematical model. This consists of relating the tolerances and the design functions to the model variables. Then, within a loop, the analysis problem is solved. In other words, for the given set of tolerance limits, the worst-case values of the design functions are found. As part of this solution, the sensitivities of the design functions on the tolerance limits are found. This relationship is then inserted into the cost equations to relate the total cost to the tolerance limits. An optimization scheme, such as Newton-Raphson, is then used to vary the tolerance limits to minimize the cost.

The sensitivities of the design functions on the tolerance limits that result from solving the analysis problem are only valid within a certain range of limits, however. The new limits must be analyzed again, to see if the old sensitivities are still valid. Thus the procedure loops until it converges to an answer.

CONCLUSIONS

In defining a tolerance analysis strategy, one needs to answer two questions:

1. What is a tolerance?
2. How do we analyze it?

This paper attempts to address these issues. It presents an overview of the GEOS tolerancing system, and it includes references to other papers which go into more detail.

The two primary definitions of tolerances are parametric and geometric. Geometric tolerances tend to represent real-world considerations better than parametric tolerances, although the latter tend to be easier to analyze. The two definitions are not equivalent, and converting from geometric to parametric tolerances can be difficult or impossible without approximation.

The feasibility space approach, based on a variational modeling strategy, provides an efficient and accurate way to analyze the tolerances and their effects in three dimensions.

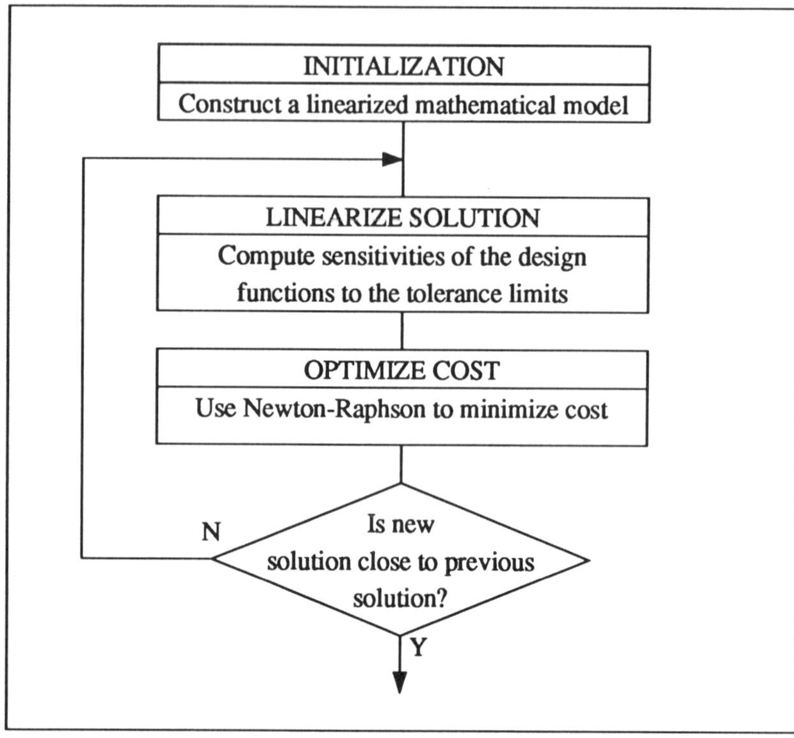

Figure 7: GEOS Logic Flow for Tolerance Synthesis

Once it is possible to analyze a given tolerancing problem, it then becomes possible to use the analysis tool within a loop to synthesize tolerance limits. By considering any variation of the design functions from nominal as a cost associated with the quality loss, the tolerance synthesis problem can be phrased as an unconstrained cost minimization problem.

This analysis and synthesis strategy form the basis for the GEOS tolerancing system, developed at RPI.

An area of future work is *tolerance allocation* (tolerance strategy). This is the process of selecting the tolerances themselves, rather than just finding values for the tolerance limits.

Given insufficient tolerances, there are some model variations that are unconstrained. Heuristics can be used to specify tolerances which control these variations. This can be taken to an extreme to actually allocate all of the tolerances from a given set of design functions. Tolerance synthesis could then be used to determine the tolerance limits. While such a completely automated scheme would probably not replace a designer, it could offer a starting point which the designer would then modify.

ACKNOWLEDGMENTS

Support for this research program is being provided by the National Science Foundation under Grant Numbers DDM-8908160, DDM-8914172, and DDM-9057598, by the New York State Science and Technology foundation through a grant to the Center for Advanced Technology in Automation and Robotics, and by the industrial associates of the Rensselaer Design Research Center. Any opinions, findings, and conclusions or recommendations expressed are those of the authors and do not necessarily reflect the views of the sponsoring agencies.

REFERENCES

American Society of Mechanical Engineers, New York, N.Y., 1982, *Dimensioning and Tolerancing*. ANSI Y14.5M-1982.

G. Bennett and L. C. Gupta, 1969, "Least-cost tolerances – I", *The International Journal of Production Research*, 8(1):65–74.

D. C. Gossard, R. P. Zuffante, and H. Sakurai, 1988, "Representing dimensions, tolerances, and features in MCAE systems", *IEEE Computer Graphics and Applications*, 8(2):51–59, March.

James Guilford, Mani Sethi, and Joshua Turner, 1992, "Worst case and statistical tolerance analysis of daughter card assembly", In *Proceedings of the 1992 ASME Computers in Engineering Conference*, volume 1, pages 343–350, San Francisco, August 2-6.

Suvajit Gupta, 1991, "Variational solid modeling for tolerance analysis", Master's thesis, Rensselaer Polytechnic Institute, Troy, NY, August. Rensselaer Design Research Center, Technical Report Number 91018.

Suvajit Gupta and Joshua U. Turner, 1991, "Variational solid modeling for tolerance analysis", In *Proceedings of the 1991 ASME Computers in Engineering Conference*, Santa Clara, CA, August 18-22.

Suvajit Gupta and Joshua U. Turner, 1993, "Variational solid modeling for tolerance analysis", *IEEE Computer Graphics and Applications*. Accepted for publication.

R. C. Hillyard and I. C. Braid, 1978, "Analysis of dimensions and tolerances in computer-aided mechanical design", *Computer-Aided Design*, 10(3):161–166.

R. Jayaraman and V. Srinivasan, 1989, "Geometric tolerancing I: Virtual boundary requirements", *IBM Journal of Research and Development*, 33(2):90–104, March.

R. Light and D. Gossard, July 1982, "Modification of geometric models through variational geometry", *Computer-Aided Design*, 14(4):209–214.

V. C. Lin, D. C. Gossard, and R. A. Light, 1981, "Variational geometry in computer-aided design", In *SIGGRAPH '81 Conference Proceedings*, pages 171–177, August 3-7. Computer Graphics, Volume 15, Number 3.

P. M. Martino, 1988, *Application of Variational Geometry to the Analysis of Tolerances in Solid Models*, PhD thesis, Rensselaer Polytechnic Institute, May.

Madhav S. Phadke, 1989, *Quality Engineering Using Robust Design*. Prentice Hall, Englewood Cliffs, New Jersey.

A. A. G. Requicha, 1983a, "Representation of tolerances in solid modeling: Issues and alternative approaches", Technical Report 41, Production Automation Project, University of Rochester, August.

A. A. G. Requicha, 1983b, "Toward a theory of geometric tolerancing", Technical Report 40, Production Automation Project, University of Rochester, March.

A. A. G. Requicha and S. C. Chan, 1986, "Representation of geometric features, tolerances, and attributes in solid modelers based on constructive geometry", *IEEE Journal of Robotics and Automation*, RA-2(3):156–166, September.

J. R. Rossignac and A. A. G. Requicha, 1985, "Offsetting operations in solid modelling", Technical Report 53, Production Automation Project, University of Rochester, June.

V. Srinivasan and R. Jayaraman, 1985, "Issues in conditional tolerances for CAD systems", In *IEEE International Conference on Robotics and Automation*, St. Louis, Missouri, March 25-58.

V. Srinivasan and R. Jayaraman, 1989, "Geometric tolerancing II: Conditional tolerances", *IBM Journal of Research and Development*, 33(2):105–124, March.

J. U. Turner, 1987a. "A vector space approach to the mathematical representation of tolerances". SIAM Conference on Applied Geometry, July 20-24. Albany, New York.

Joshua U. Turner, 1987b, *Tolerances in Computer-Aided Geometric Design*, PhD thesis, Rensselaer Polytechnic Institute, May.

Joshua U. Turner, 1988. "A conceptual framework for tolerances in CAD". IFIP/NSF Workshop on Geometric Modeling, September 18-22. Rensselaerville, New York.

Joshua U. Turner, 1993, "A feasibility space approach for automated tolerancing", *ASME Journal of Engineering for Industry*. Accepted for publication.

G. P. Turner and D. C. Anderson, 1988, "An object-oriented approach to interactive, feature-based design for quick turnaround manufacturing", In *Proceedings of the 1988 ASME Computers in Engineering Conference*, San Francisco, CA, July 31 – August 4.

Joshua U. Turner and Alex Beascoechea Gangoiti, 1991a, "Commercial software for tolerance analysis", In *Proceedings of the 1991 ASME Computers in Engineering Conference*, Santa Clara, CA, August.

Joshua U. Turner and Alex Beascoechea Gangoiti, 1991b, "Tolerance analysis approaches in commercial software", *Concurrent Engineering*, 1(2):11–23, March/April.

Joshua U. Turner and Michael J. Wozny, 1988, "A mathematical theory of tolerances", In M. Wozny, J. Encarnacao, and H. McLaughlin, editors, *Geometric Modeling for CAD Applications*. North-Holland.

Joshua U. Turner and Michael J. Wozny, 1990, "The M-space theory of tolerances", In *Proceedings of the 1990 ASME Design Automation Conference*, Chicago, Ill., September 17-19.

Nanxin Wang and Tulga M. Ozsoy, 1991, "A scheme to represent features, dimensions, and tolerances in geometric modeling", *Journal of Manufacturing Sysems*, 10(3):233–240.

Zhang Wu, W. H. ElMaraghy, and H. A. ElMaraghy, 1988, "Evaluation of cost-tolerance algorithms for design tolerance analysis and synthesis", *Manufacturing Review*, 1(3):168–179, October.

CRTD-Vol. 27, International Forum
on Dimensional Tolerancing and Metrology
ASME 1993

STATISTICAL TOLERANCING: A CASE STUDY

Roberto E. Altschul and Friedrich W. Scholz
Department of Research and Technology
Boeing Computer Services
Seattle, Washington

ABSTRACT

This case study applies statistical tolerancing to the problem of attaching a cargo door to an airplane body through a series of ten hinges of two different types. The problem is treated only in the hinge axis dimension and at issue is whether the gaps and lugs of the hinges mate successfully when we allow for variation in lug widths and pitches and in the positioning of the hinges on door and body of the airplane. The problem is first attacked in the context of an unfastened hinge pair and then generalized to the whole sequence of fastened hinges, by viewing the two sequences as giant hinge halves. The hinge problem exhibits the feature that the usual linearization paradigm does not apply to the derived clearance criterion C, since it is not differentiable at the nominal dimensions. However, it is easy enough to simulate the distribution of C while properly modeling the various sources of variation.

INTRODUCTION AND PROBLEM DESCRIPTION

A cargo door is to be attached to an airplane body with two types of hinges, namely with a sequence of eight "piano hinges" flanked by one "main hinge" at each end of the sequence. Figures 1 and 2 show stylized versions of both types of hinge sets. Note that in Figure 1 the top half of the hinge has the same design as the bottom half. We address here some tolerancing aspects that need to be considered for proper fitting of door to body. The tolerancing problem will only be treated in a one-dimensional setting, namely for variations along the hinge axis. Such variations cover the widths of lugs and gaps, their pitch (positioning of lugs and gaps), and the positioning of hinge pairs on door and body.

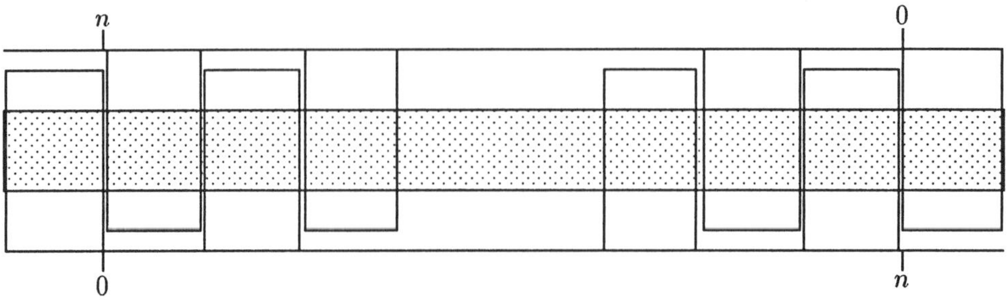

Figure 1: PIANO HINGE

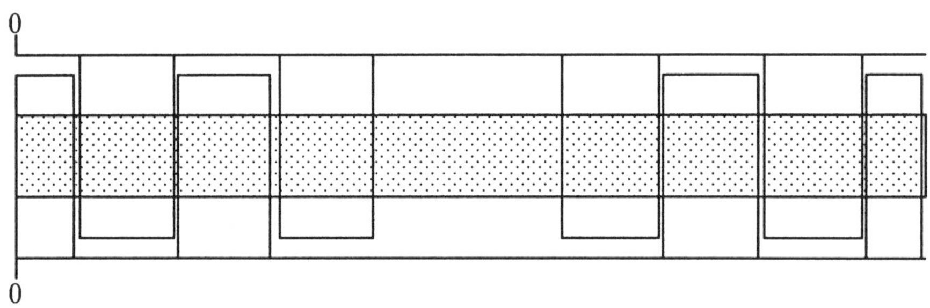

Figure 2: MAIN HINGE

The problem is first treated for a single hinge pair not fastened to door or body. Of interest is the mating clearance criterion C, which has to be positive in order for proper mating of the two hinge halves to occur. It turns out that C is a not a fully differentiable function of the widths and pitches of lugs and gaps. This prevents us from treating this as linear tolerance stacking which is usually justified by a first term Taylor series expansion.

Since this linear stacking paradigm is so pervasive in the literature (see Cox (1986), Evans (1974, 1975), and Nigam and Turner (1992)) it is of interest to mention the following example as another exception. Suppose two circles with nominal radii $r_1 > r_2$ are to have common nominal centers at the origin. The actual centers and radii are (X_i, Y_i) and R_i, $i = 1, 2$. If the second circle is to fit within the first one, then the clearance criterion

$$D = R_1 - R_2 - \sqrt{(X_1 - X_2)^2 + (Y_1 - Y_2)^2}$$

should be positive, see Figure 3. The function $D = D(R_1, R_2, X_1, Y_1, X_2, Y_2)$ is not differentiable at $(X_1, Y_1, X_2, Y_2) = (0, 0, 0, 0)$, i.e., at the nominal positions of the circle centers, and thus D cannot be linearized. For the same reason it is not possible to linearize

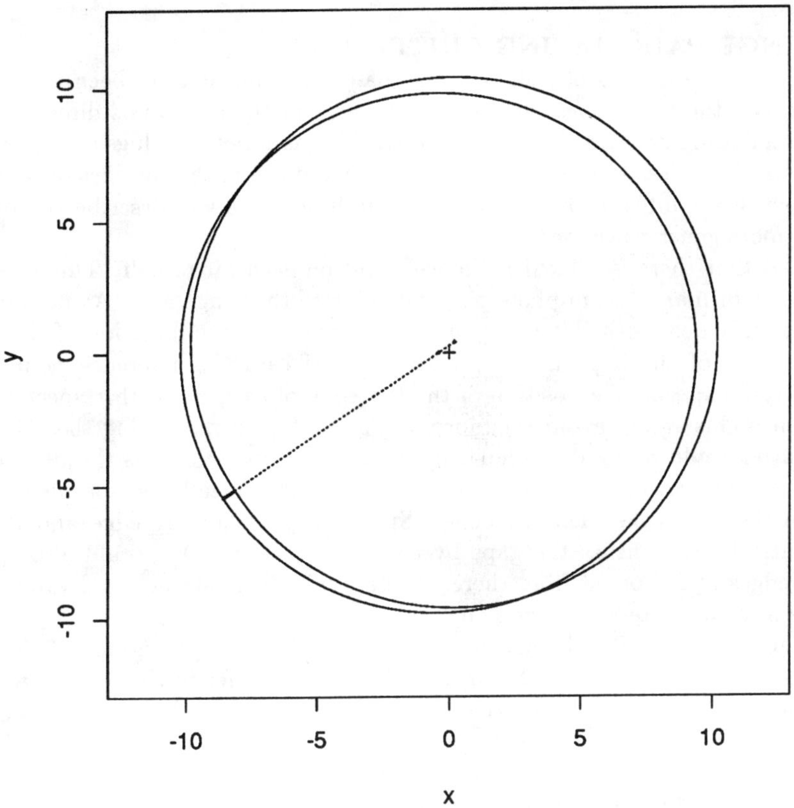

Figure 3: HOLE CLEARANCE C

the hinge clearance C and we cannot use the usual analytic avenues (Cox (1986), Evans (1975)) to evaluate the proportion of hinge pairs that will not mate. However, it is easy enough to simulate the distribution of C and thus assess the appropriateness of the design.

After resolving the hinge mating problem it is a simple conceptual step to proceed to the door to body mating problem. Here we only deal with mating the fastened hinges on the door with the fastened hinges on the body and with no other issues concerning proper fitting of door and body. The idea is to treat the door with its fastened hinges as one half of a giant hinge pair and likewise treat the body with its fastened hinges as the corresponding other half of the giant hinge pair. The remainder of the problem is mainly one of notation and of tracking the various sources of variation and dependencies, but that is easily dealt with in the simulation.

SINGLE HINGE PAIR MATING CRITERION

Here we address the mating of a single hinge pair which has not yet been fastened to door or body, i.e., the two halves can be moved freely in the hinge axis direction until lugs and corresponding gaps are aligned correctly. This assumes that lugs and gaps have proper widths and pitches. Rather than getting bogged down in the specifics of the hinge pair design, which are different for piano and main hinges, we will describe the mating criterion in a more generic fashion.

It is assumed that there is a fixed reference point on each hinge half. This reference point defines the primary datum plane perpendicular to the hinge axis. As pointed out earlier, we will only deal with this problem as a linear one in the direction of the hinge axis. All positions of the lugs can be described relative to this reference point. For example, one could specify the position of the left edge of each lug with respect to this reference point. This would avoid tolerance stack buildup in machining these left lug edges. The latter could occur if the left edge of a lug was specified as the distance to the left edge of the previous lug. The lug widths would typically be specified as the distance from the established left lug edge. Specifying the left lug edge and the lug width automatically establishes the gaps by complementation. One could also specify the right lug edges and proceed from there. The specifications are nominal values with a $\pm$ tolerance interval centered on the nominals.

In order to properly describe the mating criterion we introduce the notion of "left and right clearances." With reference to Figures 1 and 2 we will refer to the bottom and the top half of a hinge pair. It is assumed that the reference points (0 and 0 in Figure 2) or appropriate auxiliary reference points (n and 0 in Figure 1) of the two halves are aligned on top of each other. Typically this means that the hinges, if machined to nominal values, would successfully mate without interference. This alignment is only done to fix the reference framework and is not essential for proper mating. As we will see, the final mating criterion does not depend on this reference alignment since an offset in one direction is traded off by an offset in the other direction.

Usually the right facing edge of a lug on the bottom hinge half has to clear the left facing edge of the corresponding lug on the top hinge half. Let the positions of these two interacting edges be $N + X$ and $N' + X'$, respectively. Here N and N' stand for the nominal values and X and X' are random variables which express the variation from nominal. They typically would be assumed to have mean zero and their probable range should cover the specified $\pm$ tolerance or they could be composed as a sum of other random variables. The latter could occur if the relevant lug edge is established by first measuring off the other lug edge and from there the width of the lug to get to the edge in question. In that case two independent variations would be compounded in an additive fashion. The difference $L = N' + X' - (N + X)$ is referred to as a left clearance. If this clearance is positive, we could still move the upper hinge half to the left by the amount L before interference occurs at that interface. For the piano hinge in Figure 1 the first left clearance needs to be treated specially since it occurs at the reference plane. It would seem appropriate to set the error term X to zero in that case, since the physical lug edge on the lower hinge half is identified with the nominal N there. However, the corresponding X' has variation.

Similarly, the left facing edge of a lug on the bottom hinge half has to clear the right facing edge of the corresponding lug on the top hinge half. Let the positions of these two interacting edges be $M + Y$ and $M' + Y'$, respectively. Here M and M' stand for the nominal values and Y and Y' are again random variables which express the variation from nominal. The difference $R = M + Y - (M' + Y')$ is referred to as a right clearance. If this clearance is positive, we could still move the upper hinge half to the right by the amount R before interference occurs at that interface.

Usually there are many left and right clearances on a given hinge pair and they are either equal in number, say n of them as in Figure 1, or they differ by a count of one, i.e., there are n clearances of one type and $n + 1$ clearances of the other type, see Figure 2. Thus generically we have n_1 left clearances $L_1, \ldots, L_{n_1}$, with $L_i = N_i' + X_i' - (N_i + X_i)$ and n_2 right clearances $R_1, \ldots, R_{n_2}$ with $R_i = M_i + Y_i - (M_i' + Y_i')$. As long as all left clearances are positive we can still move the upper hinge half to the left by the amount

$$T_L = \min(L_1, \ldots, L_{n_1})$$

without causing interference at any of these "left" interfaces. Similarly, as long as all right clearances are positive we can still move the upper hinge half to the right by the amount

$$T_R = \min(R_1, \ldots, R_{n_2})$$

without causing interference at any of these "right" interfaces. Finally, if $C = T_L + T_R$ is positive we have enough "play," namely the amount C, in moving the top hinge half left or right, so that no interferences occur at any of the hinge interfaces. Needing C to be positive for clearance does not mean that both T_L and T_R have to be positive. For example, we could have $T_L < 0$ but T_R is sufficiently large positive. In that case we could make $T_L > 0$ as well through right motion of the upper hinge half, but in the process we also reduce T_R. Of course, it is assumed that these movements are so minute that no other interference issues arise.

In typical hinge design the nominal left and right play is constant from lug to lug, i.e., $N_i' - N_i = \Delta$ and $M_i - M_i' = \delta$, so that

$$T_L = \Delta + \min\left(X_1' - X_1, \ldots, X_{n_1}' - X_{n_1}\right)$$

and

$$T_R = \delta + \min\left(Y_1 - Y_1', \ldots, Y_{n_1} - Y_{n_1}'\right) .$$

This shows that neither T_L nor T_R is differentiable at the nominal. Also, it is not easily possible to analytically get the distribution of T_L, T_R and thus of C, since the X's and Y's are not all independent. The latter dependence is due to the carry-over effect of the error in determining one lug edge and measuring off the lug width to obtain the other edge. Thus left and right clearances can be affected by the same source of variation. However, such dependencies are easily handled when simulating the distribution of C in order to assess what proportion of simulated hinge pairs have positive clearance with the chosen design.

DOOR TO BODY MATING CRITERION

Having dealt with the single hinge pair we can now carry over this approach to the mating of the door with the ten attached hinges to the body with the attached corresponding hinges. As indicated in the introduction the problem is solved by treating door and body as two giant hinge halves, but with greater variety of lug/gap structure. The only added complication is that now the notation and analysis will have to be carried out with respect to some pair of fixed reference points on the door and body and we need to account for the additional variation from nominal in the positioning of the reference points of the individual hinges within the giant hinge structure. This time we have many more left clearances and many more right clearances, say $L_1, \ldots, L_{m_1}$ and $R_1, \ldots, R_{m_2}$, with m_1 and m_2 being roughly certain multiples of n_1 and n_2, respectively. Using the same reasoning as before we have as the overall clearance criterion

$$C_T = \min(L_1, \ldots, L_{m_1}) + \min(R_1, \ldots, R_{m_2}) \ .$$

MODELING SOURCES OF VARIATION

A position error in one particular hinge affects all the left and right clearances coming from this hinge in the same way, i.e., it provides a common random effect on those clearances. Superimposed on this effect are the variations within that hinge, namely the lug width errors and the pitch errors which are typically independent from lug to lug, but which may introduce some additional mild dependencies in the left and right clearances from that hinge. The position error of the hinges depends strongly on how the positioning is accomplished. For example, we may use a special tool that shows the positions for each of the hinges of on the body. These positions on the tool may be in error. One of these positions may serve again as reference plane and thus has no error. These body tool errors would then act as a consistent bias in positioning the hinges. However, even this error can change if the tool is recalibrated occasionally. On top of this tooling error we have to allow for the hinge positioning and fastening error relative to the tool.

In a similar fashion we may have another tool for positioning the hinges on the door. This door tool may have its hinge position indicators established with error, either relative to an absolute reference point or relative to the body tool. This error again acts as a random bias unless the tool is recalibrated every so often. Again there is the error of positioning and fastening the hinges with this door tool.

Writing down the probabilistic model for the various sources of variation entering the clearance criterion C_T is tedious, straightforward, but omitted here, but simulating the distribution of C_T poses then no difficulty.

In translating the various tolerances for statistical tolerancing one may use normal distributions with a $\pm 3\sigma$ range over the given tolerance intervals for lug width, pitch and positioning of hinges on doors and body. All these variations are assumed to be independent, but can result in significant dependence structure among the various left and right clearances. Alternatively, one can also try uniform variations over the tolerance ranges, with naturally higher likelihood of mating failure. In the particular example for which the above approach was carried out we found that with normal variation about 99.4% of the doors would have mated properly, but with uniform variation only 14.2% cleared.

Such simulations should be quite useful in deciding how to set the tolerances and the design clearances. In particular, the effect of tool errors can be studied. Since many factors can be varied here, namely hinge positioning/fastening tolerances, tool tolerances, lug pitch and width tolerances, and design clearances, it may be useful to employ experimental design techniques to sort out the best design factors. However, as the difference in clearances under normal and uniform variation already made clear, it would be of much greater value to use real data, namely how good is the positioning of the hinges, and what tolerances can be achieved in producing the hinges? It is of not much use designing something and then not being able to hold the stated tolerances. The real variation in the process ultimately drives the fallout rate.

REFERENCES

Cox, N.D. (1986). *Volume 11: How to Perform Statistical Tolerance Analysis.* American Society for Quality Control, Milwaukee, Wisconsin.

Evans, D.E. (1974). "Statistical tolerancing: The state of the art. Part I. Background" *Journal of Quality Technology* **6**, 188-195.

Evans, D.E. (1975). "Statistical tolerancing: The state of the art. Part II. Method for estimating moments." *Journal of Quality Technology* **7**, 1-12.

Nigam, S. and Turner, J. (1992). "Statistical approaches to tolerance analysis." Tech. Report No. 92021, Rensselaer Design Research Center, Rensselaer Polytechnic Institute, Troy New York 12180.

CRTD-Vol. 27, International Forum
on Dimensional Tolerancing and Metrology
ASME 1993

COMPUTATIONAL METROLOGY[1]

Theodore H. Hopp
Factory Automation Systems Division
National Institute of Standards and Technology
Gaithersburg, Maryland

ABSTRACT

Coordinate measuring systems (CMSs) assess length-based characteristics of mechanical parts by measuring points on the part surface and analyzing the point data. Data analysis software can contribute significantly to the total measurement error of a CMS. Factors affecting software performance include the choice of analysis method, the quality of the software, and characteristics of the specific measurement task. By *computational metrology* we mean the study of how data analysis factors affect the measurement uncertainty of a CMS.

This paper describes research at NIST on computational metrology. Metrological questions fall into two categories: (1) the proper choice of data analysis objective for a particular application and (2) the performance of the implementation. Our research goal is to develop the basis for a national standard in the U.S. for CMS software performance evaluation. We are implementing a Special Test service to be offered through the NIST Measurement Services Program. The service is based on a black-box model of software, in which the internal structure of the software and the choice of solution methods are assumed to be unknown. The model identifies a number of error sources for data analysis software. We are designing test methods for identifying the various components of the model and how those components relate to measurement uncertainty in inspection applications.

INTRODUCTION

Data analysis software has become increasingly important in modern dimensional measurement systems. This is particularly true of coordinate measurement systems (CMSs) such as vision systems, theodolites, photogrammetry, and coordinate measuring machines. Software computations to convert raw data to reported results can be a major source of error in a measurement system. Yet there are no standards or

---

accepted methods for evaluating the effects of software on the overall uncertainty of measurements. The phrase *computational metrology* refers to the study of the effects of data analysis computations on the performance of measurement systems. This paper identifies those aspects of computational metrology that are particular to CMSs. In particular, we will examine the objective of the computations, how these functions are carried out in software, and how software performance can be tested.

Not everyone shares the view that computational metrology is a significant area of study. For instance, an ASME standard on measurement uncertainty (ASME, 1985) states:

> Computations on raw data are done to produce output (data) in engineering units. Typical errors in this process stem from curve fits and computational resolution. These errors are often negligible.

The standard does deal with the *propagation* of errors through computations, but the above is the only mention of computations as a *source* of errors.

During the last few years, however, much evidence has been discovered that data analysis can be a significant source of errors. In the mid-1980s, Germany began a program (Porta and Wäldele, 1986) of testing coordinate measuring machine software, with the express purpose of improving what was perceived as low quality of commercial fitting algorithms. In 1988, Walker issued an advisory in which he reported the results of experiments with commercial inspection systems:

> Certain algorithms . . . are capable of stating that the measurement is worse than the actual data gathered up to an error of 37% and that the measurement is better than the actual data gathered up to an error of 50%.

And in 1989, Estler analyzed a measurement device for inspecting the casings of the solid rocket boosters for the NASA space shuttle. He reported that the data analysis software was the single largest source of error in the entire system.

There is considerable ongoing research on CMS algorithms. (See Feng and Hopp (1991) for a survey.) There is also growing interest by government standards laboratories in testing the performance of CMS software, particularly in Great Britain (Cox, 1992) and Germany (Wäldele et al., 1993). Both countries offer services to test CMS software by comparing results for test data sets to results obtained from reference software. In the U.S., NIST is developing a similar service (Diaz, 1993).

Of particular interest is the performance of software for calculating geometry that in some sense best fits a set of points measured on a part surface. Fitting is at the core of most measurements made by CMSs.[2] From a metrological point of view, two factors determine fitting software performance: the choice of fitting objective and the quality of the software implementing that objective.

To date, testing methods have focused on the quality of implementation of software (Cox and Forbes, 1992). These methods have been designed on the basis that the internal structure of the software is a "black box." Testing is based on evaluating the fitting results for specially designed data sets.

Published testing methods have uniformly been based on what might be called an external view of the behavior of the software: whether the software is computing what it is advertised to compute. This view has two weaknesses. First, it provides few guidelines for designing a suite of data sets or for interpreting the results. Second, it provides no guidelines for evaluating the utility of different fitting objectives for an application. In this paper we outline the elements of a theory of software testing based on an operational model of sources of error in fitting software. We believe that an operational model will provide a basis for designing test methods based on sound metrological principles.

---

[2] Fitting is usually thought to involve optimizing the fit between a single geometric form—plane, cylinder, etc.—to a data set. However, so-called soft functional gaging can also be thought of as fitting a solid model of a functional gage to the data.

The next section of this paper discusses the relationship between fitting objectives, measurement error, and tolerance theory. Following that, we will examine implementation issues: the interaction of optimization methods and computing environment, handling of extreme cases, and code correctness. In the last section, we discuss the testing methods being developed at NIST. We conclude with a summary of issues that need further attention.

FITTING OBJECTIVES

Fitting can be viewed as an optimization problem: find the parameters of substitute geometry that optimize a particular fitting objective for a set of points. The fitting process is illustrated in Figure 1. As mentioned above, the choice of fitting objective is an important determinant of CMS performance. Part tolerances are generally interpreted in terms of extremal fits. That is, the fitting objective is to find the geometry that fits the extremes of the data: the largest inscribed, smallest circumscribed, or minimum separation geometry. Also, simulation of functional gages (sometimes called soft functional gaging) can be interpreted as finding the maximum clearance (or minimum interference) solid model fit to the data. Measurement practice, on the other hand, most commonly involves averaging fits, typically least-squares (orthogonal distance regression) but also objectives such as least median of squares (Rousseeuw, 1984; Mamileti et al., 1992) and other robust techniques for outlier detection.

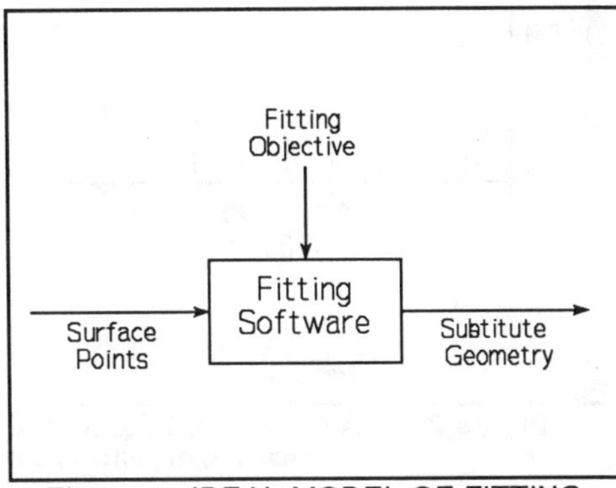

Figure 1 IDEAL MODEL OF FITTING.

In general, averaging fits commonly used in metrology are biased with respect to the extremal fit objectives suggested by tolerancing theory. That is, in the limit as the point density goes to infinity and the measurement error for each point goes to zero, averaging fits will be different from extremal fits.[3] On the other hand, it would seem that averaging fits are less sensitive to measurement errors than are extremal fits (Vecchia et al., 1993).

We used a circle-fitting problem to study these effects for least-squares and minimum circumscribed fitting objectives. We generated twenty points evenly spaced around a unit circle and then perturbed the points by moving each point in a random direction by a random distance. The directions were uniformly distributed over the unit circle and the distances were uniformly distributed between zero and an error scale factor δ. Perturbations were uncorrelated between points. We varied δ from 0.005 to 0.05 and generated 100 data sets for each value of δ. For each data set, we computed both the least-squares and minimum circumscribed fit.

The next figure shows the results of these experiments. Figure 2a shows the distance between centers and Figure 2b shows the difference in radii for the two objectives. The median distance between centers for the two fit objectives varies approximately linearly with the perturbation scale δ, with a slope of about 0.5. Similarly, the difference in radii between the two objectives varies linearly with a slope of about 1. If the perturbations are viewed as form errors in the feature, these results suggest that the bias of least-squares fitting is about half of the form error. The perturbations can also be viewed as errors in measuring individual points; a similar analysis, comparing the fits to the unperturbed circle, then shows that minimum circumscribed fitting is about twice as sensitive to measurement error as is least-squares fitting. More generally, extremal fits will propagate more of the point measurement error than will least-squares fitting.

---

[3] Depending on one's viewpoint, bias might be considered to be an error in the model. We reserve the term *model error*, however, to indicate the error resulting from picking a particular geometric form to fit to the data.

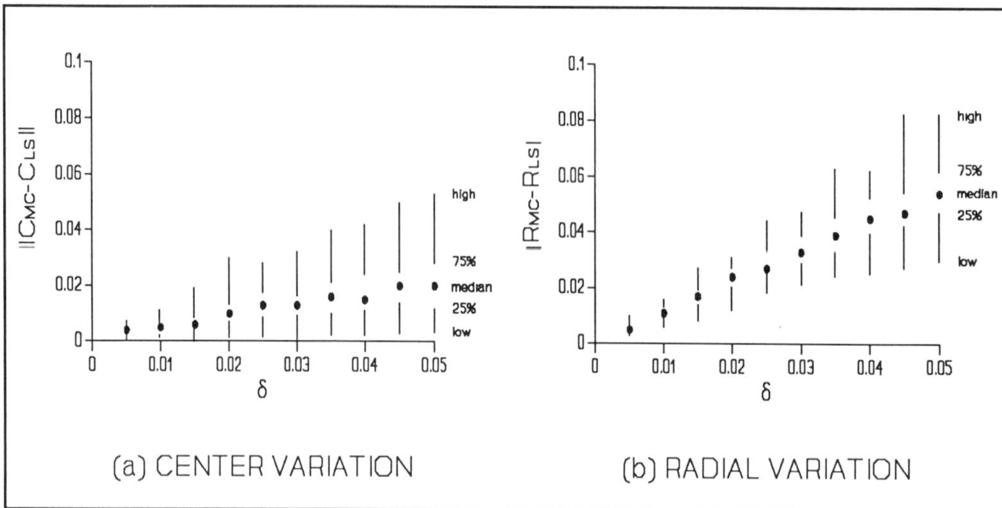

Figure 2 QUARTILE PLOTS COMPARING 100 LEAST SQUARES AND MINIMUM CIRCUMSCRIBED FITS FOR EACH VALUE OF δ.

Many fitting criteria can be expressed as special cases of a general criterion called *L_p-norm estimation.* This is a mathematically convenient formulation for which there are many results (Gonin and Money, 1989). The L_p-norm estimation problem is to find the fit parameters that minimize the L_p norm:

$$L_p = \left[\frac{1}{n}\sum_i |r_i|^p\right]^{1/p}$$

where r_i is the ith residual and the sum is over n data points. Least-squares fitting corresponds to the case p=2. The limit of L_p as p goes to infinity is the largest magnitude residual, so the L_∞ problem is to minimize the maximum magnitude residual—that is, to find the minimum zone fit. Generally, the bias and sensitivity errors of the fit will vary with p. This relationship is shown in Figure 3. As p increases, the sensitivity of the fit to point measurement error increases, but the bias with respect to the fit prescribed by tolerance theory decreases. The precise shape of the curves depends on the distribution of residuals, the geometry of the particular fitting problem, the configuration of measured points, and the measurement uncertainty of the points.

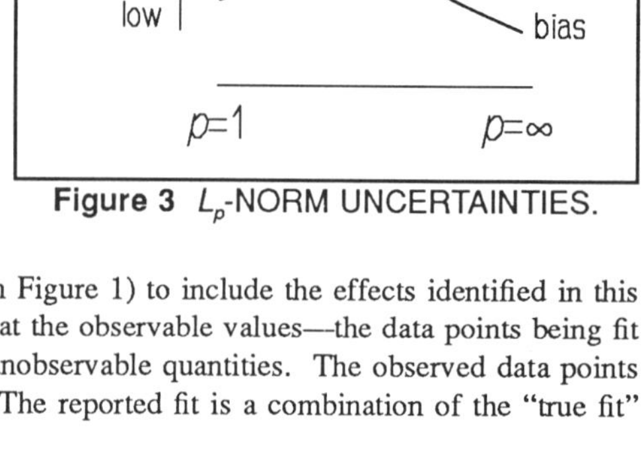

Figure 3 L_p-NORM UNCERTAINTIES.

We see, therefore, that it is very difficult to develop general guidelines for the proper choice of fitting objective for a practical CMS. Least-squares fitting is widely used and widely debated, with many claiming that extremal fitting is better because it "conforms" to tolerance theory. In fact, the best choice of fitting objective is that which produces the smallest combined uncertainty in the result. How to make the best choice is not at all clear.

We can modify the ideal model of fitting (shown in Figure 1) to include the effects identified in this section. The new model, drawn in Figure 4, shows that the observable values—the data points being fit and the reported fit—are combinations of inherently unobservable quantities. The observed data points always include some element of measurement error. The reported fit is a combination of the "true fit"

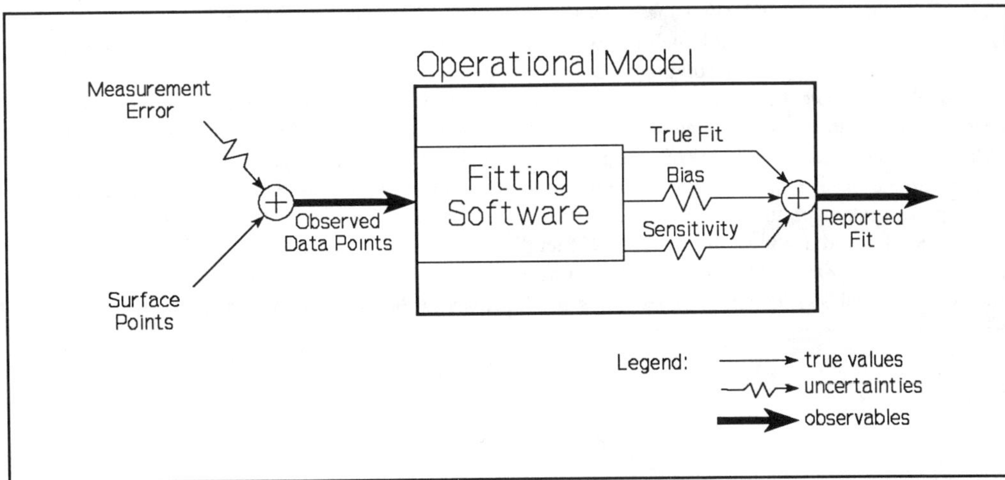

Figure 4 MODEL OF FITTING SOFTWARE, WITH BIAS AND SENSITIVITY ERRORS.

(the fit to the surface points) and two additional errors: the bias introduced by the choice of fitting objective and the sensitivity of the fit to the measurement error.

The sensitivity and bias results reported herein are highly specific. We need to develop more general results on these aspects of commonly used fitting objectives. These general results can then form the basis for practical criteria for selecting the best objective function for a particular application. Such criteria would be of great benefit in inspection planning.

Work is needed to develop objectives with smaller combined uncertainty. We would like to use non-ideal geometry models, to reduce the model error of fitting, say, perfect circles to features that are not circular. Going further, we would like to draw on pattern recognition techniques to develop methods for recognizing and classifying part form deviations. Finally, we would like to have fitting objectives that can make use of models of the production process. It is common knowledge in practice that knowledge of how a part was manufactured will help an experienced inspector do a better job. The inspector can draw on internal models of the form deviations expected from the manufacturing process to develop better inspection plans and data analysis strategies. Bayesian methods might be the basis for developing inspection systems with similar capabilities.

IMPLEMENTATION ISSUES

This section discusses the second aspect of performance of data analysis software: the implementation of the fitting objective in software. Four factors determine the quality of the implementation: the optimization method chosen to compute the fit; the computing environment (word length, rounding method, dynamic range, etc.); handling of extreme cases; and code correctness. We will discuss optimization methods and computing environment in this section. Handling of extreme cases and code correctness are discussed in the following section on testing issues.

The choice of optimization method greatly affects how the computing environment impacts measurement uncertainty. In particular, the effects of rounding due to machine precision can be much greater for some optimization methods than for others, even for the same objective function. Analyses of rounding errors are often done in terms of a quantity called the *unit roundoff error*, denoted here by **u**. **u** is the largest number that, when added to 1 in floating point arithmetic, produces an answer of 1. (Typically **u** is about 10^{-7} for single precision and about 10^{-15} for double precision.) Some iterative algorithms rely on convergence tests to end. The effects of the convergence tests can sometimes be modeled as an increased value of **u** for the computing environment.

The study of how optimization methods and rounding errors interact can draw on results in linear algebra perturbation theory. Although fitting problems are often non-linear in nature, most optimization algorithms work by repeatedly solving a linear approximation to the fitting objective. So the results of linear algebra can potentially be applied to a much broader class of optimization methods than might be apparent. All results in this section are based on linear algebra found in Golub and Van Loan (1985).

One example of how the choice of optimization method can dramatically affect the impact of rounding errors is simple linear regression. Suppose we wish to compute a regression line, represented by a vector v of slope and intercept, to the three points (-100,-100), (0,0), and (100,100). Clearly, the regression line has a slope of 1 and a y-intercept of zero. Generally, the computed value of v (by any method) will be different from the exact solution v_{LS}. The fitting objective is to find the v that minimizes the sum of squares of residuals. This can be modeled as solving for v in the matrix equation $Av \approx y$, where

$$A = \begin{pmatrix} -100 & 1 \\ 0 & 1 \\ 100 & 1 \end{pmatrix}$$

and $y=(-100, 0, 100)^T$. The most common solution method is the use of the *normal equations* for the model, obtained by multiplying the original matrix equation on the left by A^T and solving for v using matrix inversion. The result is $v=(A^TA)^{-1}A^Ty$. If v is computed with roundoff error $\mathbf{u}$ using the *Choleski decomposition* of A^TA, then the error of the solution is $\|v-v_{LS}\| \approx 6{,}700\mathbf{u}$. (The coefficient 6,700 rises from the *condition number* of the matrix A^TA. The condition number, roughly, characterizes how hard the problem is; it is a measure of how close the matrix is to a singular matrix, which cannot be solved.)

The same objective can be evaluated using the *singular value decomposition*. A can be decomposed into the product $U^T\Sigma V$, where U and V are orthogonal matrices and Σ is a diagonal matrix. Let Σ^+ be the diagonal matrix obtained from Σ by setting each element of Σ^+ to zero if the corresponding element of Σ is zero, or else to the inverse of the corresponding element of Σ. Then $v=V\Sigma^+U^T$ is the least-squares solution. If v is obtained by evaluating the singular value decomposition with roundoff error $\mathbf{u}$, then the error of the solution is $\|v-v_{LS}\| \approx 82\mathbf{u}$. Thus, for this very simple problem, the effects of rounding can be changed by nearly two orders of magnitude by the choice of optimization method. (In this case, the measure of difficulty is the condition number of A, a much smaller number.)

The above is an example of *forward error analysis*. It says that the computed fit is the exact fit plus some error. This is illustrated graphically in Figure 5. General results for forward error analysis can be

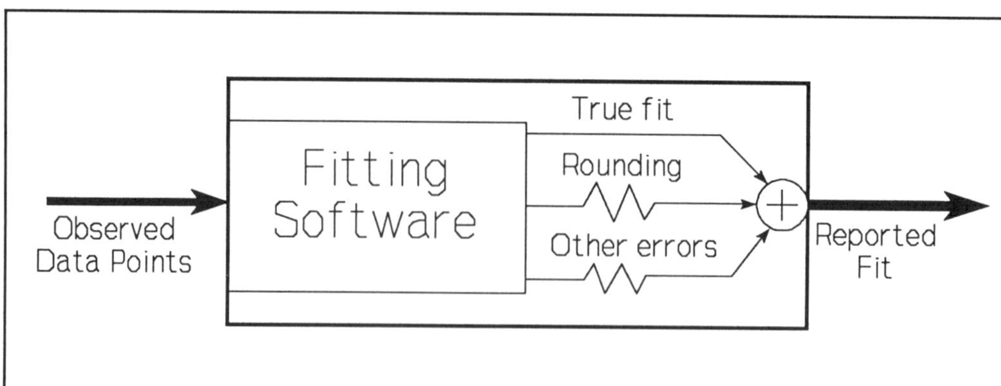

Figure 5 THE FORWARD ERROR ANALYSIS MODEL.

obtained for some algorithms. For instance, orthogonal distance regression of a plane to a set of points can be solved using the singular value decomposition of the data matrix D. This algorithm has a relative error of about $\mathbf{u}\|D\|_2/(\lambda-\mu)$, where $\|D\|_2$ is the largest variance of the data in any direction, λ is the smallest variance of the data in the plane of the fit and μ is the variance of the residuals. Forward error analysis has the shortcoming that the relative error can only be characterized in terms of the same quantities (λ and μ) that are used to compute the fit. Thus, it is very difficult to write software for estimating the fit uncertainty that will work reliably in the same computing environment that is being used to calculate the fit.

A different approach, called *inverse error analysis*, is to view the computed solution as the exact solution to a "nearby" problem. Consider again orthogonal distance regression computed by the singular value decomposition of the data matrix D. The computed fit is the exact fit to some other data set $D+\Delta$. Here, now, it is easy to characterize the effects of rounding; we have $\|\Delta\|_2 \approx c\mathbf{u}\|D\|_2$, for some constant c near one. Inverse error analysis is illustrated in Figure 6. The effects of rounding are modeled as

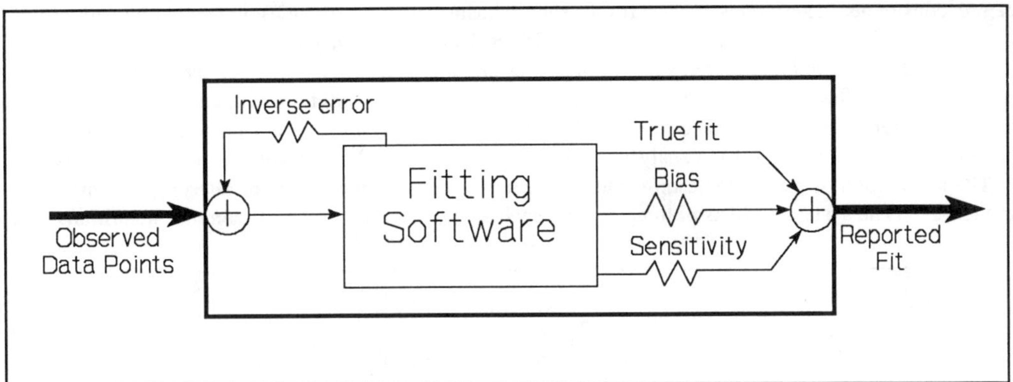

Figure 6 THE INVERSE ERROR ANALYSIS MODEL.

adding additional noise to the data *before* the fit is computed. The computation, however, is then viewed as being done in exact arithmetic. The effects of the inverse error on the reported fit are determined by the sensitivity of the objective function.

Two methods, forward and inverse error analysis, can be used to study the effects of optimization method and rounding on the uncertainty of reported fits. Only inverse error analysis results are known for many linear algebra procedures. This is because many of the factors that affect sensitivity also affect the forward error analysis, but need not be considered for inverse error analysis.

In principle, error analyses can readily be done for linear, least-squares and total-least-squares problems solved by common methods. However, much more needs to be done. Error analysis results need to be developed for all commonly used algorithms. The analysis methods also need to be extended to non-linear problems. (These are often solved using linear approximations, so the extensions should be relatively direct). Harder, but equally important, is to extend the methods to non-differentiable objectives (such as extremal fits). Often, combinatorial methods are used for these objectives instead of iterative linear approximations. The ultimate objective for this work is to develop practical algorithms for estimating the uncertainty of a reported fit.

TESTING ISSUES

The previous two sections outlined a theory of computational metrology. The purpose of testing is to assess the performance parameters of data analysis software in relation to this theory. Performance parameters include the error models identified above, handling of extreme cases, and coding errors. The results in this section are being used to develop a software testing service to be offered by NIST.

Relatively little work is being done in software testing. As mentioned in the introduction, Germany and Great Britain offer a test service for least-squares fitting routines. A joint British/German project is aimed at extending these services to other fitting objectives. In the U.S., ASME Working Group B89.4.10 on CMS Software Performance is developing a standard for characterizing and testing the performance of data analysis software. The test service to be offered by NIST will directly support this emerging standard. We do not mean to suggest that this is the only work on inspection algorithms. Most of the work, however, is aimed at developing new kinds of software or developing usage guidelines.

Our work on software testing is based on three assumptions. The first is that tests should be designed and interpreted according to a well-defined error model. The models presented above identify four error sources: bias, sensitivity, rounding error in the fit, and induced error in the data from rounding. The two other sources of error, handling of extreme cases and code correctness, will be discussed below.

The second assumption is that test results must be interpreted and reported in terms directly related to inspection tasks. Part tolerances generally require conformance of the part to tolerance zones. These zones are volumes or areas for which the part features or substitute geometry elements must observe certain constraints. This means that test results should quantify the uncertainties of geometric relationships computed by the software. Representation-specific results (e.g., parameter values) should be avoided.

The third assumption is that the software to be tested is a "black box." That is, the internal structure of the software—the optimization method, code structure, computing environment, etc.—is unobservable. Thus, the testing method is limited to supplying the software with fitting problems and analyzing the fit results. The problem set and the analysis are designed to identify the elements of the error models.

The assumption of black-box testing leads to the now-common architecture of a testing system shown in Figure 7. The system consists of three components: a data generator, a set of reference algorithms, and

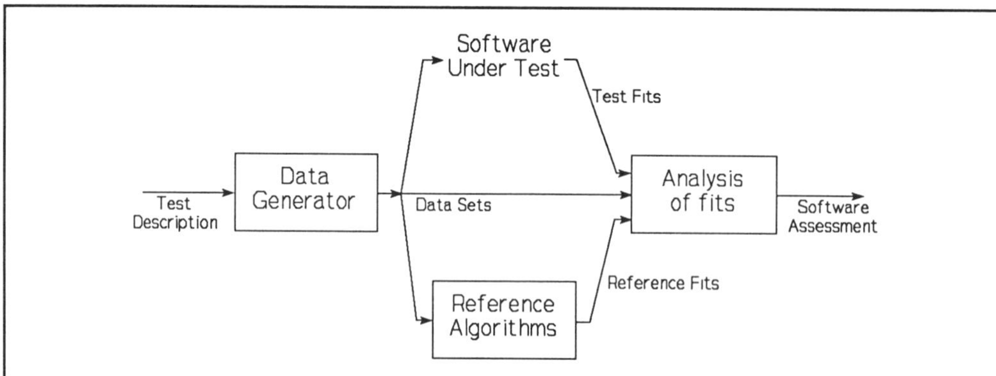

Figure 7 ARCHITECTURE OF THE NIST SOFTWARE TESTING SYSTEM.

fit analysis. The data generator is driven by a test description, a high-level specification of the collection of data sets to be generated. The data sets are processed by the software under test to generate test fits. At the same time, the data sets may be processed by fitting software supplied with the testing system. The data sets, test fits, and reference fits are used to assess performance measures for the software under test. In the remainder of this section, we will discuss the contents of a test description and how fit results are analyzed in the NIST system.

A data set is generated from four classes of information: the nominal (ideal) geometry of the feature, the form error(s) of the feature being simulated; the sampling plan (distribution of points on the feature) for the data set; and the (random) measurement error distribution for the points. A test description consists of ranges of instance values for each of these information classes. This provides the flexibility to study the behavior of the software in a controlled manner. For instance, a test description that varies the simulated measurement error alone can be used to assess the sensitivity errors of the software for the particular combination of other factors. Similarly, variation of other factors can be used to study other aspects of the error model.

A comprehensive test will check the behavior of the software for extreme cases. These cases fall into three categories, pathologies, degeneracies, and extreme values. Pathologies are data sets for which no solution is possible. These include data sets with too few points and problems with unbounded solutions (such as fitting a circle to collinear points). Degeneracies are configurations that should pose no problem but often do. These include data sets with zero residuals (perfect geometry), "vertical" slopes, and similar configurations. Extreme values include number of points (minimum required; maximum of the implementation), distance from the origin, narrow geometry bounds (small arcs, etc.), and large residuals.

Code errors are difficult to test for. Code errors can only be found if a test case exercises the relevant code. This is done in the analysis by comparing test fits to corresponding reference fits. This, in fact, is a primary purpose of computing reference fits. However, it is essentially impossible to design tests that will find all code errors in a black box. The best one can do is to generate data that are representative of the intended application.

As mentioned before, fits are analyzed in terms of geometric differences among fits, not parameter values. In our system, all fit geometries are trimmed by the projection of the data onto the fit geometry. Comparison of test fits to reference fits is used to detect code errors and to identify systematic bias. Bias can arise when an implementation optimizes an approximation to the stated fitting objective. (Commercial software sometimes uses linear approximations in place of non-linear objectives to gain a speed advantage.)

The main tool for evaluating software is measures of variation in fit geometry in response to varying test conditions. These measures are independent of reference fits. A different analysis is used for each type of geometry. For instance, cylinder fits are analyzed as follows. The axis of each cylinder is trimmed by the orthogonal projection of its corresponding data set onto the axis.[4] The smallest cylinder that encloses the axis segments is then computed. The diameter of this enclosing cylinder is a measure of dispersion for tolerances such as position, orientation, and parallelism. Depending on which parameters of the test description are involved, this dispersion is an estimate of sensitivity, bias, or effects of rounding errors. Similarly, the spread of radii of the cylinder fits measures dispersion for size. The spread of angles between the axes is measured as the angle of the narrowest cone that encloses all the direction vectors of the axes. This measure does not relate directly to common tolerance applications, but can be used for diagnostic purposes.

Comparison of fits to a reference is done on a data set-by-data set basis. (That is, a different reference is used for each data set.) Differences between test fits and the reference for each data set are computed in a way similar to that described above. Thus, for instance, the maximum deviation of axis segments from the reference axis measures axis location differences. Difference measures can be accumulated across data sets to study the sensitivity of the differences to test parameters. A detailed description of analysis methods used at NIST will be reported elsewhere (Hopp, 1993b).

The theory of testing data analysis software is not well developed. The approach presented in this paper is a start, but several research issues still need to be addressed. Software performance is affected by many factors. It is a daunting task to define tests that represent all possible variations of these factors. Also, tests should be representative of the applications for which the software is to be used. Statistical design of experiments can be used to greatly reduce the number of tests, but appropriate models of the factors and their interactions need to be developed. Even a tightly designed test suite is likely to have too many cases to examine individually. A significant research challenge is to develop methods for summarizing and visualizing test results in ways that relate to the application and do not hide important information. The measures discussed above for dispersion and difference from reference fits are one form of summarization. They are useful for developing general conclusions, but there is no assurance that all important aspects of performance will be represented in the summary results. More work is needed in this area as well. Finally, the relationship of testing to the error models has only been developed in outline.

---

4 Fits from more than one data set may be analyzed together.

Much work remains in understanding how to design and interpret tests that provide an accurate picture of software performance.

SUMMARY AND DISCUSSION

This paper has discussed three aspects of computational metrology: fitting objectives, implementation issues, and testing issues. In regard to fitting objectives, clearly, different measurement applications can differ on measurement goals. For instance, part inspection for tolerance conformance should be based on geometric tolerancing semantics (Hopp, 1993a). Part measurement for process control, however, might better be based on vectorial tolerancing semantics (Wirtz, 1991). We described a number of research issues regarding fitting objectives.

Any assessment of measurement uncertainty must be based, foremost, on a definition of the measurand. Only then can the bias of the fitting objective be assessed. The ASME Working Group Y14.5.1 on Mathematical Principles of Dimensioning and Tolerancing is developing semantics of design tolerances. This standard, expected now to be released in early 1994, will define the measurand for inspection applications.

A second aspect of computational metrology is the classification and study of implementation issues. The challenge is to develop effective operational models of software that support practical test methods. This paper has presented a model that appears to be effective for certain kinds of fitting. More work needs to be done in developing models for other fitting criteria. Accepted practice is that a measurement is incomplete without a statement of measurement uncertainty. Much remains to be done in developing tools for assessing measurement uncertainty in practical applications.

Data analysis software, like any tool, can be misused. Guidelines for proper use of such software are as important as models of software performance. A discussion of usage has been beyond the scope of this paper. However, the ASME Working Group B89.3.2 on Dimensional Measurement Methods is developing a standard that addresses this issue.

The third aspect of computational metrology is testing methods. We have described activity at NIST and elsewhere in software testing. The ASME B89.4.10 Working Group on CMS Software Performance is developing a standard for software testing. The planned NIST testing service will be based on, and will support, this standard. Today, there are no mechanisms to validate the results of data analysis with respect to international standards of length. As research issues are addressed and testing technology matures, we believe that concepts of traceability of software results will finally become possible.

Inspection software performance has become an important issue in the last few years. Software problems, or their avoidance, have become a significant expense for many companies. Theories, tools, and standards in computational metrology are now beginning to emerge. It is our hope that with these advances, the utility of CMS software will greatly improve.

REFERENCES

ASME, 1985, ANSI/ASME Standard PTC 19.1-1985, *Measurement Uncertainty*, American Society of Mechanical Engineers, New York.

Cox, M.G., 1992, "Improving CMM Software Quality," NPL Report DITC 194/92, National Physical Laboratory, Middlesex, U.K.

Cox, M.G., and Forbes, A.B., 1992, "Strategies for Testing Form Assessment Software," NPL Report DITC 211/92, National Physical Laboratory, Middlesex, U.K.

Diaz, C., 1993, "Concept for an Algorithm Testing and Evaluation Program at NIST," National Institute of Standards and Technology, Gaithersburg, MD, *in preparation.*

Estler, T., 1989, "Accuracy Analysis of the Space Shuttle Solid Rocket Motor Profile Measuring Device," NISTIR-89/4171, National Institute of Standards and Technology, Gaithersburg, MD.

Feng, S.C., and Hopp, T.H., 1991, "A Review of Current Geometric Tolerancing Theories and Inspection Data Analysis Algorithms," NISTIR 4509, National Institute of Standards and Technology, Gaithersburg, MD.

Golub, G.H., and Van Loan, C.F., 1985, *Matrix Computations*, Johns Hopkins University Press, Baltimore, MD.

Gonin, R., and Money, A., 1989, *Nonlinear L_p-Norm Estimation*, Marcel Dekker, New York.

Hopp, T.H., 1993a, "The Language of Tolerances," *Quality through Engineering Design*, W. Kuo, ed., Elsevier, New York, NY, pp. 317-332.

Hopp, T.H., 1993b, "Requirements for a Production Algorithm Testing System,", National Institute of Standards and Technology, Gaithersburg, MD, *in preparation.*

Mamileti, L., Wang, C.M., Young, M., and Vecchia, D.F., 1992, "Optical Fiber Geometry by Gray Scale Analysis with Robust Regression," *Applied Optics*, Vol. 31, pp. 4182-4185.

Porta, C., and Wäldele, F., "Testing of Tree Coordinate Measuring Machine Evaluation Algorithms," Technical Report BCR EUR 10909 EN, Commission of the European Communities, Luxembourg.

Rousseeuw, P., 1984, "Least Median of Squares Regression," *J. Amer. Stat. Assoc.*, Vol. 79, pp. 871-880.

Vecchia, D.F., Wang, C.M., and Young, M., 1993, "Outlier-Resistant Fitting of Gray Scale Images Illustrated by Optical Fiber Geometry," Proc. 1993 Meas. Sc. Conf., Los Angeles, Jan. 21-22.

Wäldele, F., Bittner, B., Busch, K., Drieschner, R., and Elligsen, R., 1993, "Testing of Coordinate Measuring Machine Software," *Precision Engineering*, Vol. 15, pp. 121-123.

Walker, R., 1988, *GIDEP Alert No. X1-A-88-01*, Government-Industry Data Exchange Program, Washington, DC.

Wirtz, A., 1991, "Vectorial Tolerancing for Production Quality Control and Functional Analysis in Design," *CIRP International Working Seminar on Computer-Aided Tolerancing*, pp. 77-84.

CRTD-Vol. 27, International Forum
on Dimensional Tolerancing and Metrology
ASME 1993

AUTOMATIC PLANNING FOR DIMENSIONAL INSPECTION

Antonia J. Spyridi and Aristides A. G. Requicha
Department of Computer Science
Institute for Robotics and Intelligent Systems
University of Southern California
Los Angeles, California

ABSTRACT

This paper describes an on-going research project that seeks to develop a planner capable of generating automatically the instructions needed to drive a Coordinate Measuring Machine (CMM), so as to inspect a set of specified surface features of a part, and determine if the part satisfies its tolerance specifications.

We divide the planning activity into high-level and low-level planning. The high-level planner produces a collection of setups, each associated with surface features to be inspected, probes for inspecting each feature, and the relative orientations of the part, CMM and probes. The low-level planner addresses primarily the issues of measurement point selection (sampling) and collision-free path generation. The focus of the paper is on high-level planning. The architecture of the high-level planner is discussed, and experimental results are presented.

DIMENSIONAL INSPECTION

Dimensional inspection is a quality control task that consists of measuring a mechanical part so as to determine whether it meets its nominal and tolerancing specifications. Coordinate Measuring Machines (CMMs), which, in essence, are very accurate 3-D digitizers, are versatile and well suited for automated dimensional inspection.

We are developing an inspection system with the architecture shown in figure 1. The input is a solid model of the part to be inspected, together with a set of surface features and associated tolerances. (A surface feature here is simply

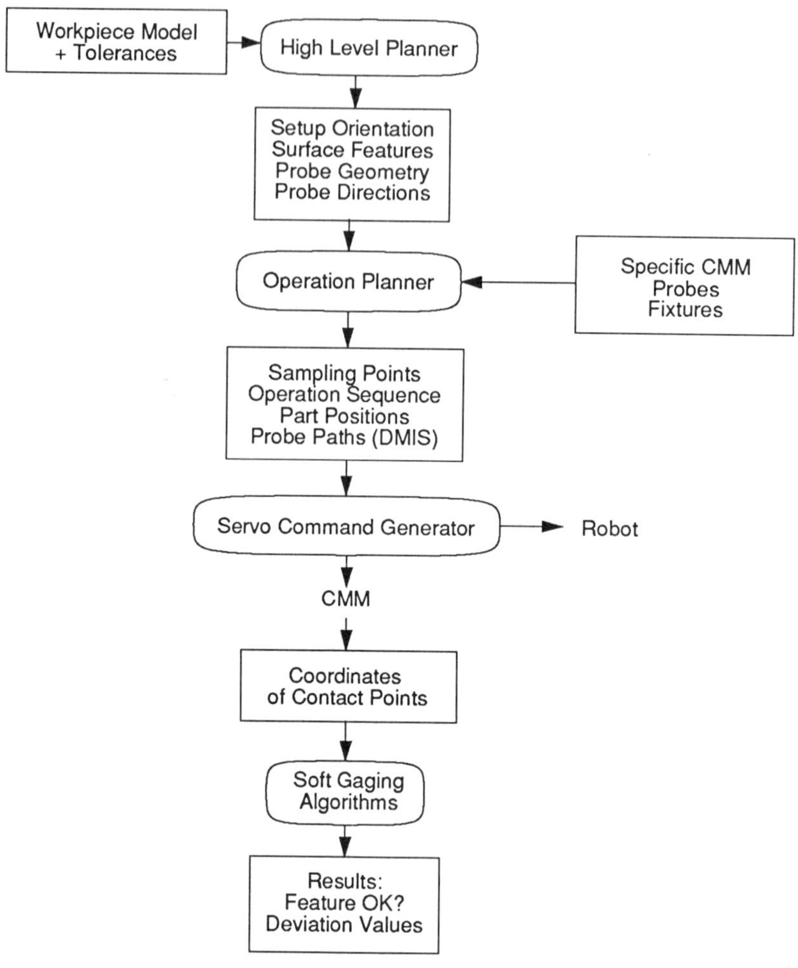

Figure 1: CMM inspection

a set of "faces" of the part.) The high-level planner produces a collection of *setups*, and a set of *operations* for each setup. A setup is characterized by a fixed workpiece orientation with respect to a CMM, and by the associated fixtures and clamping equipment. An operation essentially consists of inspecting a surface feature, and has one or more probes and probe directions associated with it. The low-level, operation planner takes into consideration the available machines, probes and fixtures, determines appropriate sampling points to inspect each feature, and produces probe paths and specific part-positioning information. Probe paths are translated into actual commands for the CMM, and part positioning information is communicated to a human operator (or, eventually, to a robot). Measured points are interpreted by a soft gaging module, which determines if a

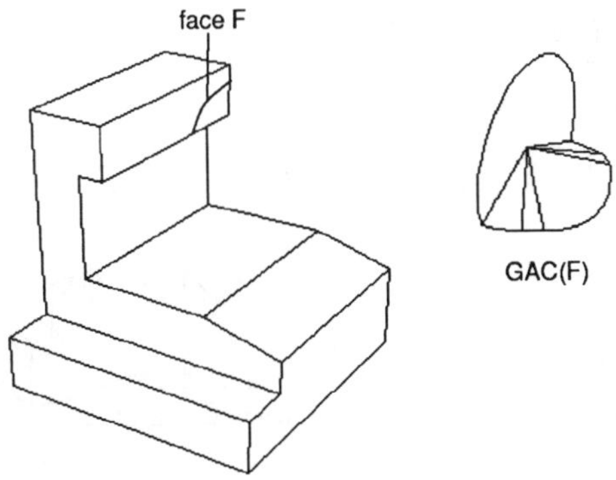

Figure 2: Global accessibility cone

part meets its tolerance specification. (Additional output information may also be useful to determine, for example, why a part is out of tolerance.)

ACCESSIBILITY ANALYSIS

We think that accessibility is a major concern in high-level planning for inspection. It strongly influences the part orientation in each setup, which features can be inspected in a setup, and the probe orientations. We developed an accessibility analysis module, whose underlying theory is based on semi-infinite probes, and on properties of *direction cones*, which represent the set of accessible directions – see [Spyridi and Requicha 1990, 1991]. The algorithm may be summarized as follows. To compute the global accessibility cone (GAC) of a face of a solid, (i) sweep a shrunk version of the face over the boundary elements (faces, edges and vertices) of the solid, (ii) compute the "silhouette" or "shadow" of the sweep as seen from a reference point in the face, (iii) union all the cones that correspond to the silhouettes, and (iv) complement the result. To compute sweeps, or Minkowski sums, it suffices to calculate the contributions from the solid's faces, edges and vertices whose normal vectors (Gaussian images) overlap those of the face (plus its bounding edges and vertices) being considered. Importantly, all of the required sweeps can be computed in the plane and produce polygons, rather than 3-D sets. Figure 2 shows the resulting GAC for the marked face of the solid.

When the intersection of the GACs of all the features is non-empty, any direction within the intersection cone may be used to inspect all the features. Usually, however, the intersection is empty, and one must arrange the cones into *clusters*, so that the cones in each cluster have a non-empty intersection. A minimal number of clusters is desirable, to ensure that a small number of probes and setups are used.

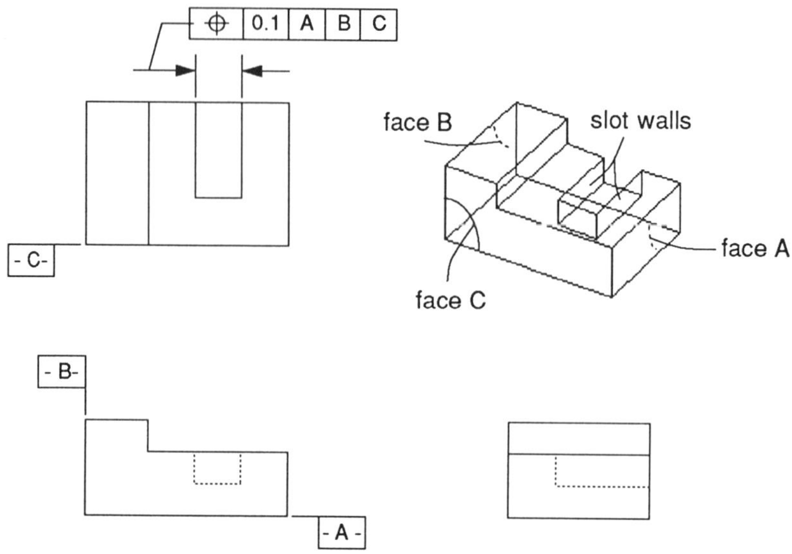

Figure 3: Toleranced part

HIGH LEVEL PLANNING

Our approach to high-level planning casts the problem as search in a space of partially-completed plans. Plans evolve towards completion through the effects of operators that map plans into plans. The state is represented by sets of 4-tuples (SU, PG, PD, SF), where SU are the the setup orientations, PG are the acceptable probe geometries, (selected from a given set of physical probes), PD are the accessible directions represented by cones, and SF is a (possibly compound) surface feature to be inspected. We follow a least-commitment strategy, and attempt to reason about all the possible solutions, gradually narrowing their range by applying constraints such as accessibility, same-setup (for related features), and so on.

We will illustrate how the planner works by means of a simple example, using only straight, non-orientable probes. The toleranced part is shown in figure 3. The task is to generate an inspection plan for verifying the positional tolerance of the slot walls.

The state (figure 4b) contains four features, F0, F1, F2 and F3, that correspond to the input tolerance features (figure 4a). (In the following we will refer to state features simply as features.) In addition to the features, the state might include constraints of type *same*, which specify that a set of features should have identical values for some of the 4-tuple fields. In this example, the features are constrained to have same setup and probe geometry.

Input tolerance

t0:
type: position
value: 0.1
f0:
part face: face A
role: primary datum
f1:
part face: face B
role: secondary datum
f2:
part face: face C
role: tertiary datum
f3:
part face: slot walls
role: feature

(a)

State

T0:

F0: (su, pg, pd, sf)

F1: (su, pg, pd, sf)

F2: (su, pg, pd, sf)

F3: (su, pg, pd, sf)

same-su(F0, F1, F2, F3)

same-pg(F0, F1, F2, F3)

(b)

Figure 4: Input tolerances and state representation

Inspection planning starts by initializing the state (figure 5a), so that all tuple variables have value *all* (in figure 5, U stands for "universe"). The problem now is to determine a sequence of operators which, when applied to the initial state, will produce a final state with valid values for the tuples of all the features.

The first operator that can be applied is called `equivalent-sf` and it computes values for the SF field of the tuple. This operator is proposed for all features, but since F0 corresponds to a primary datum, we give the operator `equivalent-sf(F0)` the highest priority and apply it to the initial state. In the resulting state (figure 5b), the SF for F0 can be either the feature itself - face A - or the CMM table. (Substituting a feature by the CMM table can only be done for a primary datum.)

Subsequent operators modify the values of PD and PG. Applying the operator `gac(F0)` to state 1 results in the PD field being filled in with the global accessibility cone of the corresponding feature (figure 6a). The operator `probe-geometry(F0)` fills in the PG field with the name of a real probe, P15 in this case (figure 6b).

After applying similar operators for the remaining features, we reach the state shown in figure 7. Note that because of the *same* constraints the values for SU and PG are shared by all the features. At this point, the probe directions can be clustered. The result of PD clustering is shown in figure 8a.

The operator `setup-from-pd` can now be applied to produce a setup such that the CMM outward surface normal is inside the PD cone. There are infinitely

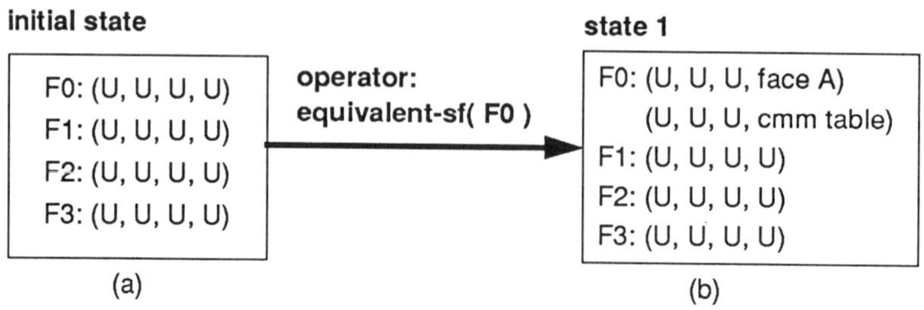

Figure 5: Initial state and application of the first operator

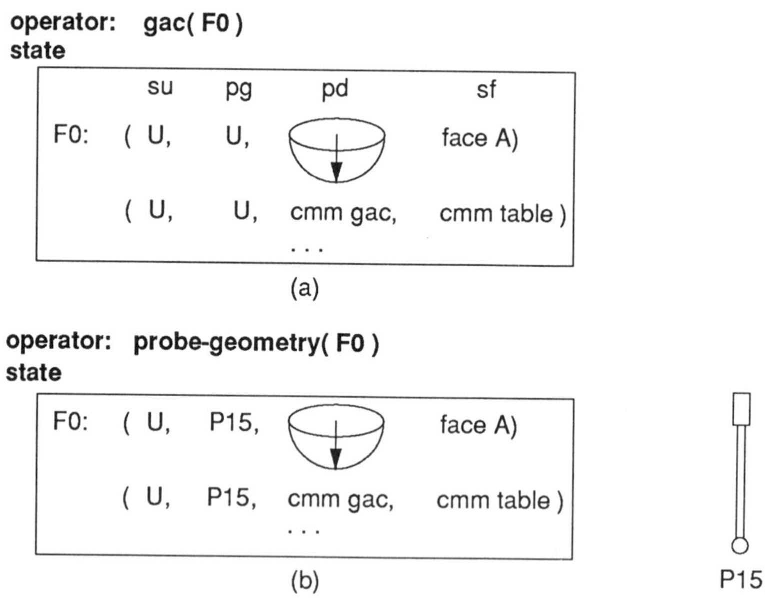

Figure 6: Applying the operators gac(F0) and probe-geometry(F0)

state	su	pg	pd	sf
F0:	U	P15	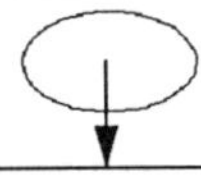	face A
			cmm gac	cmm table
F1:				face B
F2:			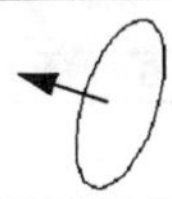	face C
F3:			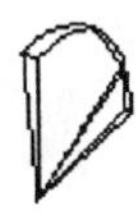	slot walls

Figure 7: Intermediate state

operator: cluster-pd
state

F0:	U	P15		cmm table
F1:				face B
F2:				face C
F3:				slot walls

(a)

operator: setup-from-pd
state

F0:	SU1	P15		cmm table
F1:				face B
F2:				face C
F3:				slot walls

(b)

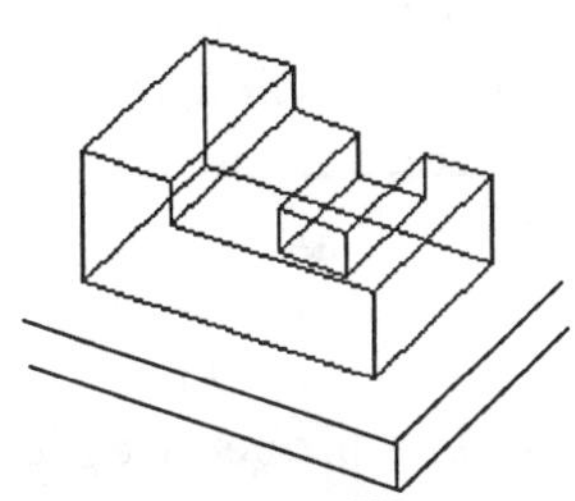

setup SU1

Figure 8: Clustering and final state

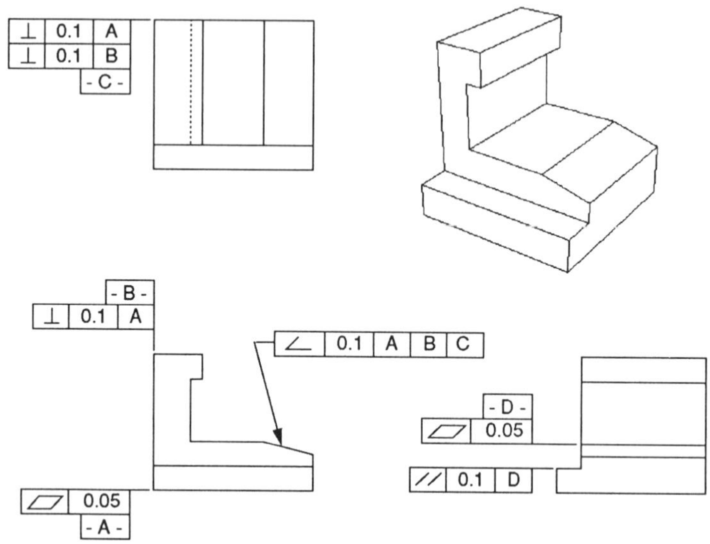

Figure 9: Part and tolerances

many such setups, but using heuristics that reflect stability considerations we can single out the setup SU1 (face A on CMM table – figure 8b), as the best setup. This plan should now be verified. In this example, the verification succeeds, hence the final inspection plan is the one shown in figure 8b.

EXPERIMENTAL RESULTS

This section presents the inspection plan produced by the planner for the part of figure 9.

The final plan (figure 10), has two setups SU1 and SU2. In SU1 the resting face of the part is face A. Cone_a gives the probe directions for the features indicated by the dashed line; probe_a is the probe selected for this inspection plan in an orientation from cone_a. Note that in this setup, face A cannot be inspected directly, but since it is a primary datum the CMM table is inspected instead. Cone_b gives the probe directions for the features indicated by the solid line and probe_b is the probe in an orientation from cone_b. In SU2 only the face A is inspected. This is necessary because of the flatness tolerance imposed on face A. The probe directions and a probe for this measurement are shown by cone_c and probe_c, respectively.

The part was defined in the Parasolids modeler. The inspection plan was generated by assuming that the setups were aligned with the principal axis directions. The setups were introduced in the state early in the planning process.

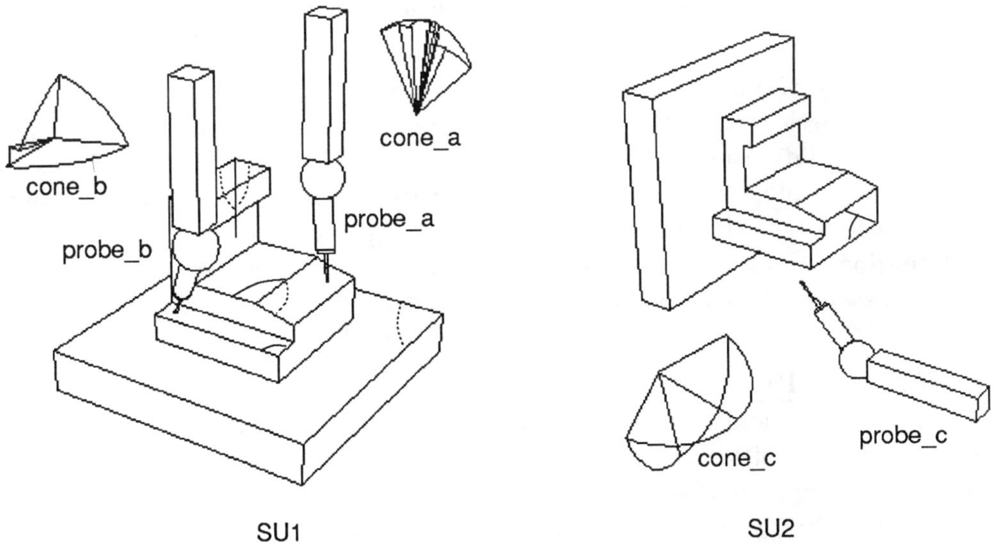

Figure 10: Inspection plan

The cones were generated and intersected by the accessibility analysis module mentioned above, and the clustering was done manually. In addition to "pure" global accessibility analysis, the cones were also shrunk in order to accommodate the width of the probe and probe head. The part and probe display were done through PADL-2 manually (the planner is not connected to PADL-2), and the cones were displayed by the accessibility analysis module. We intent to remove all the manual steps in the next version of the planner.

CONCLUSIONS

In this paper, we presented a system for automatically generating high-level inspection plans for the inspection of mechanical parts by CMMs. The system is based on the Artificial Intelligence technique of state-space search, where the state is a collection of inspection plans (some of which may be invalid), and the operators eliminate the invalid plans from the state and attempt to arrive at an optimal plan. The main advantage of the system is the flexibility in ordering the operators through the use of different search strategies. The drawback is that the state space might be very large (although it might be necessary to look through a large number of plans in order to come up with a close-to-optimal inspection plan). We are working on reducing the number of plans stored in intermediate states (1) by devising strategies for operator selection that result in better operator ordering, (2) through the use of methods that produce a restricted range of alternative plans, and (3) through the use of "lazy evaluation" techniques that postpone the expansion of operators until needed.

ACKNOWLEDGEMENTS

The research reported in this paper was supported in part by the National Science Foundation under grants DMC-87-96192, DDM-87-15404 and CDR-87-17322, by the industrial members of the Institute for Manufacturing and Automation Research (IMAR), and by the Industrial Associates of the Programmable Automation Laboratory, Institute for Robotics and Intelligent Systems (IRIS) of the University of Southern California.

BIBLIOGRAPHY

A. J. Spyridi and A. A. G. Requicha, "Accessibility Analysis for the Automatic Inspection of Mechanical Parts by Coordinate Measuring Machines", *Proc. IEEE Int'l Conf. on Robotics and Automation*, Cincinnati, OH, pp. 1284-1289, May 13-18, 1990. (Published also as IRIS Rept. No. 257, Institute for Robotics and Intelligent Systems, University of Southern California, October 1989.)

A. J. Spyridi and A. A. G. Requicha, "Accessibility Analysis for Polyhedral Objects", in S. G. Tzafestas Ed., *Engineering Systems with Intelligence: Concepts, Tools and Applications.* Dordrecht, Holland: Kluwer Academic Publishers, Inc., 1991, pp. 317-324. (Published also as IRIS Rept. No. 280, Institute for Robotics and Intelligent Systems, University of Southern California, June 1991.)

CRTD-Vol. 27, International Forum
on Dimensional Tolerancing and Metrology
ASME 1993

TOLERANCES AND ASSEMBLIES IN CAD/CAM SYSTEMS

George Allen
Unigraphics Division
Electronic Data Systems
Cypress, California

ABSTRACT

The architectures of many of the leading mechanical CAD/CAM systems have undergone significant changes in the last few years, and these changes may provide new opportunities for representing and using tolerance and assembly information.

This paper is divided into two major sections. The first describes the techniques used to represent nominal models in contemporary CAD/CAM systems. We pay particular attention to features, associativity, and other aspects of model intelligence. The second section explores some ideas for incorporating tolerance information into nominal models, and for using this information in applications algorithms.

NOMINAL MODELS

This section outlines the techniques used to represent nominal models in a typical modern CAD/CAM system such as EDS/Unigraphics. We cover geometry, topology, assemblies, and the methods used to link entities together to establish associative relationships between them.

Geometry

The most basic building blocks for representing shape information are geometric entities — points, curves, and surfaces. Points are so simple that they do not warrant any further discussion. The curve types generally available are:

- Line (i.e. straight line)
- Circle
- Conic section curves (ellipse, parabola, hyperbola)
- A spline curve, usually a non-uniform rational b-spline (NURB) curve

Similarly, the surface types typically available are

- Plane
- Circular cylinder
- Circular cone
- Sphere
- Torus
- Surface of revolution
- Extruded surface
- Offset surface
- Some sort of free-form surface, usually a NURB surface

This is roughly the set of geometric entities specified in standards like IGES and PDES, and (more importantly) is roughly the set of geometric entities supported by most popular systems. Because of this commonality, exchanging curve and surface data between systems generally works fairly well. Conversely, market demand for good basic data exchange facilities and "open" systems deters vendors from deviating too far from this data model, so it is likely to remain quite stable in years to come.

Topology

Most systems use some form of boundary representation in which curves and surfaces are combined together to describe bodies. The basic topological entities are vertices, edges, loops, faces, shells and bodies. They record various types of adjacency or connectivity relationships, as illustrated in the figure below.

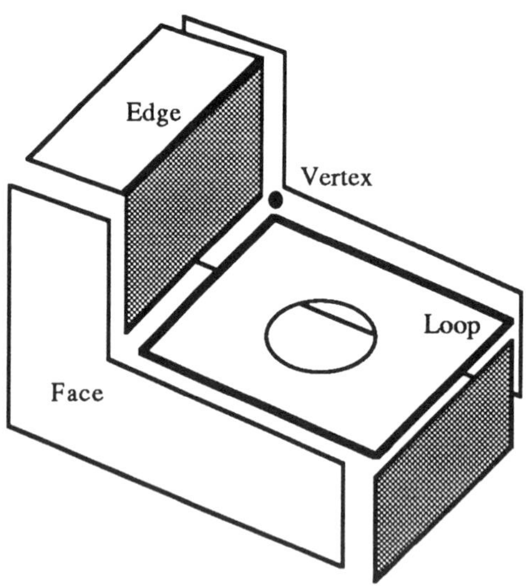

Body: A body represents a single connected piece of material. There are three types of body, namely solid, sheet, and wire. A body references one or more bounding shells.

Shell: A shell represents a single simply-connected boundary separating material from air. In a given body, one of the shells is distinguished as the exterior of the body and the other shells (if any) represent internal voids. A shell references its owning body, and its constituent faces, edges and vertices.

Face: A face represents a single connected portion of a surface. A face references its owning shell, the surface on which it resides, and its bounding loops.

Loop: A loop is a closed connected ring of edges and vertices representing some portion of the boundary of a face. A loop references its owning face, and its constituent edges and vertices. If a loop consists of a single closed edge (e.g. a circular edge), it may not contain any vertices. On a face that is roughly planar, one of the loops can be recognized as the exterior, or peripheral loop, but on highly curved faces such a distinction is not possible.

Edge: An edge represents the boundary between two faces. An edge references its two owning loops, the curve on which it resides and (optionally) its bounding vertices. Closed edges (such as complete circles) may not have any bounding vertices.

Vertex: A vertex represents the junction between two or more edges. It references a point entity which records its position in space, and the edges that meet there.

Given the boundary representation of a body, it is quite difficult to determine whether or not the body is valid (physically realisable). For example, it may well be that the body is self-penetrating, because some of the places where faces intersect are not recorded as edges. Many systems take a fairly cavalier attitude towards validity checking in an effort to obtain improved performance (Hoffman, 1993).

Many systems support only manifold bodies. Roughly speaking, these are bodies in which each edge is bordered by only two faces. The industry trend is towards more general topologies, including non-manifold bodies, cellular objects, and so on. See (Weiler, 1986) for a description of one well-known scheme for representing non-manifold bodies.

Intelligence, Parameterisation and Associativity

In modern CAD/CAM systems, most entities are intelligent, or associative. This means they remember that they were defined in terms of certain arguments (often references to other entities), and they update automatically whenever their arguments change. We think of an intelligent entity as consisting of two parts: a definition and a value. The definition records the defining data of the entity plus the rule used to generate the entity. The value is the result of using an evaluation processor to "execute" the definition and obtain an answer.

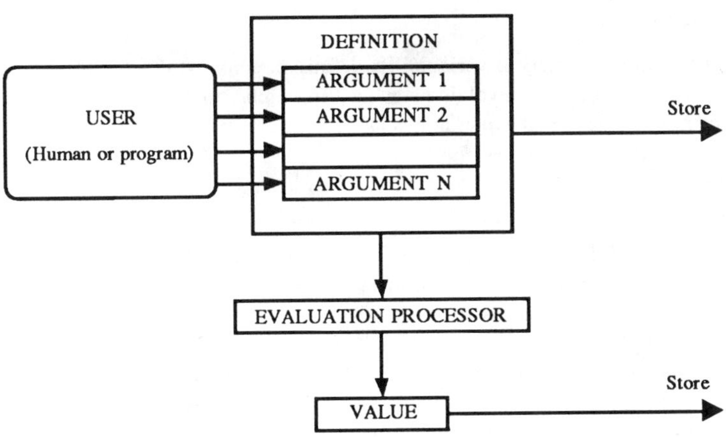

This is a somewhat abstract notion, so an example may help. Consider the case of a simple slab shape, as illustrated below.

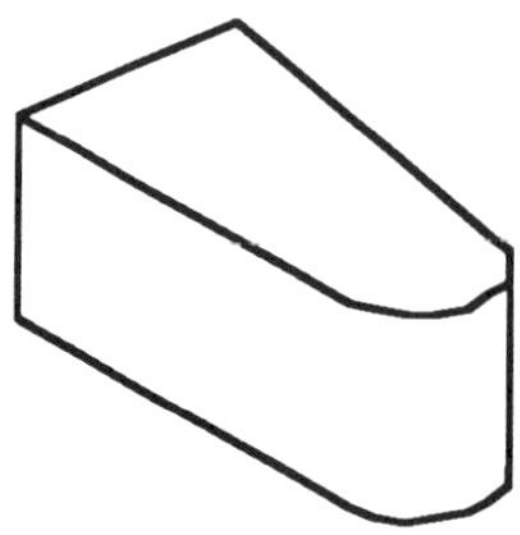

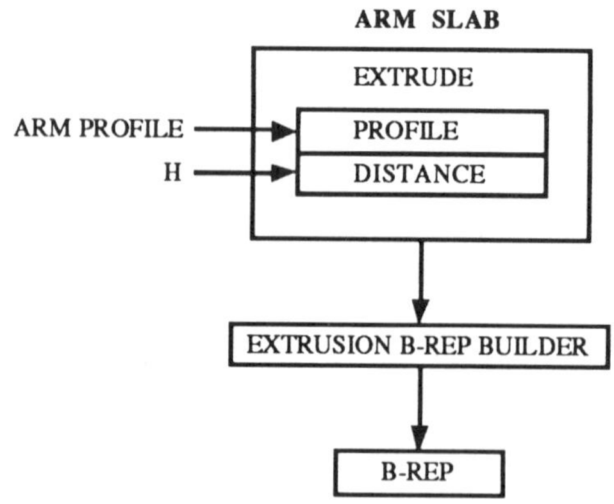

The definition of this body consists of a set of curves describing its profile, a numerical variable H specifying its height, and some positioning data (not shown in the picture). Its value is a boundary representation consisting of twelve edges and six faces. If the value of H is changed, or if any of the curves in the profile are modified, then the system re-evaluates the slab definition to obtain a new boundary representation.

If the arguments used in defining an entity are numerical (like the height H in the slab example above), they are often referred to as the parameters of the entity, and it is said to be parameterised or parametric. Alternatively, we might say that there is an associative relationship between the slab and its profile.

Expressions

An expression is just a simple numerical variable. Its definition consists of a list of references to other expressions together with the formula (involving standard arithmetic operations, trigonometric functions, etc.) used to produce it. The value of an expression is just a real number.

Relative Positioning Constraints

Most systems use a variety of simple techniques for relative positioning of two bodies. In some cases, the bodies might represent piece parts in an assembly. In another scenario, one of the bodies is a "target" and the other body represents a feature that is to be attached to it (see below). The desired positioning is specified by imposing simple constraint relationships between geometric entities that belong to the two bodies. For example, one might stipulate that a cylindrical surface on one body is to be coaxial with a conical surface on the other body. One of the bodies (the parent) is considered fixed, and the system calculates a position for the other body (the child) in which the positioning constraints are satisfied. The constraints are "remembered" so that the positioning can be associative — if the parent body changes, the system can use the constraints to calculate a new position for the child body. Some examples of commonly-supported relative positioning constraints are the following:

Mate Planes: Positions two planes so that they coincide and their normal vectors have opposite senses. This is a special case (offset distance equal to zero) of the Mate/Offset constraint described in the next paragraph.

Mate/Offset Planes: Positions two planes so that they are parallel and their normal vectors have opposite senses. The distance between the two planes (the offset distance) is also optionally specified in this constraint.

Align Planes: Positions two planes so that they coincide and their normal vectors have the same sense. This is a special case (offset distance equal to zero) of the Align/Offset constraint described in the next paragraph.

Align/Offset Planes: Positions two planes so that they are parallel and their normal vectors have the same sense. The distance between the two planes (the offset distance) is also optionally specified in this constraint.

Align Axes: Positions two axes so that they are collinear. Each of the axes may be either a datum axis or the (implied) axis of a surface of revolution. This might be used to align a pin with a hole, for example.

Planes at an Angle: Positions two planes so that the angle between them is as specified.

Sometimes it is convenient to think of a set of relative positioning constraints as an entity in itself. In this scenario, the value associated with the entity is the rigid motion that moves the child body into the desired position.

Note that using these types of positioning constraints may allow us to develop tightly-integrated applications for the study of mechanisms. A joint between two elements of a mechanism could simply be represented as a deficient set of positioning constraints.

Sketches

A sketch is a collection of curves, together with dimensions and geometric constraints that determine their size and position. Sketches are often used in the definitions of extruded solids or solids of revolution, as we saw in the slab example earlier. An example of a simple sketch is shown below:

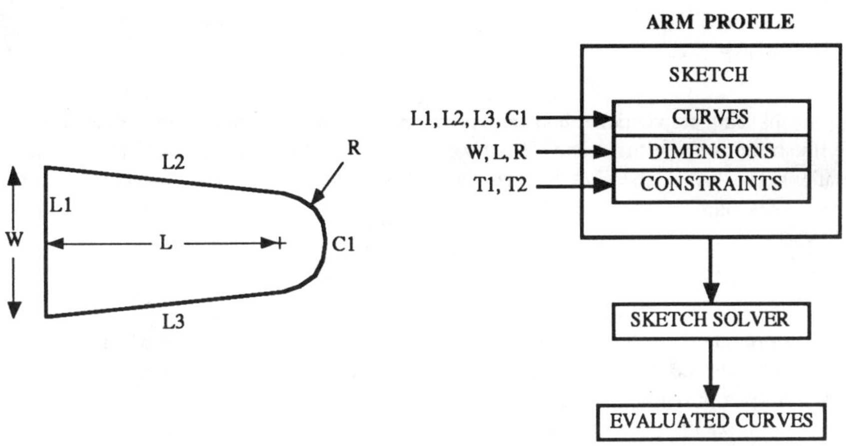

In this case the geometric constraints indicate that the lines L2 and L3 are to be tangent to the circle C1, and that the profile is symmetric. These geometric constraints may be specified explicitly by the user or may be inferred by the system as the curves are being constructed. The sketch solver modifies the curves so that they conform to the dimensions and constraints. Note that in general a sketch definition is declarative — we simply specify that certain constraints are to be satisfied, but we do not have a simple procedural recipe for producing a solution in a sequential series of steps. Solving sketches usually requires that the system somehow simultaneously solve a system of non-linear equations. There are many approaches to this problem, none of which is completely satisfactory at this point. See (Light & Gossard, 1982), (Stevenson, 1988) and (Owen, 1991) for some partial solutions.

Features

A feature can be thought of as a parameterised collection of faces and edges that is to be attached to some target body. The feature might be positive (boss, pad, rib) or negative (hole, pocket, groove). Features can be loosely classified as:

Form Features: Hole (simple, counterbored), pocket, boss, pad, etc.

Body Features: Block, extrusion, free-form bodies

Operation Features: Blend/round/chamfer, taper, offset, hollow, etc.

Enumerative Features: Circular and rectangular arrays

Feature Attachment

As noted in (Hoffman, 1993), the feature attachment operation in many systems is somewhat different from a conventional boolean union or subtraction. In fact the semantics of the operation are generally not well defined mathematically — the only real definition is the system code itself. A feature attachment operation is often computationally less expensive than a conventional boolean, and sometimes produces more intuitive results. An example is shown below — we see an end view of a thin-walled pipe, and we wish to attach a cylindrical boss to the outside of it:

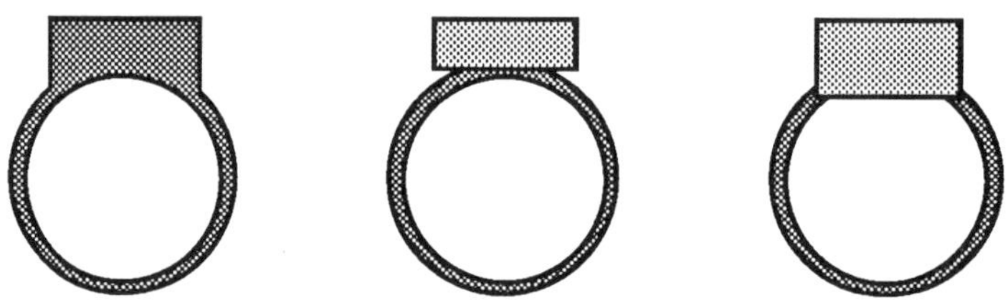

Desired result — boss on pipe | Base of boss too high | Base of boss too low

The naive approach, using conventional boolean operations, would be to unite a cylindrical solid with the pipe, but the illustrations show that this will not work. One can obtain the desired effect by using two boolean operations, but this is expensive. See (Hoffman, 1993) and (Ranta, 1993) for further discussion of feature attachment operators.

Datums

Systems typically provide three types of datum entities: datum planes, datum axes, and datum points. Datum entities can be viewed as "pseudo-geometry" — they are different from other geometric entities because they are not part of the b-rep of the model. Many different methods of defining datum entities are usually provided. For example, a datum plane might be defined through three points, through an axis, normal to a plane, etc. Datum entities are used to position and orient sketches and features.

Dimensions

A dimension entity generally references the geometric entities being dimensioned, and also includes information used to position it on a drawing, plus rendering information such as character size and font, line thicknesses, and so on. In most systems, dimensions are associative — if the referenced geometry changes, the dimension entity will be regenerated to reflect the change. Note that for historical reasons, dating back to the days of wire-frame modeling, dimensions are often attached to curves (or edges). The trend is towards attaching them to surfaces (or faces) instead. Dimension entities often include ± tolerances as additional text fields.

The Model Graph

As we have seen above, there are many situations in which model entities are defined via associative references to other ones. The chains of references form a graph structure that we call the model graph. In (Hoffman, 1992) this structure is referred to as an E-rep (editable representation). It is quite similar to a traditional CSG graph — the terminal node represents the final model, and internal nodes represent its constituents and various intermediate stages in its construction. Contemporary model graphs differ from traditional CSG graphs in several ways, however:

Traditional CSG Graph	Model Graph
Lowest-level bodies (leaf nodes) are intersections of half-spaces	Lowest-level nodes can be arbitrary b-reps
Small range of node types: usually just numerical expressions, boolean operations, and explicit rigid motions	Wide assortment of node types, including sketches, sweeps, feature attachment operations, blends, offsets, dimensions
Features combined using standard boolean operations	Features combined using specialised feature attachment operator
Used for editing purposes and also as a source of information for divide-and-conquer algorithms	Used for editing purposes only, for the most part — seldom used to drive applications algorithms.
Semantics well-defined, and validity guaranteed	Semantics not well-defined, so validity questions unanswerable

The model graph plays a crucial role in editing functions. When the user changes some entity, the links in the model graph are used to decide which other entities need to be regenerated to reflect the change. For example, if the user changes the value of some expression variable, we may have to calculate new values for several other expressions that depend on it, which may cause the sizes and/or locations of several features to change, and so on. This propagation of change can be extremely powerful, and users seem to grasp it quite readily. We generally explain the process by telling users that it is analogous to the behaviour they have seen in spreadsheets.

ADDITION OF TOLERANCES TO NOMINAL MODELS

In this section, we consider ways in which tolerance information might be added to our nominal models, and the improvements this might produce in system functions. Our aim is to understand how the functionality and usage of systems might change if they incorporated tolerance information. The focus is on short-term extensions that can be accommodated without changing the basic architecture of current systems.

General Techniques for Representing Geometric Variations

If the nominal models in a CAD system are of the type described in the previous section, then the most obvious way to describe variations in size and shape is to vary the parameters that were assigned to the model during the design phase. However, it may happen that the desired variations can not be produced simply by altering the values of the parameters used by the designer — a different parameterisation may be required. One possibility is to use the natural parameterisation inherent in any boundary representation, namely the one in which the parameters are simply the defining parameters of the surfaces of the model. This parameterisation is too rich (or too verbose) for design use, as it allows too may degrees of freedom, but it appears to be quite useful for variational studies. To escape from the world of perfect form, and allow even more variation in shape, we can also replace simple surfaces like planes and cylinders with NURB surfaces whose shape can vary as much as we please.

An example may help to clarify some of these ideas. Suppose a designer has created a simple block shape in a modern "parametric" CAD system. We would expect this block to have nine degrees of freedom — six for its position and orientation, and three for its size (length, width, and height). The range of variation of this block is quite limited: by varying length, width and height we can only produce shapes whose opposite faces are parallel. If we want more degrees of freedom, we can discard the designer's parameterisation, and think of the block as just a boundary representation. It has six planar faces, each of which has three degrees of freedom, giving us a total of 18. Now we can represent a much broader range of shapes (including some which are not physically realisable), but still they all have planar faces. We can obtain further flexibility by replacing each of the planar surfaces by a simple bicubic patch. Each patch will have 48 degrees of freedom, giving us a total of 288.

Tolerances

In most systems, tolerances are represented in a fashion that is appropriate for drafting, but little else. Limit tolerances are often stored as text strings appended to dimensions, which are in turn associated with curves or edges. A datum callout or a GD&T callout is stored as a "label" attached to a curve or edge. Sometimes the characteristic symbols and other information in a GD&T callout are represented by type codes (sequences of control characters) but in some systems they are merely stored as unstructured sets of graphical elements (lines, circles, arrows, etc.).

Some users have devised ways to represent tolerances using the general-purpose attribute mechanism provided with most systems. When represented this way, the system has no way of understanding the meaning of the tolerance information, but at least it is retained and propagated during modeling operations, and is available to drive user-written applications.

Other users are even more creative. For example (Taylor, 1993) describes a successful scheme in which tolerance zones are represented as short line segments in a CAD model.

GD&T Syntax Checking

Most systems provide facilities for creating feature control frames as annotations on drawings. For the uninitiated, it is easy to make mistakes when creating these frames, so it is useful if the system provides some sort of simple syntax checking. For example, the system might alert the user that a form tolerance includes a datum reference, and might even display the text of a standards document to explain the relevant rules. In a point-and-click style of interface, illegal combinations of symbols can be prevented by offering the user only those options that are appropriate based on prior entries in the frame. Note that this sort of checking is more difficult if the feature control frame is represented merely as an unstructured collection of graphical elements.

GD&T Functional Checking

A further level of checking is possible if the feature control frames are associated with model geometry in an intelligent way. For example, if two datums in reference frame were parallel, the system would be able to detect this fact and inform the user of the problem.

Use of Limit Tolerances in Planar Milling

If asymmetric tolerance bands are specified, then the nominal geometry may not be useful for an NC device (like a milling cutter, a laser, or a wire EDM machine). The problem is that the nominal geometry (which is the only information used by the toolpath generation algorithms) is not centrally located within the tolerance band, so is not likely to produce good results.

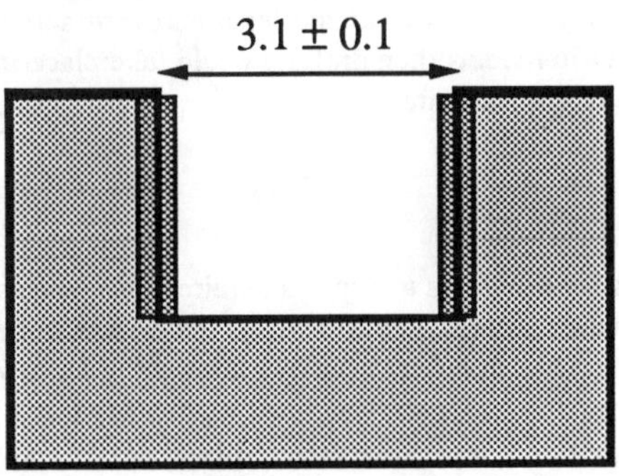

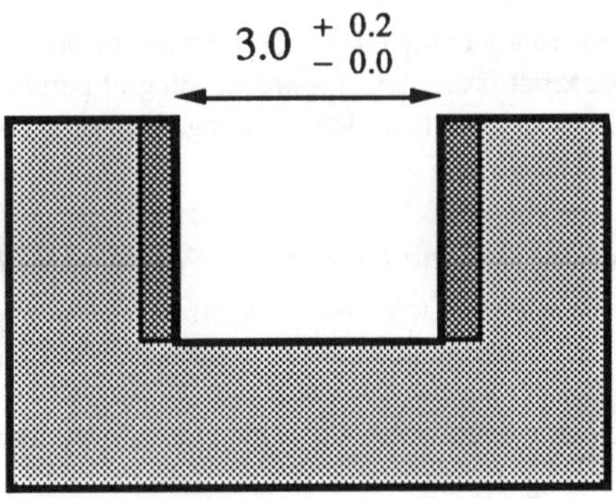

Interference and Clearance Analysis

Current systems sometimes provide two types of interference checks. Crude checks for gross collisions often use approximate (facetted) models, for performance reasons. Tolerances are not relevant in this type of checking.

More precise checks of fit in assemblies should definitely use tolerance information. For simple worst-case analysis, the user simply needs to make a temporary revision of each piece part involved, setting it to maximum material condition, and then have the system determine whether or not these maximised parts interfere.

A more realistic approach would create sample variants of the piece parts using some statistical scheme, and test these for interference. By doing this repeatedly, one could calculate the percentage of the assemblies that will fit together properly. For repeated trials of this sort, the variants of the nominal model would have to be generated automatically, perhaps by varying dimensions according to known probability density functions and regenerating the model. However, in this approach, each trial requires a regeneration of at least some portion of the model, so an experiment with several thousand trials is likely to be very time-consuming.

Process Planning

Many CAD/CAM systems provide functions for generating NC instructions, but most of them focus on the operation planning phase of the activity, leaving process planning to human beings. Information about nominal shape and required tolerances is necessary for effective process planning, but by itself it is not enough: we also require information about the precision of various manufacturing processes, so that the appropriate ones can be selected.

In some companies, tolerance assignment and process planning are actually one and the same thing. When the design engineer chooses a material and assigns tolerances, he is implicitly dictating how the part will be made, even though he may not realise this. In other cases, design engineers will deliberately stipulate the manufacturing process to be used, since this is a simple (though perhaps overly restrictive) way for them to express their design requirements.

Engineering Drawing Generation

As we have seen, handling of tolerances in CAD systems have historically been drafting-oriented. The tolerance information is placed on drawings, rather than being an integral part of the geometric model. This implies that the model itself does not provide sufficient information for many downstream activities, so drawings have to be produced, even though they may never actually be plotted on paper. Even with the best of CAD systems, layout and annotation of a drawing of a given model is quite time-consuming. Therefore the creation of the drawing can easily become a critical-path task in the overall production process. Many companies will continue to produce drawings (if only for outdated legal and contractual reasons), but if tolerances were an integral part of our geometric models then drafting could take place in parallel with manufacturing planning, rather than being a costly pre-requisite.

Comments on the Expressive Power of GD&T

Dimensioning and tolerancing schemes are touted as a means for designers to express requirements, and as a way for manufacturers to understand inspection and acceptance criteria. Unfortunately, there seem to be some deficiencies in the expressive power of the schemes that are in common use — they do not allow designers to directly say what they really mean, in some cases.

Free-form geometry seems to be especially problematic. There does not seem to be any way to specify that a curve or surface should be free from inflexions (for aesthetic or aerodynamic reasons). Similarly, there does not seem to be any way to indicate that a curve or surface should have continuous curvature, which might well be important in the design of cams, airfoils, or consumer products. Since they have no way of expressing their true requirements, designers often place huge numbers of positional constraints on free-form curves and surfaces, which increases their complexity unnecessarily.

As noted in (Parratt, 1993), problems also arise when engineering requirements can only be expressed as declarative constraints involving multiple design parameters. One simple example would be the requirement that the area of a rectangular plate (of length L and width W) should be within a tolerance E of some nominal value A. Conventional tolerancing schemes do not seem to provide a way to express the simple requirement that $A - E \leq L*W \leq A + E$. Global requirements like symmetry also seem difficult to express in any reasonable way.

REFERENCES

C. M. Hoffman and R. Juan, "EREP: An Editable High-Level Representation for Geometric Design and Analysis", Technical Report CER 92-24, Department of Computer Science, Purdue University, 1992.

C. M. Hoffman, "On the Semantics of Generative Model Representations", Technical Report CER 93-07, Department of Computer Science, Purdue University, 1993.

N. P. Juster, "Modelling and Representation of Dimensions and Tolerances: a Survey", *Computer-Aided Design*, Vol. 24, No. 1 (January 1992), pp. 3-17.

R. A. Light and D. Gossard, "Modification of Geometric Models Through Variational Geometry", *Computer-Aided Design*, Vol. 14, No. 4 (January 1982), pp. 209-215.

J. C. Owen, "Algebraic Solution for Geometry from Dimensional Constraints", *Proceedings First ACM/IEEE Symposium on Solid Modeling and Applications*, Austin, Texas, 1991.

S. W. Parratt, "A Theory of One-Dimensional Tolerancing for Assembly", Ph. D. Thesis, Cornell University, 1993.

Mervi Ranta, Masatomo Inui, Fumihiko Kimura, Martti Mantyla, "Cut and Paste Based Modeling with Boundary Features", *Proceedings Second ACM/IEEE Symposium on Solid Modeling and Applications*, Montreal, Canada, 1993.

A. A. G. Requicha, "Representation of Tolerances in Solid Modelling: Issues & Alternative Approaches", Technical Memo No. 41, Production Automation Project, University of Rochester, August 1983.

J. R. Stevenson, "Assembly Cost as a Function of Component Tolerance", M. S. Thesis, California State University, Long Beach, May 1988.

B. Taylor, "Dimensional Control Through the Tolerancing of a Solid Model", these proceedings.

J. U. Turner, "Tolerances in Computer-Aided Geometric Design", Ph. D. Thesis, Rensselaer Polytechnic Institute, May 1987.

K. J. Weiler, "Topological Structures for Geometric Modeling", Ph. D. Thesis, Rensselaer Polytechnic Institute, 1986.

CRTD-Vol. 27, International Forum
on Dimensional Tolerancing and Metrology
ASME 1993

CURRENT TOLERANCING PACKAGES

Joshua Turner
Design and Manufacturing Institute
Rensselaer Polytechnic Institute
Troy, New York

ABSTRACT

Commercially available software provides support for analysis of the effect of geometric and parametric tolerances on the functionally significant properties of parts and assemblies. Available technology still requires considerable input from the user in building the analysis models, and a thorough understanding of the assumptions and limitations of the package. Support for the ANSI/ISO geometric tolerancing standards is limited. However, when used with care and knowledge, the available products can be of considerable value to design engineers in reducing product cost and improving product quality.

This paper provides an overview of the basic capabilities for tolerance analysis of the technologies used in currently available software packages. Three major generic approaches are identified.

INTRODUCTION

Tolerances must be considered early in the design cycle in developing product specifications for quality products that can be manufactured cost-effectively. Software for tolerance analysis can play an important role in evaluating the design concept.

This paper provides an overview of the basic capabilities for tolerance analysis of the technologies used in currently available software packages. Available technology still requires considerable input from the user in building the analysis models, and a thorough understanding of the assumptions and limitations of the package. Support for the ANSI/ISO geometric tolerancing standards is limited. However, when used with care and knowledge, the available products can be of considerable value to design engineers in reducing product cost and improving product quality.

Sections 2 provides a short review of terms and concepts. Section 3 describes the two principal approaches to tolerance modeling: parametric tolerances and geometric tolerances. Section 4 highlights the major techniques used by the available software packages. Three principal approaches are identified. Section 5 suggests directions for further improvements in the available software.

TERMS AND CONCEPTS

This section briefly reviews some of the terms and concepts used in the paper.

A **design function** is some product characteristic which is subject to significant variation due to variation in the exact geometries of the component parts.

A **tolerance** is a constraint on the size, the orientation, the location, or the form of a **tolerance feature.** A design engineer specifies tolerances as an indirect way of controlling the variation in the design function.

A **tolerance feature** is a part face, a centerline, or a centerplane. For purposes of size tolerances, a pair of opposed parallel planar faces is also referred to as a tolerance feature.

Size and form tolerances are applied to the feature itself, but orientation and location tolerances are applied relative to one or more datums. A **datum** is an infinite surface of perfect form, orientation, and location. The form, orientation, and location of the datum surfaces are determined based on the *nominal* (ideal) part geometry.

In a **worst-case** tolerance analysis, the objective is to verify that the value of the design function will fall within specified limits for every possible assembly of in-tolerance parts.

In a **statistical** tolerance analysis, the objective is to determine how the design function will be distributed. Before a statistical tolerance analysis can be performed, a distribution must be determined for each significant toleranced feature, based on the manufacturing processes that will be used to produce that feature.

In this paper we are concerned with computer software that performs a worst-case or a statistical tolerance analysis of some design function.

TOLERANCE MODELING

There are two broad general approaches to modeling tolerances. Both approaches are available in one commercial package or another. Both are useful, but the user needs to be aware of the differences.

Parametric Tolerancing

In this approach, each type of tolerance is interpreted as a limit or pair of limits applied to some scalar parameter of a user-constructed parametric model.

Depending on the complexity of the problem, it may be easy or difficult for the user to construct a suitable parametric model. Results depend on how the parametric model is constructed.

The main advantage of the parametric approach is that it may be easy to develop a general understanding of the main effects of some of the sources of variation, upon the design function. A parametric approach is well-suited to a 2-D analysis performed early in the design cycle, where the analysis results may indicate major changes in the product concept.

The parametric approach supports either a worst-case or a statistical analysis. In the latter case, the user associates a distribution with each of the parameters.

Geometric Tolerancing

In this approach, each type of tolerance is interpreted as a passive constraint on the geometry of a tolerance feature. Each tolerance defines a tolerance zone within which the toleranced feature must lie.

The ANSI tolerancing standard (1) defines different types of tolerance zones for tolerances of size, orientation, location, and form.

In principle, software for geometric tolerance analysis can be developed which performs an analysis directly on the CAD model, without further model development by the user. (The GEOS package illustrates this potential (2).) Thus results are not dependent on user interpretation, but the geometric tolerances specified by the user must be correct, and must be functionally meaningful.

The main advantage of the geometric approach is that it supports a detailed understanding of the impact of the geometric tolerances used on the product functionality. The design engineer can develop an assurance that if the processing planning is consistent with the geometric tolerances used, then the product functionality will fall within design requirements.

The geometric approach is best suited to worst-case analysis. While some forms of statistical analysis are possible, there is only an indirect relationship between the tolerance zone sizes used on the drawings, and the probability distributions of the tolerance features that arise from the process plans. Additional effort may be required of the user to model how the tolerance features will actually vary when the parts are made.

TECHNIQUES USED BY COMMERCIAL PACKAGES

Our review of the available packages indicates that there are three principal techniques in use: techniques based on procedural models, techniques based on dimension-driven models, and special-case techniques. This section highlights some of the main features of each of this basic techniques.

Procedural Models

Techniques based on procedural models analyze a parametric model in which each model parameter (or a small set of parameters) determines the location of one of the tolerance features. In some cases the procedural model is explicitly constructed by the user, in other cases the software is capable of constructing the procedural model automatically.

The most interesting commercially available procedural modeling technique is based on a set of 3-D points. The VSA package (3; 4) employs this technique. The procedural model specifies how each of the points is to be constructed.

If the procedural model is constructed by the user, then the user must identify the key points of each of the parts, and determine how to construct each point. Parameters are used in the construction procedure specified for each of the points. The software analyzes the influence of each of these parameters on the design function.

The user must be careful in selecting the points to be modeled, and the construction procedures to be used in modeling the points. Figure 1 illustrates a user-constructed model which is probably a defective one.

In this figure,

1. P_1 is constructed at $(0, 0)$
2. P_2 is constructed at $P_1 + (a, 0)$

Figure 1: User-Constructed Procedural Model

3. P_3 is constructed at $P_1 + (0, b)$
4. P_4 is constructed at $P_3 + (c, 0)$

a, b, and c are the parameters. Note that because of the way the model has been constructed, the 90 degree angle at the lower left corner of the part cannot vary, and the top and bottom edges of the part will remain parallel. Thus it is likely that significant sources of variation have been overlooked.

In most cases the procedural model is constructed by an experienced user or a consultant, from the toleranced drawings and from a knowledge of the anticipated behavior of the manufacturing processes to be used. It is possible to develop a procedural model that reflects either a parametric tolerancing scheme, or a geometric tolerancing scheme. The parametric approach may require less effort. If geometric tolerances are to be analyzed, then the user must take care to construct a procedural model that accurately reflects the definition of each of the geometric tolerances as given in the standard.

A new version of VSA scheduled for release this summer, employs artificial intelligence techniques to try to automatically construct the procedural model. Effectively, the package incorporates a tolerancing expert which attempts to construct a correct procedural model from the geometric tolerance specifications and the manufacturing process specifications. It remains to be seen whether the new software is able to take the place of an experienced user in constructing the procedural models automatically.

Dimension-Driven Models

Techniques based on dimension-driven models analyze a 2-D dimensioned sketch. If the sketch dimensions are varied, the software package is able to recompute the sketch geometry so that the geometry is consistent with the new dimensions. A tolerance analysis is performed by analyzing the effects of variations in the sketch dimensions.

Cognition's Mechanical Advantage package (5; 6) and Analytix (7) both perform sensitivity was the first commercial tolerancing package based on the dimension-driven technique, and is also the most well-established. Saltire's Analytix package (7) has also been available for some time. More recently, SDRC and Intergraph also announced dimension-driven capabilities for use within their CAD systems.

The basic operation of a dimension-driven package is as follows:

1. First, the user enters a 2-D sketch, and the software creates a geometric model consisting of vertices and lines or curves.
2. Next, the user adds dimensions, which are used to define algebraic equations on the (x, y) coordinates of the vertices.
3. Next, the software solves these equations to determine the values of the vertex coordinates.
4. A tolerance analysis is performed by varying each of the dimensions and determining the effect of these variations on the design function.

Figure 2 illustrates the dimension-driven approach.

The following equations are generated based on the dimensions given in the sketch:

$$X_1 = 0$$

$$Y_1 = 0$$

$$Y_2 = 0$$

$$D_1 = tan^{-1}\left(\frac{Y_3 - Y_1}{X_3 - X_1}\right) - tan^{-1}\left(\frac{Y_2 - Y_1}{X_2 - X_1}\right)$$

$$D_2 = tan^{-1}\left(\frac{Y_1 - Y_2}{X_1 - X_2}\right) - tan^{-1}\left(\frac{Y_4 - Y_2}{X_4 - X_2}\right)$$

$$D_3 = \sqrt{(X_2 - X_1)^2 + (Y_2 - Y_1)^2}$$

$$D_4 = \sqrt{(X_3 - X_1)^2 + (Y_3 - Y_1)^2}$$

$$D_5 = \sqrt{(X_4 - X_2)^2 + (Y_4 - Y_2)^2}$$

$$D_6 = \sqrt{(X_3 - X_5)^2 + (Y_3 - Y_5)^2}$$

$$D_6 = \sqrt{(X_4 - X_5)^2 + (Y_4 - Y_5)^2}$$

The D's are given by the user. The X's and Y's are unknowns. In total, the software generates 10 equations in 10 unknowns.

The dimension-driven approach is ideally suited to the analysis of parametric tolerances: the parameters are the dimensions, and the tolerances give plus-minus limits on the dimensions. At an early stage in product design, it is extremely useful to sketch up the principal parts and analyze the main effects of the dimensional variations. When an analysis is performed at this stage, it may serve to identify critical problems with the product concept.

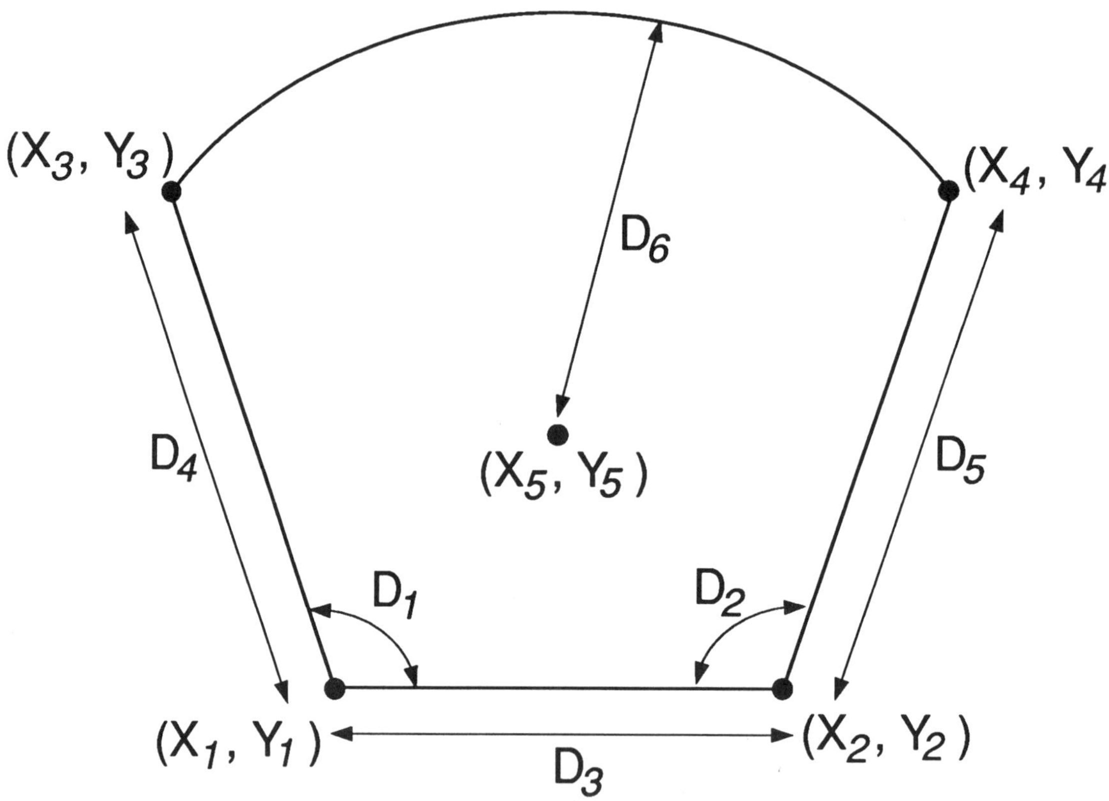

Figure 2: Dimension-Driven Model

However, the dimension-driven approach becomes unwieldy when applied to 3-D problems, and it is not well-suited to the analysis of complex geometric tolerances. Most of the packages provide only limited approach for the geometric tolerancing standard.

Special-Case Techniques

Several packages provide software designed to facilitate the analysis of a few common special-case situations. We will focus on two such situations — one-dimensional analysis, and mating pins and holes.

One-dimensional tolerance analysis refers to a class of problems where all the relevant dimensions stack up in a straight line. When parametric tolerances are used, both worst-case and statistical tolerance analysis is straightforward. Design engineers have performed such analyses by hand for decades. Recently Texas Instruments has introduced a 1-D analysis capability (TI/TOL) for use with PRO/Engineer.

Part mating situations involving the assembly of two parts with mating hole patterns and floating fasteners, or with mating pins and holes, are extremely common. Both size and location tolerances are usually involved. Design texts give rules for the the assignment of tolerances to achieve a desired class and grade of fit. The VALISYS package (8; 9; 10) provides capabilities for tolerance assignment and analysis of such situations. Geometric locating tolerances are supported. The Texas Instruments package apparently contains similar support.

DIRECTIONS FOR FUTURE GROWTH

The current capabilities of the tolerancing packages are largely restricted to parametric tolerances. Some limited support for geometric tolerances is built in. The user may be able to construct analysis models that use parametric tolerances to approximate the semantics of the geometric tolerances. The VSA package, in particular, seems to be making serious efforts to reduce the user efforts required to model the geometric tolerances.

Different packages may be suitable based on the tolerance problem to be solved. The dimension-driven approaches appear to be best suited to the analysis of design concepts at early stages in the design process. The procedural modeling approach appears to be best suited to the analysis of detailed designs, where assembly sequence plans, fixturing requirements, and process distributions have already been determined. The special-purpose approaches may be most efficient in cases where the analysis problem fits the package capabilities.

None of the current packages can really act as a substitute for human insight. If used without understanding, serious tolerancing problems can be overlooked. But a manual analysis has the same drawback. Even if only used to provide assistance with the manual tolerance calculations, the available software can be well worth the purchase price.

Although truly automated tolerance analysis software is not commercially available, the existing packages are continuing to grow and evolve. With increased usage, we expect that trend to continue.

References

[1] American Society of Mechanical Engineers, New York, N.Y., *Dimensioning and Tolerancing*, 1983. ANSI Y14.5M-1982.

[2] J. U. Turner and J. D. Guilford, "Research in automated tolerancing at RPI," in *Proceedings of the 1993 NSF Design and Manufacturing Grantees Conference*, (Charlotte, NC), pp. 709–714, January 6-8, 1993. Technical Report No. 92026, Rensselaer Design Research Center.

[3] Applied Computer Solutions, St. Clair Shores, Michigan, *VSA User Guide*, 1990.

[4] Applied Computer Solutions, St. Clair Shores, Michigan, *VSA Language Reference*, 1990.

[5] Cognition, Inc., Billerica, MA, *Mechanical Advantage Reference Manual*, November 1989.

[6] Cognition, Inc., Billerica, MA, *Mechanical Advantage Training Manual*, December 1989.

[7] Saltire Software, Beaverton, OR, *Analytix Tutorial and Reference Manual*, 1989.

[8] IBM Corp., Kingston, New York, *VALISYS General Information Manual*, March 1990. Manual Number GH52-0097.

[9] IBM Corp., Kingston, New York, *VALISYS Training Manual*, March 1990. Manual Number GH52-0089.

[10] IBM Corp., Kingston, New York, *VALISYS Reference Manual*, March 1990. Manual Number GH52-0090.

CRTD-Vol. 27, International Forum
on Dimensional Tolerancing and Metrology
ASME 1993

FEATURE-BASED TOLERANCING FOR INTELLIGENT INSPECTION PROCESS DEFINITION

Curtis W. Brown
Department of Mechanical Quality Engineering
AlliedSignal Incorporated
Kansas City, Missouri

ABSTRACT

This paper describes a feature-based tolerancing capability that complements a geometric solid model with an explicit representation of conventional and geometric tolerances. This capability is focused on supporting an intelligent inspection process definition system. The feature-based tolerance model's benefits include advancing complete product definition initiatives (e.g., STEP - Standard for Exchange of Product model data), suppling computer-integrated manufacturing applications (e.g., generative process planning and automated part programming) with product definition information, and assisting in the solution of measurement performance issues. A feature-based tolerance information model was developed based upon the notion of a feature's toleranceable aspects and describes an object-oriented scheme for representing and relating tolerance features, tolerances, and datum reference frames. For easy incorporation, the tolerance feature entities are interconnected with STEP solid model entities. This schema will explicitly represent the tolerance specification for mechanical products, support advanced dimensional measurement applications, and assist in tolerance-related methods divergence issues.

INTRODUCTION

Industry today faces new challenges as it pursues precision manufacturing, higher quality products, and greater competitiveness. Many strive toward achieving these goals by implementing computer-integrated manufacturing. Efforts toward addressing these challenges have produced many technological advances, such as computer-aided design (CAD) systems and computer-controlled production systems. Unfortunately, these advances have created islands of automation in which integration among these areas is still a time-consuming and labor-intensive task. However, an automated process definition link that bridges the product definition to

* Operated for the United States Department of Energy under Contract No. DE-AC04-76-DP00613.

production machines will provide the critical flowtime reduction component required for obtaining rapid response manufacturing.

DIMENSIONAL MEASUREMENT TECHNOLOGY CHALLENGES

Because of industry's acceptance of computer-controlled coordinate measuring machines (CMMs) as viable inspection devices, the automated definition link for dimensional measurement process is a rapid, cost-effective generation of inspection process plans, CMM measurement programs, and inspection support documents. In conjunction with this need, specific challenges have been identified that are attributed to industry's migration from traditional inspection methods to computerized, sample-point measurement equipment (e.g., CMM). These activities have resulted in the recognition of measurement performance issues and underlying technological voids. Therefore, primary challenges in advancing dimensional measurement are not only to overcome the costs of process definition flowtime, but to concurrently address the issues of methods divergence. As a result, the need for a complete and unambiguous representation of tolerances crosses three separate but interwoven domains. These domains consist of intelligent inspection process definition, product definition modeling, and measurement performance.

Intelligent Inspection Process Definition

The first domain of interests involves automated definition activities. AlliedSignal Inc., Kansas City Division (KCD) is developing an automated inspection process definition system. Process definition involves the preproduction activities required for production. Dimensional measurement process definition involves the generation of process plans, CMM part programs, and any support documentation. This function is typically performed by one or more inspection process planners and CMM part programmers. KCD's IPPEX (Inspection Planning & Programming EXpert), (Brown, 1991) system, as illustrated in Figure 1, is an automated tool that applies artificial intelligence techniques, resource information, and product definition to generate inspection process definitions. Through the experience gained by developing an automated inspection process definition prototype, a number of technology risks were identified. A major requirement involves the scope of this paper. It is the representation of tolerances as an integrated part of the product definition. This information is paramount to dimensional inspection because any inspection scope must be defined by a list of tolerances requiring verification. Therefore, any automated inspection process definition system must provide a means for representing and identifying tolerances. Furthermore, tolerance representations are not unique to inspection applications but are needed for design, process planning, assembly, and many other manufacturing applications.

Product Definition Modeling

As stated earlier, a major requirement for the successful development and deployment of advanced computer-integrated manufacturing applications is the need for a product definition. This product definition must not only provide a complete and unambiguous description of a product's nominal shape but must also contain non-geometric property attributes, such as tolerance specifications and functional descriptions. This information is required to successfully support advanced fabrication applications throughout the product life-cycle. To support these applications, non-geometric property attributes must actually be associated with solid model entities and/or features of the solid model. Furthermore, for a given workpiece tolerance representation, the tolerance modeler must perform feature evaluation, feature-tolerance evaluation, and tolerance model checking. Finally, for functional use, the product definition's

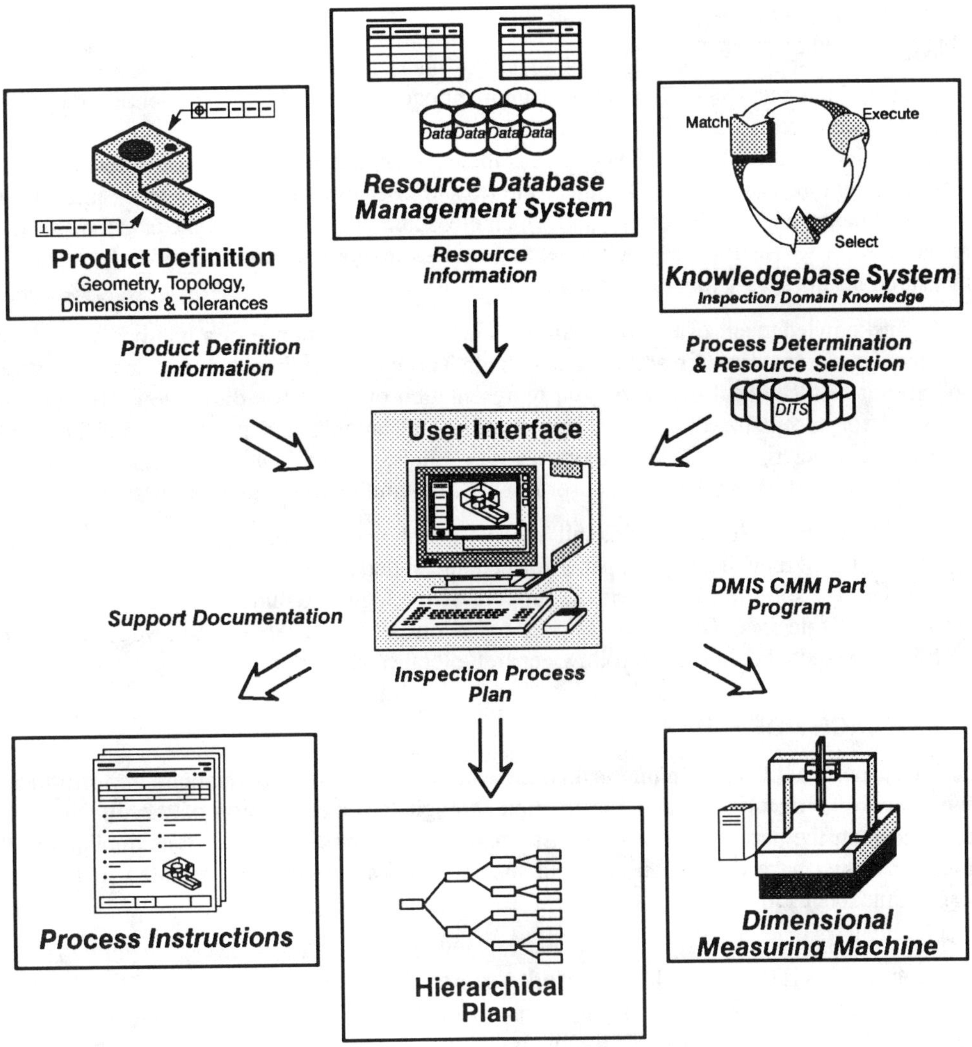

Figure 1. IPPEX System Architecture

features, geometry, topology, tolerance data, and any modeler manipulative routines must be fully accessible through a robust application programmers interface.

It is apparent that modeler developers have succeeded in representing the shape of an object accurately and reliably using solid modeling technology. Unfortunately, most solid modeling systems today do not allow tolerances to be directly associated with the geometry. Although some solid modeling systems allow dimensional attribute associations and many systems allow tolerances as text, few actually associated conventional and geometrical tolerances to the geometric entities they control, and still fewer associated these tolerances in an accessible and intelligent manner. As a result, many CIM applications must augment their geometry model with dimensions and tolerances or not consider tolerances at all. Fortunately, the STEP community has recently devoted attention to this need with work on Shape Tolerances.

Measurement Performance Issues

The third domain of interests is dimensional measurement. The emergence and increased reliance on automated sample-point measurement equipment has focused attention on measurement performance issues. The primary challenge in this domain is to overcome the issue of methods divergence. Methods divergence is the realization that dissimilar measurement methods yield different measured results for the same feature. Because of this divergence, questions such as which measured value is correct, and which measured value represents the actual condition are raised. As a result, todays practitioners of this technology are not confidently assured that measured results satisfy functional intent.

The acknowledgment of this problem has identified the existence of a number of root causes that requires further research and understanding. This expanded knowledge includes the need for an explicit mathematical definition and representation of geometric dimensions and tolerances, the need for acceptable dimensional measurement techniques and practices, and the need for standard, characterized substitute geometry algorithms. Several national standards committees (B89, ANSI Y14), industries, and consortiums are committed to working on these areas.

A look at these three domains all reveal a real need to further the understanding and representation of tolerances. This paper furthers the understanding for representing feature-based tolerances in a solid model system in support of intelligent manufacturing applications. The feature-based tolerance model was designed to be consistent with ANSI Y14.5M-1982, integrated with STEP standard entities, and follow general tolerance abstractions.

TOLERANCE MODELING

Today, process planners, numerical control analyses and part programmers perform their job functions by extracting implicit information through the interpretation of the dimensions and tolerances on the part drawing. Likewise, many computer-aided applications duplicating these functions must have dimension and tolerance semantics explicitly represented in a computer understandable form.

Many researchers have recognized the technological void of representing dimensions and tolerances. This is represented by key work performed by Kimura et al. (1986), Requicha (1986), Gossard et al. (1988), Wickens (1988), Shah and Miller (1989), and Jararaman and Srinivasan (1989). These approaches vary as to how much they attempt to use traditional approaches, how independent they are from related solid modeling systems, what geometric coverage they have, and whether they emphasize tolerance analysis, user interaction, or manufacturing.

Additional work of note for associating tolerances to solid geometric models includes Burkett's (1985) work. Burkett defined the principle information necessary to communicate ANSI Y14.5M tolerances in a Boundary Representation (BREP) solid model. He based his conceptual modeling on the following assumptions:

1) A boundary representation method of geometric modelling. The connectivity of the dimensions and tolerances must correspond to the topology (faces, edges, and vertices) of the geometric model.

2) The model defines the theoretically exact or nominal shape of the object.

3) The ability to reference functional geometry not referenced to topological entities. This occurs with:

a) adjunct geometry that contributes to the definition of something (e.g., hole centerlines);

b) surface geometry on surfaces (e.g., point, curve, or sub-region);
c) derivation geometry that are not intimately related to the shape (e.g., planar offsets);
d) loose geometry used as information only (e.g., major reference plane of a mating object).

4) A feature capability allows the capability to address a collection of entities to which a tolerance applies as if it were a single entity.

With the understanding that geometric tolerance information primarily references analytical geometric forms, the Consortium for Advanced Manufacturing - International (CAM-I) contracted Johnson (1985) to define a computer data structure for a Dimensions and Tolerance (D&T) model. These data structures represent the dimension and tolerance data for a part or an assembly in association with a BREP geometric model. This study was significant because the data structure could be created, modified, and interrogated through an application programmers interface. It also associated tolerances to the feature classes identified in ANSI Y14.5M-1982.

Ranyak and Fridshal (1988) furthered Johnson's preliminary design by expanding, implementing, and demonstrating a Dimension and Tolerance modeler (CAM-I, 1986). The D&T modeler implementation was designed to unambiguously represent the variational model of a part, thereby complementing the nominal solid model. Furthermore, the tolerance model had its own application programmatic interface. This provided the modeler with the functionality of creating, deleting, and interrogating dimensions and tolerance information. Application programs requiring both tolerance data and geometric data in a computer-intelligible form can interface both the solid geometric and tolerance modelers through subroutine calls.

FEATURE-BASED TOLERANCING

The KCD Feature-Based Tolerance Modeler is an object-oriented schema for representing conventional and geometric tolerances compatible with ANSI Y14.5M-1982. It is founded upon the success of past work, specifically Johnson, Ranyak, and Fridshal. The basic foundation of the model is the triangular entity relationship that was represented in the CAM-I D&T Model. As illustrated in the Information Analysis diagram in Figure 2, the modeler is based upon the interrelationship between three high-level entity node types: features, tolerances, and datum reference frames (DRFs). Reading counterclockwise from the bottom right, a tolerance controls one or many tolerance features, where a tolerance feature helps define zero, one, or many datum reference frames, and each DRF is referenced by one or many tolerances. In the clockwise direction, a tolerance references zero or only one datum reference frame, a DRF is defined by one or many features, and each feature is controlled by one or many tolerances. Conversely, this structure allows one tolerance node to tolerance many features and one feature node to act as a datum feature for many tolerances. Finally, the connectivity of the feature to the geometric model is through one or more solid model entities.

Tolerance Features

The Tolerance Feature class taxonomy contained in the KCD Feature-Based Tolerance Modeler is illustrated in Figure 3. Tolerance features are classified as simple or compound tolerance features. The simple features are of subtype feature-of-size (FOS) or non-feature-of-size. FOS tolerance features maybe internal or external cylindrical features. These are typified by a hole or circular boss, respectively. Another FOS feature is an opposite parallel plane feature. Typical inner opposite parallel plane features are slots, while external opposite parallel plane features are described as blocks or tabs. Additional FOS features include internal and external spheres. Non-FOS tolerance features are planar faces and cylindrical arc segments. These arc segments

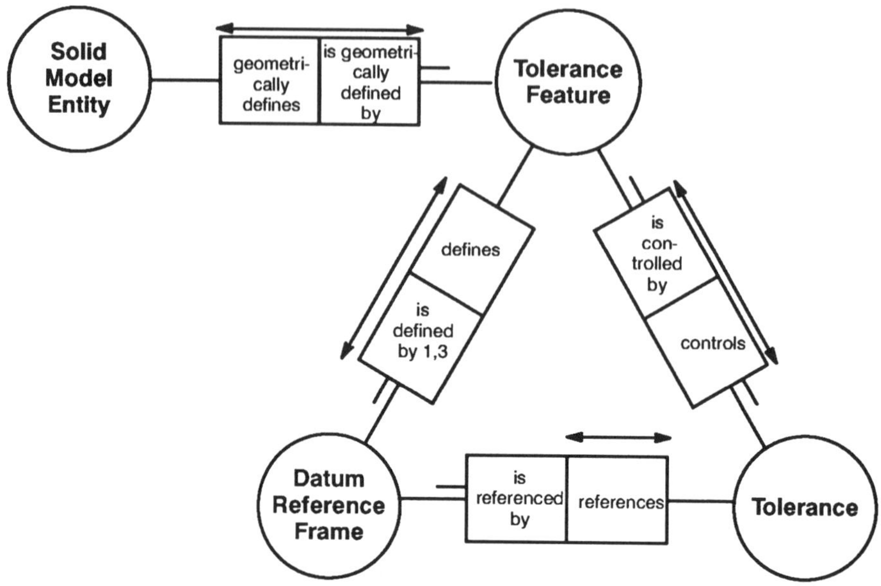

Figure 2. Tolerance Feature, Tolerance and Datum Reference Frame Interrelationship.

consist of less than 180° of the circumferential surface. Compound features consist of simple feature patterns, compound simple features, profile groups, revolute and linear swept features, and complex sculpture features. Simple feature patterns consist of two or many identical types of FOS features that as a group resolves about a center-line. Examples of simple feature patterns include bolt hole patterns and patterns of tabs. Compound simple features consist of two or more geometrically identical simple features. Examples of compound simple features are two planar faces divided by a slot that together represent a datum or have a common tolerance constraint. A profile group feature consists of two or more arc-wise connected simple tolerance components. A profile group is usually associated with a profile tolerance. The revolute swept feature corresponds to a surface of revolution. The linear swept tolerance feature consists of a surface of linear extrusion. The complex sculptures tolerance feature consists of a bounded b-spline surface.

As illustrated in Figure 4, the KCD Feature-Based Tolerance Modeler is based upon the notion that all features must be constrained by the toleranceable aspects of location, orientation, form, and surface finish, and if the feature is a feature-of-size, a size toleranceable aspect is required. For example, a hole must have a set of tolerances that controls all five tolerance aspects. A location and a size tolerance are always given. This results in implementing the location and size toleranceable aspects for the hole feature. Interestingly, the location tolerance provides a certain degree of orientation tolerance aspect. However, if more orientation control is required, then orientation tolerance may be added. Likewise, the size tolerance provides the feature with a default form tolerance aspect. However, if more form control is needed then a specific form tolerance is added. The surface finish aspect is usually applied by an overall part default unless specifically called out.

The KCD Feature-Based Tolerance Modeler is implemented by connecting tolerance features with the entities in STEP Part 42. Each tolerance feature references one or more tolerance component entities. Tolerance component entities can be either another tolerance feature or a STEP solid model entity consisting of either a manifold solid brep body, face, sub-face, edge,

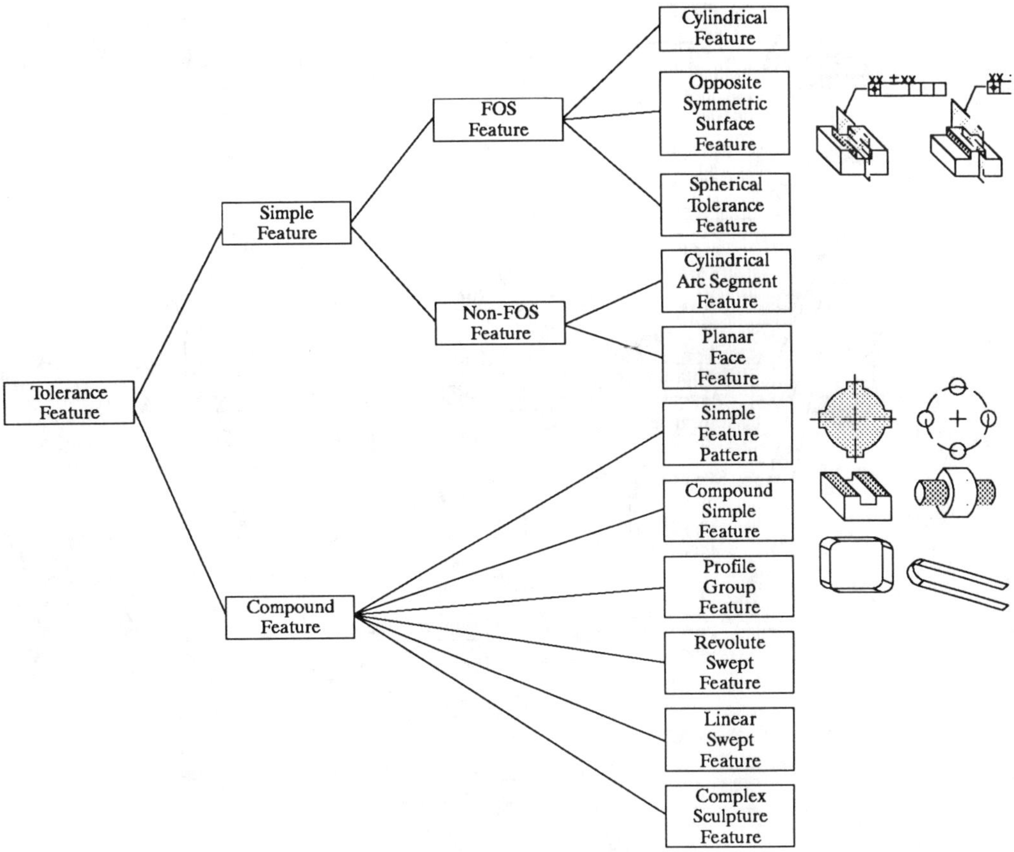

Figure 3. Tolerance Feature Hierarchy

curve on surface, vertex, or point on surface. Also, general property attributes such as thread specifications, material type, edge breaks, or cosmetic attributes can be assigned to a tolerance component entity. Furthermore, if a feature is used as an explicit datum feature, the feature's resolvables are used as arguments for the datum reference frame's transformation matrix.

Tolerances

Tolerances are created and assigned to formally constrain tolerance features. The model's tolerance classes are angle, distance, form, orientation, position, profile, radial, runout, size, and surface finish tolerances. Figure 4 shows the tolerance aspects and the associated tolerance classes that help control it. Although these tolerances are different than those explicitly defined in ANSI Y14.5M, they are fully compatible. Since the ANSI Y14.5M standard is drawing-based, its tolerance classes must also convey geometric information, whereas the Feature-Based Tolerance Modeler implements a more theoretical approach for tolerance abstraction and permits the geometric modeler to provide the geometric information. As a result, the ANSI class for any tolerance-feature combination is derivable. For example, parallelism and perpendicularity are special cases of orientation which have a specific geometric angle relationship of 0° and 90°, respectively. Therefore, if the geometric relationship between two planar faces is 90° and they are related by an orientation tolerance, then we can easily deduce the ANSI perpendicularity tolerance.

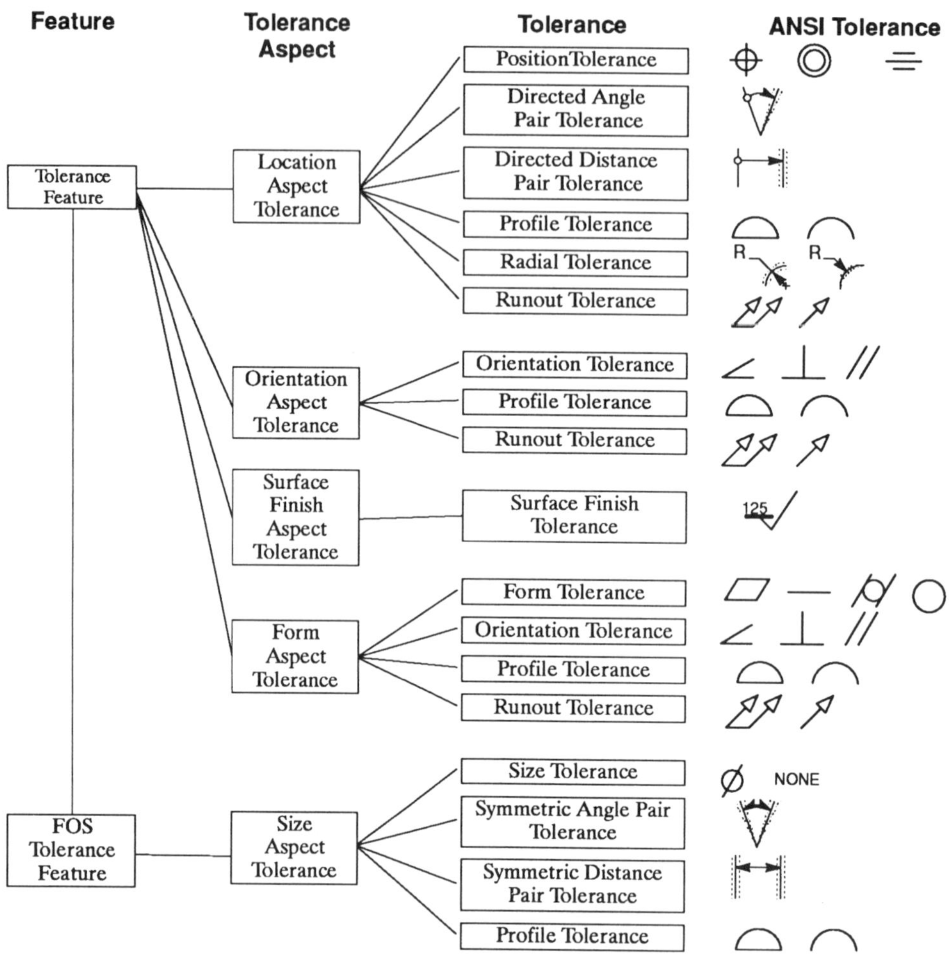

Figure 4. Tolerance Aspect Implementation

Datum Reference Frames

Each datum reference frame (DRF) may be defined by one, two, or three existing tolerance features that have been designated as an explicit datum feature. For dimensional measurement, each DRF actually represents an inspection set-up. As a result, the completed DRF must explicitly define a coordinate system with an explicit origin location and axis directions. The classifications of DRFs follow the CAM-I datum reference frame classes. The DRF's datum feature resolvables explicitly define a cartesian coordinate system. A datum feature of a DRF typically defines an axis direction and/or one or many origin coordinates. The determination of the direction and origin is influenced by the DRF's datum precedence. One DRF datum feature rule is that the primary datum's resolvable defines the Z-axis component of the resulting DRF.

Tolerance Modeler Functions

For a tolerance modeler to be implemented and useful, the modeler must have the functions to validate and verify the tolerance model during creation and modification. Furthermore, upon completion of a tolerance model, the overall scheme of the model must be checked. The following are tolerance modeler functions needed to help verify the tolerance model.

Feature Evaluation. The tolerance model must evaluate each tolerance feature by checking the correctness of its associated solid model entities. For example, this capability allows planar face tolerance features to be only assigned to planar face solid model entities.

Feature-Tolerance Evaluation. The tolerance model must validate the tolerance attributes, correctness of the tolerance's assigned feature, and correctness of each DRF. Furthermore, the tolerance model must be able to derive ANSI Y14.5M tolerance classes from the feature-tolerance relationship. To incorporate this, many toleranced features must have construct qualifiers to help control specific aspects. Usually, these qualifiers help resolve the feature to a lower dimensionality to incorporate specific 2D tolerances (e.g., a cylindrical tolerance feature toleranced by a form tolerance with a qualifier would result in a ANSI circularity tolerance). Qualifiers include feature resolution, intersect/section, and cross-section qualifiers. Next, the tolerance model must perform distance evaluation for determining dimension requirements. This is easily performed by the solid geometric modeler's capabilities. Additionally, the basic dimensions are obtained from the nominal geometric solid model.

Datum Reference Frame Evaluation. The tolerance model must evaluate the defined DRF by checking for valid datum features and by creating a transformation matrix that mathematically represents a coordinate system. Datum features for a DRF must be resolvable to a point, line, or plane. Furthermore, if a datum feature is a FOS, then it must have a material condition modifier assigned to it.

Tolerance Model Checking. For a part to be unambiguous and fully toleranced, the location of every feature should be fully controlled and every FOS feature must have a size tolerance. The tolerance modeler shall perform locating tolerance checking. To facilitate this capability, the modeler must be able to determine the feature's locating dimensionality, determine the part's overall datum reference frame, determine location of features with respect to DRF or implicit datums, and check for size tolerance for any FOS features.

Geometric Reconciliation. The tolerance capability shall have a mechanism to reconcile tolerance features upon any changes to the geometry in the solid geometric model. If the topology remains intact, reconciliation is usually possible. However, if the model is unreconcilable, then any detached tolerance features must be identified for user disposal or reconciliation.

Tolerance Modeler Programmers Interface

For a tolerance modeler to successfully support advance manufacturing applications, it must provide a programmatic interface. The Feature-Based Tolerance Modeler capability provides a programming interface through procedure calls or through direct access to C++ objects. The programmatic interface will allow the capability of extracting all tolerance features, tolerances, datum reference frames, and topology connectivity information. The interface is designed to provide, at a minimum, the following:

- Enquire and create default tolerances and part DRF.
- Enquire and create tolerance, feature, and DRF attributes.
- Add and release tolerances and model entities to/from features.
- Add and release a datum to/from DRFs.
- Set a material condition for datum features.
- Query features of assigned tolerances.
- Query DRFs of referencing tolerances.
- Query tolerances of a feature.

- Query tolerances of an implicit datum feature.
- Query DRFs using an explicit datum feature.
- Query solid model entities of a feature.
- Query the parent feature of a feature.

SUMMARY

The Feature-Based Tolerance Modeler defines a scheme for representing conventional and geometrical tolerances. The model is based on the interrelation of tolerance features, tolerances and datum reference frames, while the tolerance features are associated to STEP solid model entitles. A tolerance feature hierarchy necessary to represent ANSI Y14.5M tolerances was discussed. Also the fundamental notion that every tolerance feature has toleranceable aspects that must be constrained is conveyed. The original tolerance model was developed in Pascal and interfaced with a number of solid modelers. Currently, the tolerance model data has been logically designed graphically and is captured in the EXPRESS information modeling language. The EXPRESS language definitions are parsed into C++ objects. Methods to the C++ data model objects are presently being defined. The Feature-Based Tolerance Modeler will complement a solid geometric model with tolerance information that will support advanced manufacturing applications such as process planning, inspection planning, and CMM part program generation.

REFERENCES

Brown, C.W., "IPPEX: An Automated Planning System for Dimensional Inspection," *Proceedings of the 22nd CIRP International Seminar on Manufacturing Systems: Computer Aided Process Planning*, Vol., 20, No. 2, 1991.

Requicha, A. A. G., "Representation of Geometric Features, Tolerances, and Attributes in Solid Modelers Based on Constructive Geometry," *IEEE Journal of Robotics and Automation*. Vol. RA-2, No. 3, Sept. 1986, pp. 156-166.

Shah, J.J., and Miller, D. "A Structure for Integrating Geometric Tolerances with Form Features and Geometric Models," In *Proceedings ASME Computers in Engineering Conference*, August 1989, San Francisco, CA., pp. 395-402.

Jayaraman, R., and Srinivasan, V., "Geometric tolerancing: I. Virtual boundary requirements, Geometric tolerancing: II. Conditional tolerances," IBM J. Res. Develop. 33, No. 2, March 1989, pp. 90-124.

Burkett, W. C., "PDDI Approach to Dimensioning and Tolerancing a Solid Model," In *Proceedings in Dimensioning and Tolerancing Workshop at Seattle, WA.*, Computer Aided Manufacturing - International, Inc., Arlington, TX., P-85-ASPP-02, July 10, 1985.

Johnson, R. H., "Dimensioning and Tolerancing Final Report," Computer Aided Manufacturing - International, Inc., Arlington, TX., R-84-GM02.2, May 1985.

Ranyak, P. S., and Fridshal R. "Features for Tolerancing a Solid Model," In *Proceedings of 1988 ASME International Computers in Engineering Conference and Exhibition at San Francisco, CA.*, July 1988, pp. 275-280.

"CAM-I D&T Modeler Version 1.0 - Dimensioning and Tolerancing Feasibility Demonstration Final Report," Computer Aided Manufacturing - International, Inc., Arlington, TX., PS-86-ANC/GM-01, Nov. 1986.

CRTD-Vol. 27, International Forum
on Dimensional Tolerancing and Metrology
ASME 1993

AN INDUSTRIAL USER'S PERSPECTIVE OF CAD

Archie R. Anderson
Trikon Corporation
Auburn Hills, Michigan

ABSTRACT

In order to make this report have more meaning, it is the result of interviews with individuals and groups in several large companies using CAD.

Subjects that are investigated include: user friendliness, time to proficiency, down stream use of math data, application of GD&T to components and assemblies, CAD system conformance to ANSI Y14.5M-1982 and methods of tolerance analysis, manual, CAD based or PC based,

The groups and individuals interviewed represented producers of machines components and formed sheet metal components. Formed sheet metal components have unique tolerancing opportunities as the components are not rigid. The report on tolerance analysis will focus on the methods used for sheet metal components.

REFERENCES

Chrysler Corporation
Ford Motor Company
General Motors Corporation
Herman Miller, Inc.
Saturn Corporation

CRTD-Vol. 27, International Forum
on Dimensional Tolerancing and Metrology
ASME 1993

SESSION 8
QUESTIONS AND DISCUSSION

Moderators: **Thomas Charlton**
Brown & Sharpe Manufacturing Company
Richard Walker
Westinghouse Electric Maine Division

Reporters: **Steffen Parratt**
Cornell University
Dean Robinson
Cornell University

INTRODUCTION

This section provides a summary transcript of questions and discussion directed at the presentations in Sessions 5–7. The talks in these sessions were tightly scheduled, and only brief questions for clarification were admitted after each talk. Substantive questions were deferred to this session (#8), and were submitted in signed, written form. The session Moderators ordered the questions and managed the ensuing discussion.

The Reporters produced this record from written notes and tape recordings; it has been edited lightly by Herb Voelcker, acting for the *Proceedings* Editor, Vijay Srinivasan. The questions, responses, and remarks attributed to individuals have not been checked by those individuals because of the compressed publication schedule for the *Proceedings*. The Reporters and Editor apologize for any misinterpretations.

QUESTIONS AND DISCUSSION

Douglas Sutphin (to Richard Walker): It appears that more questions have been raised and more uncertainty introduced by the forum than we have solved. What should the average user do now to obtain answers?

Walker: I think it is good that we raised questions, and maybe there is uncertainty. From the first time we met several years ago a lot of work has been accomplished. More CMM users and manufacturers understand GD&T. Five years ago a Y14.5 meeting had twenty attendees; now we have 60 or 70 people. CAD vendors are beginning to integrate Y14.5 into their products. I think a lot of good things have been done.

Jahir Pabon (to James Guilford): How do you compute the "geometric sensitivities" in the intermediate block of your flow diagram?

Guilford: We are looking at the worst case value of the design function over the feasible region. Since we are dealing with small variations in the model we can linearize the equations. We use a linear program to find the worst case value and the Revised Simplex algorithm provides sensitivities of the final values on the limits.

Bill Grant (to Ted Hopp): Do you think it important to develop test data sets ("soft artifacts") from actual measured data? These data have non-statistical deviations (such as outliers, differences between real surface data and probed data, surfaces with regular features in the probed surfaces, turning marks, lobes) that cause algorithms to do weird things that you don't see with data sets generated from random number generators.

Hopp: It is important to be able to simulate form errors and that sort of thing in test data sets. The system we are developing at NIST does have systematic error simulation capability.

Mark Nasson (to Curtis Brown): Would you comment on the recent patent covering the use of DMIS?

Brown: I think it will be protested. It doesn't cost much to protest a patent.

Walker: I heard from Paul McKim that it has already been nullified.

Guilford (to Georg Henzold): You seemed to indicate that under ISO, the position tolerance controls a bent axis, whereas under vectorial tolerancing, a derived straight axis is controlled. Under ANSI Y14.5, an RFS position tolerance controls the derived straight axis. Is the ISO standard inconsistent with Y14.5 in this particular case?

Henzold: That is correct. Vectorial tolerancing only deals with substitute elements. It eliminates form error and then deals with position, orientation, etc. In this case, vectorial tolerancing can only deal with straight substitute axes.

Tom Charlton: There were several related questions on how to handle functional fitting of components using vectorial tolerancing.

Henzold: For this case the conventional method of tolerancing is appropriate. However, we can benefit by using both vectorial and conventional tolerancing depending on the application. Each method is appropriate for certain applications.

Paul Wreede (to Henzold): Do you feel that it is appropriate to submit the issue of vectorial tolerancing to ISO TR10 360-1 prior to acceptance of this technique (for features) as a new work item by ISO TC10 SC5?

Henzold: The ISO procedure has already been started.

Wreede: The new work item as proposed disagrees with the Abstract paragraph four. The new work item noted that the method of vectorial tolerancing does not adequately address size, orientation, and location. Is there a conflict between the Abstract and the proposed work item?

Henzold: I don't see any conflict.

Nasson (to Henzold): You seemed to indicate near the beginning of the presentation that technological advances in metrology would make the advantages of vectorial tolerancing less significant. Is this the case?

Henzold: You must have misunderstood me. Just the contrary, I think that vectorial dimensioning and tolerancing will become more important because of the advantages I have just mentioned, e.g., SPC. I am convinced that SPC can only be managed with vectorial tolerancing.

Walker (to Hopp): An individual here at the conference has several coordinate measuring systems. He stated that using your algorithm testing system the numbers come out different on these systems. Are you aware of this? What is the magnitude of the difference? What is the source of the difference?

Hopp: I was aware of this. The differences were between software results from their systems and the NIST references algorithms, and also between different measurement systems in their own plant. I am not familiar with any of the details.

Charlton: Does this person want to comment further?

[unknown]: We are using the NIST software to generate data sets and are running them through the software packages we have in-house. We can't seem to find any that agree as far as measuring sizes and positions of circles.

Charlton: What size circles are you measuring and what size errors are you seeing?

[unknown]: We're measuring a 20 unit circle and seeing a couple thousandths of positional error.

Charlton: How many sample points are you taking?

[unknown]: 16, 21, and 29 points, sampled uniformly.

Charlton: The PTB and MPL have formal programs for software certification, and NIST is developing a formal program. If we develop a new measurement, such as the measurement of gage blocks, it would be anticipated that there would be an international round robin among the national laboratories to compare the results of the certification procedure. Has there been any discussion of doing this among these laboratories?

Hopp: There has been some discussion between NIST and PTB, and between NIST and NPL. I don't know what PTB and MPL have discussed between themselves. The results of our efforts have been disappointing. We sent our results to the PTB and NPL. They promised to exchange their results with us, but we haven't heard from them in close to a year now. We would like to see more cooperation.

Pat McCuistion (to Ben Taylor): Where do you find designers with the knowledge required to carry out your process (i.e., design, manufacturing, tool, fixture, tolerance, and inspection engineering) all in the same one or two persons?

Taylor: Actually, we don't. But, we have started putting all of these types of engineers in a central location so that they work closely together. More importantly, we also have about 35 sponsored co-op students that spend time in each of these disciplines. We don't have engineers with all of this knowledge, but in a few years we hope to have engineers that have most of it.

Bruno Brazauskas (to Alvin Neumann/Rich Walker): Has there been any discussion, within the

Y14.5 or Y14.5.1 Committees, to expand the list of modifying symbols to include the desired algorithm which should be used to most closely approximate the functionality? For example, a hole size and position measurement requiring a minimum inscribed circle algorithm vs. a least squares algorithm.

Walker: I think that the closest thing we have to this is the new statistical tolerancing symbol. I don't think there is going to be any move to create a new symbol.

Charlton: I think that it is a tricky issue because, on the one hand, it involves process, where the process is inspecting the part, which I am sure the Y14.5 Committee would say is forbidden. On the other hand, since it is part of the designer's understanding of what he or she means when putting the symbol on the paper, it could be part of the design intent.

Sam Levy: On some projects we put the inspection process on the drawing telling the inspector exactly what we want. Sometimes you know exactly how the part is to be measured and you cannot get it by just leaving it to chance. I believe that the time will come when this type of information will be acceptable for drawing use.

Henzold: I think that this will be discussed in the near future within ISO as well. The designer may want to specify whether only the outer points of the feature shall be taken into account, or whether he wants every point of the feature to be taken into account. The default case should be standardized, and the other case should be indicated explicitly.

Ken Sheehan: I have heard that the Y14.5 Committee resists new symbols in the interest of avoiding clutter on drawings. It may not be an issue in the future when the drawing is in electronic form, but on paper it could be an issue.

Charlton: I think that is correct. There is also a very strong philosophical point of view to keep process related information off the drawing.

Sutphin (to Hopp): Does the average CMM user understand that they are certifying parts with algorithms that have not been certified? How much of an impact could this have on industry? Are CMM/software vendors open to a certification process?

Hopp: I think the answers are "yes", "a lot", and "yes". I think that some of the CMM users are aware that their software isn't certified because there is no certification available. I also believe that many people think that software does not need to be certified.

Levy: Who would do the certifying?

Hopp: What we are planning to do with our test service, similar to what the PTB does, is to provide some measure of the variability that we can expect from the software for specific ranges of test cases.

Levy: Will NIST be responsible for certified software?

Hopp: Probably no one will "certify" software; we will test and evaluate software.

Charlton: Let me comment on the last issue of whether CMM vendors are open to a certification process. All the major CMM vendors have already been through this process because it is necessary to do business in Germany. We don't really see this as an issue.

Henrik Neilsen: I don't think that the average CMM user knows he or she is certifying parts with uncertified software. The average user thinks that when a person comes in once a year and calibrates the machine then the machine is fine, including the software. The average user wouldn't think that the software could make a mistake. I think it is a problem that is invisible to the average user.

Sheehan: The PTB test takes a set of software results and compares them to what they believe are the true results and categorizes the software into one of several different classes. The best class says the software can contribute no error greater than, say, a tenth of micron. My understanding is that anybody that didn't get into that class on the first try, dug into their software and made modifications so that they made it into that first class on the next try. I think everybody, as far as the PTB is concerned, is on a level playing field for the basic geometric algorithms for a specific set of data. The question is, whether algorithms that have been tuned and optimized for one set of data work well for other sets of data. This may be why NIST has taken a different approach to certification than PTB.

Going back to the second question about certifying software and product liability issues: you can't certify software in such a way that you know that your measurements are good. You can misuse software, misuse a CMM, or misuse any other type of gage. There are very specific cases where the CMM, a part program, and a software system, as a unit, have been certified. But, in the general case, you can't be assured that the CMM and software are going to be used properly.

Charlton: I might also comment that Ken said that everybody is certified at the highest level. But, of the 10 or 12 twelve software packages out there, only one of them made it through the first time. Which means there were probably a lot of products shipped that had fitting bugs in them, particularly a lot of the older software packages. I'm sure they were not tested. So there is some need for caution when saying that all that software out there is certified.

Neilsen: Isn't it true that PTB certification is only for least squares and simple geometry?

Charlton: Does anybody know of efforts to certify software using methods other than least squares?

Henzold: I think the PTB is going this way.

[unknown]: I believe that PTB and NPL are working on certifying min/max and circumscribed circle algorithms.

Paul Drake (to Guilford): The tolerance synthesis process depends on manufacturing cost parameters that are input by the user. Are you (or anyone else) collecting this kind of data? What is the difference between "tolerance synthesis" and "tolerance allocation"?

Guilford: First, there is very little data available to validate the cost models. We are doing a little work in this area. Second, tolerance allocation is the process of finding a set of tolerances from scratch, so to speak. Synthesis is the process of finding values for a given set of tolerances.

Roberto Altschul (to Grant): Why use a Taguchi approach that disregards interactions?

Grant: As I understand it, the Taguchi method primarily is non-interactive, but it can be used for interactions. The idea of Taguchi is to reduce Design of Experiments to a non-interacting set of parameters to reduce the size of the problem. We use the non-interactive parameterization approximation, and even this is time consuming in our application.

Ari Requicha (to Hopp): Suppose we did an *approximate* min/max fit by first doing least squares and then "moving" the surface to encompass all the points. How far is the result from the true min/max fitting?

Hopp: That procedure wouldn't affect the difference in computed centers at all. The center variation would still be there, which can be a significant difference. What I was suggesting was to take the least squares and then expand the radius to cover all points. I suspect the overestimation in the radius is going to be on the order of the error in centerline distances.

Barry Lawler (to Hopp): If the purpose of measurements is either to control the process or to predict the conformance of the product to requirements from the dimensional samples, what work has been done on *adaptive sampling algorithms* which determine sampling strategies based on known sample values and measurement objectives? How should test descriptions be changed dynamically based on measurement results?

Hopp: That's an interesting approach. It means sampling and saying "It looks like there may be a problem here" and going back to locally sample some more. I haven't seen any applications of this for CMMs.

Charlton: Adaptive sampling with scanning probes is commercially available from more than one vendor because it is necessary to follow the surface. If the machine goes at constant speed and the surface changes quickly, the probe can't follow the surface. So that's a reason for adaptive sampling.

Wreede (to Taylor): What is your opinion of imbedded intelligence in the CAD model such as *attributes* associated with the individual features of the model? Attributes can specify target information, tolerances, finishes, inspection points, etc., but do not specify process, typically. Specifications of specific process cannot be used due to limitation and exclusion of manufacturers with some contract work.

Taylor: I still think that if you are developing the part and you haven't decided how it is going to be made, then you probably haven't optimized the design. The design isn't just for part function. You also design it to reflect how you are going to make it. Because we use the process as part of the tolerance, we see nothing wrong with calling the process part of the specification. We can make it work, but we can't get subcontractors to make our parts – that's the price we pay.

Henzold: I want to make an observation about traceability using CMMs. I know that the PTB is working on a solution to that problem. The idea is that once the CMM is delivered it goes through an acceptance inspection. In doing this acceptance inspection, the behavior and accuracy of the CMM is detected and the data is stored. This data will be used to simulate, for a certain measurement, the measurement uncertainty. At first I thought is was impossible to take all of these different influences into account. But, I believe this method will work. We will wait a half a year and see what comes out.

Charlton: The general name of this work is "virtual machine studies". I think the six month estimate is optimistic, because the great difficulty in accomplishing traceability is modeling the sources of repeatability of machines. We have good methods and estimates for accuracy, but we don't have good models for repeatability, and you must have both.

[unknown]: CMMs certified using a least squares algorithm may give large errors when used for applications that require outer point measurements.

Hopp: You're right. The test is designed to verify that the software works as advertised. It does not verify that the software works for your particular application. Those are two separate issues.

Herb Voelcker (to all): I want to pose a general question to the group. Is tolerance design so well understood that it is not worthy of discussion? Jim Guilford was the only person who talked on the subject, and it was but one topic of several in his talk. Is the issue solved and trivial? Is it too hard to talk about? Or, is it not interesting to this group because the group views itself as a "service agent" that explains what tolerances mean and how to measure them, but not how to use them?

Norman Cutler: Where do tolerances come from? They come from the design engineer who works with the designers and checkers who do tolerance studies and who have knowledge of the processes that you plan to use.

Voelcker: Then the question is: is this understood in a scientific sense? One could make analogous comments about metrology thirty years ago. Yet here we are today talking about theories of metrology, and systematic analyses of metrology algorithms. Is tolerance design understood in a similar manner, or is it black art?

Cutler: It is not terribly well understood.

CRTD-Vol. 27, International Forum
on Dimensional Tolerancing and Metrology
ASME 1993

CHAIN OF STANDARDS: A NEW CONCEPT IN TOLERANCING AND ENGINEERING DRAWING GPS-STANDARDS – GEOMETRICAL PRODUCT SPECIFICATION STANDARDS

Per Bennich
PB Metrology Consulting
Vaerloese, Denmark

ABSTRACT

Chains of standards is a new concept in tolerancing and engineering drawing standards. Chains of standards is a new way to link Geometrical Product Specification (GPS) standards together. Each chain of standards has six links - all six chain links shall be covered by standards (definitions) to form an unambiguous link between the drawing indication (design intent) and the result of assessment of a workpiece in e.g. SI-units. 49 chains of standards are defined in the area of General GPS chains of standards. It makes 6 x 49 = 294 matrix cells in the General GPS matrix. Only 48% of these cells are at the moment covered by ISO standards! In many cases the existing "covering" are outdated and insufficient. Only two of the 49 defined chains of standards are complete with standards covering all 6 links.

Consequently it can be demonstrated that only extreme few ISO GPS drawing indications are unambiguous. Major general gaps exist. There are even contradictions between the existing ISO GPS standards produced by ISO/TC 3, TC10/SC5 and TC57. The situation is almost the same in most national GPS standards-systems. A need for a close coordination between the three ISO/TC's is demonstrated in order to solve the observed problems.

INTRODUCTION

The term "Geometrical Product Specification" - abbreviated GPS - is used as a collective name for all sorts of specifications on an engineering drawing used to define the "geometric" features of a workpiece and their interrelations, i.e. specifications of: Size, distance, form, orientation, location, surface roughness, etc.

An engineering drawing consists of two main parts: the symbolic "picture(s)" and the GPS(s) (geometrical characteristics: Size, distance, form, surface roughness, etc. and their related tolerances). The purpose of the drawing is to show and communicate the design intent.

The drawing pictures consists in principle of a "selection" of ideal geometrical features (straight lines, flat surfaces, correct angles, perfect circles, perfect cylinders, and even more complex geometric perfect features). This is probably - by experience - also the "picture" which exists in the mind of the designer!

The geometrical characteristics or GPS are basically related (defined) in only this geometrical ideal "world". The majority of the geometrical characteristics will lose its meaning completely, when the geometrical features involved are not in an ideal condition.

In the real world - in the workshop - a workpiece consisting only of ideal features have never been seen.

Workpiece features are only approximations to the ideal geometrical features shown/indicated on the drawing. Therefore the GPS indicated on the drawing will have no exact or unambiguous meaning if only the theoretical geometric definitions are available.

This problem is well known in the workshop area and especially by the measuring people in industry. It is also a well known problem in standardization. A number of standards exist, which contains the necessary supplementary definitions which "translate" (extend) the definitions of the ideal geometrical characteristics to be valued also in the real - non-ideal geometry - world. The oldest of these supplementary definitions is the Taylor criterion (ISO/R 1938 et al) equivalent with the US "Principle no. 1" covering the problem of extending the definition of the diameter of a ideal circle or cylinder to a "non"-circle or -cylinder. A collection of similar supplementary definitions exist for various other geometrical characteristics (i.e., definition of roundness, straightness, flatness, cylindricity in ISO 1101 and surface roughness in ISO 4288). In this paper these supplementary definitions are designated: Definitions for Actual Feature characteristic or parameter (abbreviated DAF).

These DAFs shall be derived from the **functional needs** of the features or the workpiece in the product. This is also the case with the Taylor principle/Principle no. 1 which is derived from the functional needs of two mating parts in a fit.

Having defined a DAF for an ideal geometrical characteristic it is now quite clear how the theoretical physical ("geometrical") property is to be understood in the real world - on the workpiece. It is then possible to set up general requirements for verifying that the geometrical characteristic lies within the tolerance, and consequently it is also possible to set up general requirements for the measurement or gauging equipment, so that the DAF is fulfilled. Standards dealing with these matters also exist.

From these general requirements concerning verification methods and measuring equipment it is possible to go into further detail, setting up very concrete requirements for specific measuring equipment, so that the specific equipment will be able to fulfil the DAF at least to a certain (known) extent. Such standards and standardised measuring equipment are common (i.e., dial gauges, micrometers, roundness and surface roughness measuring instruments, etc.).

The last problem is to set up requirements and procedures for verifying that the specific measuring equipment fulfils its stated requirements. These procedure(s) is known as the calibration of the measuring equipment. The calibration requires in most cases one or more calibration standards, which for other reasons (quality assurance - ISO 9000/ISO 10012-1) shall be traceable to internationally accepted standards and definitions of the (SI-) units concerned. These calibration procedures and calibration standards are in several cases also covered by ISO standards.

CHAINS OF STANDARDS

For a specific geometrical characteristic the above mentioned problems and the related standards forms some sort of a "family" - dealing with details in the process of connecting the design intent indicated on the drawing as an ideal geometrical characteristic with the practical verification of the specification traceable to the SI-units concerned.

In the following clauses of this paper these families will be dealt with and set up in a special format "**chains of standards**". Chains of standards were first introduced to ISO/TC 3, 10 and 57 in the spring of 1992. Chains of standards have been in use by the Danish Standards for more than ten years with a remarkable effect on the organization of work, planning and production of standards as well as the understanding for standardization by the authorities.

A chain of standards is formed by **all** the related standards concerning the same geometrical characteristics (GPS). A (main) chain of standards can in many cases be subdivided into sub-chains, each dealing with particular types of features - but handling the same geometric characteristic. A chain of standards is characterized by the fact that each single standard, which is a part of (a link in) the chain, affects the other standards, so that the full understanding and application of each standard requires the knowledge of the other standards in the chain of standards.

When considering the title of a standard or the subject of the standard, mentioned in the subtitles, each single standard in a chain of standards is often aimed towards a limited or special group of users. It is a common opinion - and misunderstanding - by the users of standards, that specific standards belong either to design, manufacturing or to quality assurance (measurement).

The rationale is, however, that each link of the chain of standards is absolutely necessary for all parties (i.e., design, production, quality control, measuring, etc.) in order to make it possible to understand the content and interpretation of each single standard and the definitions mentioned in each single standard in the chain. All standards in a chain shall be known by all the mentioned parties - not least the design people! While the later standards in the chain often alters the meaning of the theoretically defined characteristic.

A chain of standards consists of six chain links (numbered 1 - 6). Each link in the chain of standards has a very special task of its own - as indicated below. A specific standard with a narrow scope is attached to only one chain link - or even only a part of a chain-link. A standard with a wider scope may cover more than one chain link. In special and rare cases, one standard covers all six chain links

The task of the total "chain of standards" is to connect - **unambiguously** - the indication on the drawing (i.e. the surface roughness symbol) with the SI unit of length, the meter (more accurate micrometer), in a way so that the tolerance limits (upper, lower or both) are defined in every case possible - irrespective of the deviations from ideal geometry of the toleranced features, or other deviations from theoretical correct conditions - whatever they might be.

The six chain links

The objective of each of the single chain links is in detail discussed in the following clauses.

Chain link no 1 - Product documentation indication - Codification. Standards concerning the drawing indication of the specification of the workpiece are located in this chain link. The indication is often in a sort of "coded" symbol - a symbolic representation of the geometrical characteristic. These standards define the symbols , how to use the symbol and the associated rules of "grammar", the small differences in the symbol, which cause a major shift in meaning.

Chain link no 2 - Definition of tolerances - Theoretical definition and values. Standards concerning definitions of the numerical values related to the "code"-symbols are placed in this chain link. These standards define the mathematical rules of translating from the code to "understandable" values of e.g. the size - and vice versa. These standards deal with (define) the ideal geometric characteristics (the tolerance) of the features of the workpieces.

Chain link no 3 - Definitions of actual feature- characteristic or parameter - DAF. This third chain-link is probably the most important of the six. In links no 1 and no 2 the standards deal only with the ideal geometry of the features concerned. The standards in link no 3 have the purpose to introduce the supplementary definitions to extend the meaning of the ideal geometry, so that also the non-ideal real world geometry (actual feature characteristic) always is defined in relation to the theoretical tolerance.

Chain link no 4 - Assessment of the deviations of the workpiece - Comparison with tolerance limits. The standards in chain link no 4 set up the general requirements for the assessment of the deviations of the workpiece. The standards shall also give the detailed rules for comparing the measurement results with the tolerance limits to be able to accept/reject the workpiece with respect to the indicated geometric characteristics and their associated tolerance on the drawing.

In several cases - in the ISO standards - these requirements have the consequence that the theoretical tolerance zone defined in chain link 1 and 2 is either increased or decreased.

Chain link no 5 - Measurement equipment requirements. The standards in link no 5 describe very specific measuring equipment, based on, and in harmony with the general requirements in link no 3 (but not always in full accordance with the link no 1). Definitions of actual characteristics of the feature or parameter. The most valuable task of this chain link is probably the harmonizing effect on the market of measuring equipment including integrated software.

Chain link no 6 - Calibration requirements - Calibration standards. The standards of link no 6 describe/define the calibration standards and the calibration procedures to be used, verifying the accuracy and functional requirements of the specific measuring equipment in chain link number 5, with traceability to the definition of e.g. the meter.

Consequences of chains of standards

The chosen order of the links of the chain of standards - in this paper - is an attempt to mention them in a logic order. The chain of standards can be seen from different angles, which each of them will require a special order. The order used reflects the situation: Using and "reading" of a finalised drawing. allowing the user to:

- Observe the codes on the drawing
- Use the theoretical tolerance definition and the tolerance values
- Translate these definitions and values and combine them with the definitions of the actual feature
- Assess the toleranced features of the workpiece accordingly - compare with the tolerance limits of the actual features
- Assessment is done by choosing/using measuring equipment which fulfils the actual requirements
- Calibration of the measuring equipment performed according to the actual calibration requirements and with the prescribed calibration standards

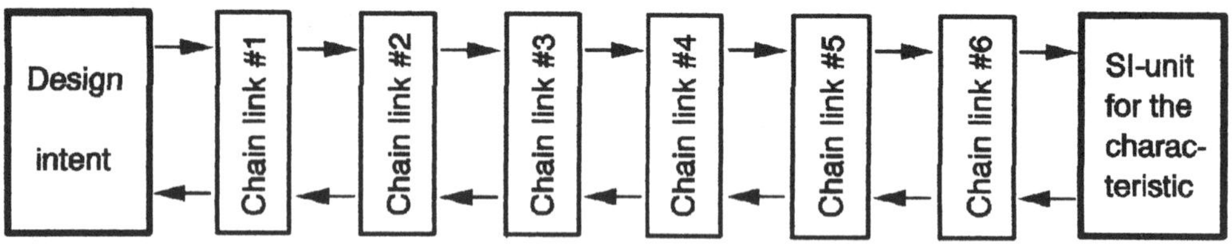

FIGURE 1

When, but only when, all six chain links are fully operational, i.e. standards exists in each of the six chain links, are updated and the necessary definitions are included, it is evident, that the chain of standards will guarantee that a drawing indication (design intent) will have one and only one unambiguous geometric solution in the workpiece. All arrows at figure 1 are present.

And vice versa, when a chain of standards is not complete i.e., standards in one or more chain-link are missing or are inadequate, the drawing indication is ambiguous. There will be more than one geometric solution for the workpiece. Boxes or arrows on figure 1 is missing. This may cause a lot of troubles, and it often cost a lot of money. As a result, quality assurance of the workpiece will be difficult - or in many cases impossible to handle.

THE CHAINS OF STANDARDS MATRIX MODEL

The various chains of standards each covering a specific geometric characteristic is only part of a complete structure - "The chain of standards matrix model". The complete structure consist of four different parts as indicated on figure 2:

- GPS chain of standards fundamental principles
- GPS global standards

The global GPS standards

GPS or related standards which deal with or influence several or all GPS chains of standards

The GPS chains of standards fundamental principles

I. The principle of independency
II. The principle of unambiguity
III. The principle of totality

General GPS chains of standards

1. The **Size** chain of standards
2. The **Distance** chain of standards
3. The **Radius** chain of standards
4. The **Angle** chain of standards (tolerance in degrees)
5. The **Form of a line** (independent of a datum) chain of standards
6. The **Form of a line** (dependent of a datum) chain of standards
7. The **Form of a surface** (independent of a datum) chain of standards
8. The **Form of a surface** (dependent of a datum) chain of standards
9. The **Orientation** chain of standards
10. The **Location** chain of standards
11. The **Circular run-out** chain of standards
12. The **Total run-out** Chain of standards
13. The **Datums** chain of standards
14. The **Surface roughness** chain of standards
15. The **Surface waviness** chain of standards
16. The **Surface defects** chain of standards
17. The **Edges** chain of standards

Special GPS chains of standards

A. The **Screw thread** chains of standards
B. The **Gears** chains of standards
C. The **Splines** chains of standards
D. The **Coating thickness** chains of standards

FIGURE 2

- General GPS chains of standards
- Special GPS chains of standards

GPS Chains of Standards Fundamental Principles

To make the individual standard, the chain of standards and the GPS-matrix work as a homogeneous system safeguarding unambiguous relations between the drawing and the workpiece and vice versa, a set of fundamental principles have to be formulated. These fundamental principles shall be included in a supervising standard or guideline valid for all GPS. The necessary fundamental principles are:

- The principle of independency
- The principle of unambiguity
- The principle of totality

The fundamental principles must in the future be used as the "constitution" to be obeyed when deciding on the contents and definitions in new standards and in revisions of existing standards.

The principle of independency. The principle of independency is already formulated and included -

but hidden - in a standard (ISO 8015 Technical drawings - Fundamental tolerancing principles) with a very special status. ISO 8015 shall be mentioned on the drawing by number to be in force. It is not quite clear to everybody - because of the special status - that the principle of independency is - and necessary must be - a universal principle.

The principle of independency reads:

Each specified dimensional or geometrical requirement on a drawing shall be met independently, unless a particular relationship is specified.

The principle of independency is absolutely necessary when several geometrical characteristics (chain of standards) are required for the same feature - and in practice this is always the case. The principle of independency **shall be defined more strictly and clearly** in the future and shall be in force for all chains of standards. It shall be deleted from ISO 8015, which defines only an alternative to the original fundamental gauging principle (Taylor principle/US principle no 1).

The principle of unambiguity. The principle of unambiguity - which is not yet included in any ISO standard - is the fundamentals of the chain of standards approach itself. It will safeguard an unambiguous relation between the drawing indication and the geometrical characteristic of the actual workpiece - and vice versa - independently of other geometrical deviations for an ideal condition.

If the principle of unambiguity is not in force in a chain of standards, the drawing indication and the chain of standards is rather useless and will definitely cause trouble for the user.

The principle of totality. Each of the chains of standards shall be complementary to each other. If not - the principle of independence will be violated or destroyed. The definitions in each of the chains **shall be formulated** in such a way **that no interference** occurs between the requirements of the various chains of standards.

The GPS system or the GPS-matrix shall offer to the user the possibility of indicating on the product specification all necessary geometrical characteristics. It shall be a total system.

When making new and expectedly rather narrow and restrictive definitions in each of the chains of standards, the user will be left in a very unhappy situation if the system is not a total one, where he can express all his necessary geometric requirements, derived from functional requirements of the part concerned.

That means: It will be necessary to investigate the needs of geometrical characteristics to fulfil all the user's requirements and, if required, formulate new chains of standards or sub chains in the already existing chains.

GPS global Standards

Not all GPS standards belong to only one single GPS chain of standards or fit into the chain of standards approach. Such standards have a **global** or special role in the GPS-matrix.

Some of these global GPS standards are of a type which influences the meaning of the indication for more than one chain, often several. It is a sort of grammatical addendum which can be used in special occasions or for special workpieces. Examples are ISO 10578 and 10579.

Other global GPS standards are of a type linking the tolerances of two chains of standards together - suspending often the principle of independency. An example is ISO 2692.

The last type of general GPS standards deal with or influence all GPS chains of standards, sometimes only one or two chain-links in all chains. They are not always really genuine GPS-standards, but are related standards. Examples are ISO 1, 10205, 10012-1 and a future standard defining conformance with specification taking into account the uncertainty of measurement and uncertainty of sampling.

General GPS Chains of Standards

One part of the GPS-matrix includes the (main) general chains of standards. General chains of standards deal with the general geometric characteristics, which are used to tolerance general features i.e. features which are not special ones (Special geometries like: Gears, splines, screw-threads, coating thickness, etc.).

At the moment, seventeen (17) general (main) chains of standards are identified (see figure 2). Most of the chains consist of several "sub chain of standards".

The number of general chains of standards is a matter of definition and discussion. Some or even all the sub-chains could be identified as (main) chains of standards on their own. The sub-chains - in several cases - have only the name in common with the other sub-chains under the present chain of standards. They often deal with geometrically and functionally very different features and geometry. Sub-chains often need separate definitions of actual feature or parameter. That is why it is beneficial and necessary to sub-divide into sub-chains. The gabs and needs for new definitions and standards become very clear.

In the following four examples of general chains of standards are dealt with in order to underline the urgent needs of improvement. For the full story see Bennich (1993).

Size chain of standards (Chain no 1) - see figure 3. This is one of the few general GPS chains of standards which is more or less complete. Anyhow improvements and a better coordination between the individual standards are necessary. Especially ISO 1938 needs modernization to reflect the gauge and measurement instrument technology of today.

<table>
<tr><th>Chain</th><th>Link #1</th><th>Link #2</th><th>Link #3</th><th colspan="2">Link #4</th><th>Link #5</th><th>Link #6</th></tr>
<tr><td rowspan="2">Size</td><td rowspan="2">ISO 129, 406, 286-1, 2768-1, 8015</td><td rowspan="2">ISO 286-1, 286-2, 1829, 2768-1 8015</td><td rowspan="2">ISO 1938, 8015</td><td>Limit gauges</td><td>ISO 1938 clause 3</td><td>ISO 1938 clause 3</td><td>ISO 3670</td></tr>
<tr><td>Indicating measuring instruments</td><td>ISO 1938 clause 4</td><td>ISO 463, 3599, 3611, 9121, 6906, 9493, 10360-1, 13365</td><td>ISO 3670, 7863</td></tr>
</table>

FIGURE 3

Distance chain of standards (Chain no 2) - see figure 4. As indicated it is necessary to distinguish between the case with two real features, and the case where one or both features are derived features. No definitions for actual features exist, and consequently any assessment will be by chance as intended by the designer. Measurement equipment is dependant of the DAF and needs standards too. This is one of several examples of indications used on nearly every engineering drawing, which is not defined in ISO (and national) standards. Derived features needs a global definition!

Sub chain	Link #1	Link #2	Link #3	Link #4	Link #5	Link #6
"Step" distance (height) between real features	ISO 129, 406, 2768-1	ISO 2768-1				
Distance between real or derived feature and derived feature	ISO 129, 406, 2768-1	ISO 2768-1				

FIGURE 4

Form of any line (independent of a datum) chain of standards (Chain no 5) - see figure 5. This example shows the need to subdivide chain no 5 into six sub-chains in order to come down to the missing definitions and standards. Figure 5 shows clearly the need for a global definition of derived features. What is not indicated is, that ISO 1101 is not dealing with the very practical problem of, which wavelengths are part of the form deviation. A closer definition of form in relation to waviness and roughness is clearly needed. This is a typical case, where more than one ISO/TC is involved: TC10 and

TC 57. Even TC3 is involved because of the responsibility for co-ordinate metrology and the related software.

Sub chain		Link #1	Link #2	Link #3	Link #4	Link #5	Link #6
Real feature	Profile any line	ISO 1101, 1660	ISO 1101, 1660	ISO 1101			
	Straightness	ISO 1101, 2768-2	ISO 1101, 2768-2	ISO 1101			
	Roundness	ISO 1101, 2768-2	ISO 1101, 2768-2, 6318	ISO 1101		ISO 4291, 4292	
Derived feature	Profile any line	ISO 1101, 1660	ISO 1101				
	Straightness	ISO 1101	ISO 1101				
	Roundness	ISO 1101	ISO 1101				

FIGURE 5

Location chain of standards (Chain no 10) - see figure 6. This example shows - once again! - the need for a definition of derived features. This is a very critical miss, while location most often are used on engineering drawings for derived features and relative seldom for real features.

Sub chain		Link #1	Link #2	Link #3	Link #4	Link #5	Link #6
Real feature	Position	ISO 1101	ISO 1101, 5458	ISO 1101			
Derived feature	Position	ISO 1101	ISO 1101, 5458				
	Coaxiality	ISO 1101	ISO 1101, 5458				
	Concentricity	ISO 1101	ISO 1101, 5458				
	Symmetry	ISO 1101	ISO 1101, 5458				

FIGURE 6

The four examples chosen out of seventeen general GPS chains of standards indicates very well the total situation. There is a lot of general gaps to be filled in.

Special GPS Chains of standards

The other part of the GPS-matrix includes the special chains of standards (see figure 2). Special chains of standards deal with tolerancing of special geometric features. These special geometrical features are the responsibility of other TC's than the three responsible for the GPS general chains of standards (ISO/TC 3, 10 and 57). The special chains identified at the moment are the:

A. **Screw thread** chains of standards. It is in reality at least 5-6 individual chains of standards (ISO-M, ISO-S, ISO-Inch, ISO-R, ISO-G, ISO-TR, etc.).
B. **Gears** chains of standards. It is in reality several chains of standards each of them dealing with the different types of gears.
C. **Splines** chains of standards. At least two individual chains (straightsided and involute splines)
D. **Coating thickness** chains of standards. It is in reality several different types of coating thickness each with its own chain of standards.

Also in this case coordination problems exist. Standards for thread are produced by more than three ISO/TC's (and national committees) using different tolerancing and measuring principles (general purpose-, tube-, gas-thread, etc.). Most of the thread chains of standards are incomplete. This list of items is definitely not complete.

GAPS AND CONTRADICTION IN THE GPS MATRIX MODEL

As already demonstrated the existing ISO (and national!) GPS chains of standards has a lot of gaps. The seventeen GPS general chains of standards can be expanded to 49 sub-chains each with six chain links. This makes a 49 x 6 matrix with 294 cells. For the moment only 48% of these cells are covered by ISO standards!

Looking at the 49 sub-chains the situation looks even worth. Only 2 - 3 of these chains of standards are complete. And in all the 2 - 3 chains complete, there are several "key" standards, which urgently needs updating.

Gaps

The geometry and the GPSs indicated on an engineering drawing are not the final and intended purpose of the designer. The real purpose of the drawing specifications is to safeguard the **function** of the part or the product. All GPS's on the drawing are the designers' "translations" of the functional requirements - using more or less known relations and company experience - into Geometrical Product Specifications.

The GPSs are thus derived specifications and not primary ones. This must be taken into account when defining all sorts of definitions related to chains of standards. Especially the definitions of:

- Geometrical characters
- Tolerance definitions and their values
- Actual features (DAFs)
- Assessment of deviations

shall primarily reflect the **functional requirements** of the workpiece or product and only secondly other technical needs related to e.g. production processes, assessment and measuring procedures or even convenience.

The problem of letting the functional requirements control the definitions of GPS is, that there is not only one single functional requirement related to one single geometric characteristic or combination of geometrical characteristics. Therefore one single set of definitions for actual features in a chain of standards is not sufficient in all cases.

At the moment there is at the most one single set of definitions related to a chain of standards - reflecting only one single **chosen** functional requirement. In most cases the necessary number of detailed definitions are not even there. In many cases, there are no definitions at all! It could be foreseen that it will be necessary - in the future - to develop the "GPS-language" and to add the necessary definitions especially on the "grammatical" side, to make it possible to indicate on the drawing more distinct and varied definitions for the same geometrical characteristic, to reflect more and not only one single requirement.

Contradictions

Even contradictions exist between the content, the rules and the definitions given in the existing ISO (and national) standards, mostly caused by the fact, that the coordination between the three ISO/TCs involved have not been of a formal and systematic nature. Many chains of standards are the responsibility of two and in few cases three TCs.

UNCERTAINTY OF MEASUREMENT - CONFORMANCE/NON CONFORMANCE WITH SPECIFICATION

At the moment only two of the 49 general GPS sub-chains mentions the "problem" uncertainty of measurement. This is in case of size and surface roughness measurements. By experience uncertainty of measurement is the no 1 source to make difficulties in the relations between contractor and subcontractor. There is an urgent need to define the terms "conformance with specification" and "non conformance with specification", taking into account the uncertainty of measurement and the limits of tolerance defined in the chain links 1 through 4.

A draft proposal for a standard concerning this matters will be drafted as a new work item proposal in ISO/TC3 in a couple of months. It is the hope that this standard could serve as a global standard for the GPS structure as a whole. The estimation of uncertainty of measurement in this document is based on the principles drafted by ISO/TAG4/WG3.

QUALITY ASSURANCE - MEASUREMENT AND CALIBRATION SYSTEM

Specifications of workpieces and products, measurements and uncertainty of measurement are closely connected in the documentation and operation of a certified ISO 9000 quality assurance system. By experience the measurement equipment is very specific and linked to only a few different sub-chains in the GPS system.

By experience the sub-chains, and even more subdivided than the 49 chains indicated, is a valid tool when a company's "measurement and calibration system" is planned and evaluated in the ISO 9000 certification process. It is on this basis possible to make a direct link between the drawings and the derived needs one side with the measurement equipment and the capacity and capability on the other side.

CONCLUSION

By using the chain of standards concept it is demonstrated, that the existing ISO (and consequently national) standards needs completion and improvement. It is at the moment not possible - in most cases - to make an unambiguous drawing indication.

This statement is not in accordance with the daily life experience in industry. In many cases there are no observations of problems like these.

The reason could be one of or a combination of the following: The tolerances on the drawings are not correct representations of the functional requirements, the tolerances used are too tight, the drawing is not the only requirements and information given to the producer, common work shop practice (craftsmanship) is adding to the drawing requirements.

By experience the gaps in the existing standards will result in an increasing amount of problems in the future.

REFERENCES

Bennich, P., ISO/TC 3 N 516. "Introduction of a new concept related to engineering drawing specification standards": "Chains of standards". April 1992. 4 pages.

Bennich, P., "Gaps and Contradictions in ISO Geometrical Product Specifications (GPS) Standards and ISO GPS chains of standards", 1. edition - February 1993. ISO/JHG (Joint Harmonization Group - ISO/TC 3, ISO/TC 10 and ISO/TC 57) N 3. 70 pages.

Bennich, P., "Gaps and Contradictions in ISO Geometrical Product Specifications (GPS) Standards and ISO GPS chains of standards", 3. edition - February 1993. CEN/TC 290/WG 1 N 3 - Revised, 3rd Edition. 72 pages.

ISO/TAG 4/WG3: Draft January 1993. Guide to the expression of uncertainty in measurement. 108 pages.

ISO/TC3/TG1: Proposal June 1993. Inspection by measurement of workpieces and measuring instruments - Decision rules for proving conformance or non conformance with specification. 10 pages.

CRTD-Vol. 27, International Forum
on Dimensional Tolerancing and Metrology
ASME 1993

A PERSPECTIVE ON ENGINEERING DESIGN EDUCATION: WHERE DO TOLERANCES FIT IN?

Gary A. Gabriele
Department of Mechanical Engineering,
Aeronautical Engineering and Mechanics
Rensselaer Polytechnic Institute
Troy, New York

ABSTRACT

The mechanical engineering curriculum at Rensselaer, along with many others, under went a significant evolution between 1950 and 1980 as the curriculum shifted from practice oriented course work to more theoretical, science based courses. As might be expected, much of this change is the result of the emergence of new technologies and engineering methods but also in response to perceived inadequacies in our engineering graduates. The area of engineering design was one of the bigger casualties in this evolution and long with it basic design engineering skills such as dimensioning and tolerancing.

This paper looks at this evolution and at the more recent changes in the mechanical engineering curriculum that have come about in response to new perceived needs about the inability of engineering graduates to produce quality designs. We will look at a proposed set of engineering design fundamentals and assess how well the current curriculum addresses them. We will also look at how the topics of dimensioning and tolerancing fit into these fundamentals and how well they are addressed. We conclude that while the current curriculum is moving in the right direction in addressing engineering design, the specific topics of dimensioning and tolerancing are usually only superficially addressed.

INTRODUCTION

The emergence of new technologies and the maturing of engineering sciences places a heavy demand on the modern engineering curriculum. The engineering curriculumof forty years ago had a heavy emphasis on the practical aspects of engineering. Students of that time would have encountered many 'hands-on' courses where they were required to be familiar with the machinery and practices being used in the field. Among these were the design courses, which often included a heavy dose of technical drawing. It was in technical drawing as well as the manufacturing process courses that the principles of dimensioning and tolerancing were often encountered.

In more recent times, the emphasis in engineering education has seen these 'hands-on' courses yield to the more theoretical courses of engineering science. Design courses, being included in this group of 'hands-on' courses, began to fade from the curriculum. The technical drawing courses were a casualty in this evolution particularly when CAD software began to make an impact in the marketplace. The growing acceptance of CAD as a means for producing design documentation seem to signal to some engineering educators that there was longer any need to teach technical drawing because the computer was going to automate it.

Fortunately, the teaching of design has become 'fashionable' again in recent years due in part to the accreditation boards and the recognition by industry of the important role that design plays in the competitiveness of a company. As we begin to struggle with restoring design teaching to the engineering curriculum we must address ourselves to the issue of what are the fundamentals of design and how do we teach them.

This paper will describe the current undergraduate mechanical engineering curriculum at Rensselaer as an example of a modern curriculum. We will look briefly at its evolution over the last 40 years and assess where it has been and where it is going. We will concentrate on the design and manufacturing portion of the curriculum and assess its content against a proposed set of engineering design education fundamentals. From this examination we will attempt to identify where these fundamentals are taught, and the problems associated with teaching them.

THE MECHANICAL ENGINEERING CURRICULUM AT RENSSELEAR

It is widely agreed that the principle purpose of the undergraduate engineering education is to provide students with an understanding of the fundamentals of engineering. To some extent these fundamentals are outlined by the criteria set forth by the Accreditation Board for Engineering and Technology (ABET) which reviews and accredits undergraduate engineering curriculums. In general, the engineering topics within the curriculum fall into two categories, engineering science and engineering design. The engineering sciences have their roots in mathematics and basic sciences but carry knowledge further toward practical applications. An examination of almost any accredited mechanical engineering program would find the following required engineering science related courses: statics and dynamics (mechanics), strength of materials, thermodynamics (introductory and advanced), fluid mechanics and heat transfer.

We define engineering design as the application of scientific knowledge to the solution of technical problems within given material, technological and economic constraints. An examination of typical mechanical engineering curriculums for engineering design related courses would find less consensus about what to teach than for the engineering sciences. The two most commonly found courses would be a machine elements course and a senior level 'capstone' course, the latter due to recent changes in the ABET criteria.

In general, the engineering schools have done a far better job of education in the engineering sciences than in engineering design. Part of the reason can be attributed to the long held belief that engineering design is more art than science and therefore difficult to teach. Another reason maybe the increased emphasis on improving engineering science education that was begun in the 1950's in response to perceived national needs in the defense and space programs.

It is interesting to see how the engineering curriculum has evolved over the past 40 years in response to these perceived needs for engineering undergraduates. Tables 1 - 3 list the mechanical engineering curriculum at Rensselaer for 1950, 1980, and 1993 respectively. Looking to 1950 (Table 1) which is before the emphasis on engineering science had taken hold, we can identify 8 courses that have a direct relationship to design and manufacturing. We also see that the courses are rather evenly distributed throughout the four years. As mentioned earlier, the course descriptions also indicate that these courses tended to be more practice oriented.

Moving ahead to 1980 (Table 2), we see what may have been the low point for of engineering design and manufacturing related courses in the curriculum. Gone are the courses in engineering graphics and drawing, and the number of required courses in design and manufacturing has shrunken to 2, both of which occur in the senior year. It is evident here that teaching engineering design had very low priority, and in fact many of those qualified to teach these topics were having a hard time justifying themselves in promotion and tenure reviews. It is also clear to see why industry was beginning to complain that they were getting good analysts in their new hires but not people who understood how to design. In the early 1980's ABET began to emphasize in the accreditation criteria the need for design related courses, in particular the 'capstone' design experience. This was meant to be a senior level course that would required students to use the engineering background acquired during the previous three years in a significant design experience.

The 1993 curriculum (Table 3) shows the effect of the ABET changes and the increased awareness by

TABLE 1 MECHANICAL ENGINEERING CURRICULUM OF 1950

First Year		Second Year	
Chemistry	Chemistry II	English II	Economic and Social Principles
Engineering Drawing	Historical Background of the Modern World	Calculus III	Engineering Math
English I	Descriptive Geometry	Shop Processes	Metal Casting and Forming
Introduction to Engineering and Science	Calculus II	Physical Training III	Physical Training IV
Calculus I	Physical Training II	Physics III (Electricity and Magnetism)	Physics IV (Acoustics, Optics, and Electronics)
Physical Training I and Hygiene	Physics II (Mechanics and Heat)	Engineering Mechanics	Dynamics of Liquids
Physics I (Mechanics)			
Third Year		**Fourth Year**	
Heat Engineering	Electrical Engineering II	Heat Transfer	Engineering Thermodynamics
Engineering Metallurgy	Applied Composition	Machine Design I	Machine Design II
Economics and Social Problems	Heat Engineering II	Industrial Engineering	Manufacturing Engineering
Electrical Engineering I	Strength of Materials	Technical Elective(2)	Technical Elective (2)
Non-technical elective	Non-technical Elective		

TABLE 2 MECHANICAL ENGINEERING CURRICULUM OF 1980

First Year		Second Year	
Computing Fundamentals I	Computing Fundamentals II	Mathematics III	Intro. to Diff. Eqns.
Mathematics I	Mathematics II	Physics III	Thermodynamics I
Chemistry I	Physics II	Strength of Materials	Engineering I
Physics I	Statics	Dynamics	Mechanics II
H&SS	H&SS	H&SS	Engineering Seminar.
Phys. Ed. I	Phys. Ed. II	Phys. Ed. III	H&SS
Third Year		**Fourth Year**	
Engineering Lab I	Engineering Lab II	Engineering Economics	Feedback Systems
Lumped Parameter Systems	Intro to Heat Transfer	Elements of Mech. Des.	Experimental Methods
Thermodynamics II	Materials I	Dynamics Elective	Design Elective
Fluid Mechanics I	Electives (2)	Technical Elective	Technical Elective
Elective	H&SS	H&SS	H&SS
H&SS			

industry and academics of the role engineering design plays in the national competitiveness. The pendulum of change has begun to swing back toward a more balanced approach between engineering science and engineering design. Engineering graphics and drawing has returned, albeit using the new technology of solid modeling, and there are now 4 required design and manufacturing related courses in the curriculum.

Although the above review points to some gains made by design and manufacturing related topics in the curriculum, there still remains the question of what should be taught. Recognizing that this is an undergraduate curriculum, what are the fundamentals of engineering design that should be covered.

DESIGN EDUCATION FUNDAMENTALS

Although there seems to be a consensus about the set of topics which cover the fundamentals of engineering science, a similar consensus of engineering fundamentals for teaching engineering design does not exist. This was pointed out by Dixon (1991a) in his assessment of the current state of engineering design education. In that paper, and his follow-on (Dixon, 1991b), he argues for the need for research to establish a 'science' of engineering design. In his follow-up paper, Dixon outlines a set of proposed fundamentals of

TABLE 3 Mechanical Engineering Curriculum of 1993

First Year		Second Year	
Intro to Engineering Analysis	Engineering Processes	Intro to Engineering Design	Engineering Mechanics
Engineering Graphics and CAD	Chemistry of Materials II	Intro to Differential Equations	Engineering Thermodynamics
Chemistry of Materials I	Mathematics II	Computer Science I	Statics and Strength of Materials
Mathematics I	Physics I	Physics II	Physics III
H&SS	Engineering Seminar	H&SS	H&SS
Phys. Ed.	H&SS	Phys. Ed.	
	Phys. Ed.		
Third Year		**Fourth Year**	
Lumped Parameter Systems	Lab. Intro to Embedded Control	Intro to Automatic Control	Engineering Economics
Fluid Mechanics I	Intro to Heat Transfer	Machine Dynamics	Technical Electives(2)
Modeling and Analysis of Uncertainty	Electronics and Instrumentation	Design of Mechanical Systems	Elective
Applications of Thermodynamics	Elements of Mechanical Design	Experimental Methods Elective	H&SS
Elective	Elective	H*SS	
H&SS	H&SS		

engineering design science derived from current best industry practices and research results. These are listed in Table 4

TABLE 4 Proposed Fundamentals of Engineering Design Science (Dixon, 1991b).

1.The Business Context of Engineering Design. Topics: Product realization process and its relation to other business functions, competitive benchmarking processes, quality cost and time to market.

2.Concurrent Engineering and Team Participation Principles. Topics: Concurrent design process concepts, practice effectively on a cross cultural team, present effective reports.

3.Manufacturing. Topics: Manufacturing processes and their economics, practices and tolerances, knowledge of DFM/A and SPC.

4.Analysis and Prototyping. Topics: Computational prototyping methods, analytical modeling processes, development of idealized design models, rapid physical prototyping methods, common process simulations.

5.Statistics. Topics: probability, statistics, decision theory, and design of experiments.

6.Systems. Topics: Systems design and integration principles, conceptual, configuration, and parametric design of systems that integrate mechanical, electronic, optical, and computer elements.

7.Computer-Aided Design. Topics: CAD, CIM and solid modeling systems.

8.Design Theory and Methodology. Topics: Descriptive, prescriptive, and computer-based models of design processes, Pahl and Beitz methodology.

9.Component Design and Optimization. Topics: Design, evaluate, and redesign mechanical components, optimization formulations and methods.

10.Assembly Design and Tolerances. Topics: design, evaluate and redesign mechanical assemblies with respect to manufacturing, cost, tolerances, computer assembly modeler, 1-D tolerance analysis.

11.New Information and Learning. Topics: Finding new information, research methods, self-learning.

Both these papers have excited considerable debate among engineering educators about the merits of establishing a science of engineering design. A recent ASME conference on engineering design education (Fisher, et. al., 1993) found no consensus on the issue of an engineering design science or on the issue of the usefulness of teaching design methodologies.

However, putting aside the debate on the merits of establishing an engineering design science, I believe that the list of fundamentals provided by Dixon can serve as an metric for assessing the design content of a curriculum. The fundamentals are derived from current best practices in industry and we can expect that a student versed in these fundamentals should be able to contribute to any modern product realization effort in industry.

In the following section we will apply this list to the current mechanical engineering curriculum at Rensselaer as a representative modern curriculum.

UNDERGRADUATE DESIGN SEQUENCE AT RENSSELAER

The four year curriculum outlined in Table 3 was begun as a pilot program during the 87/88 academic year and was adopted as the official curriculum in the 1992/1993 academic year. Although there were changes along several fronts, we will concentrate on those courses that have a close relationship to the design and manufacturing area and address the fundamentals outlined in Table 4.

The courses described below that occur during the freshman and sophomore years are part of the Pre-engineering curriculum at Rensselaer. All engineering students, regardless of discipline, take these courses. A short description of each course is provided and as well as a brief discussion of the fundamentals addressed. Table 5 provides a reference map between the courses described and the fundamentals listed in Table 4.

TABLE 5 Coverage of Engineering Design Fundamentals

Course	1	2	3	4	5	6	7	8	9	10	11
EG/CAD							4				
ENGP			4								
IED		4					4	4			4
EMD				4					4		
DMS	4	4	4	4				4			4
DD/CAD							4			4	
AML		4	4							4	

Freshman Year

Engineering Graphics and CAD. This course provides an introduction to engineering drawing and CAD using the solid modeling package, ProEngineer. The objectives of the course are to: (1) enable students to produce freehand sketches, (2) develop skills in the visualization of physical objects when they are represented as engineering drawings, and (3) enable the student to develop solid models and engineering drawings of objects and assemblies.

The topics covered in this course include freehand sketching, orthographic projection, auxiliary views, visualization, introduction to solid modeling and computer aided design.

This course addresses some of the basic needs outlined in Fundamental 7. Although CIM is not covered, students do become versed in the basics of CAD and solid modeling

Engineering Processes. This course covers the use of basic machine tools such as lathes, milling machines, drill presses, band saws, and grinders, micrometers, vernier calipers, and other devices of use in a machine shop or laboratory. Some welding and tool making are sometimes covered if time permits. The students must actually fabricate and build an assembly (typically a toy cannon) based on an engineering drawing provided to them during the course.

This course, much like the CAD course, provides some basic manufacturing process instruction that can be built on in later courses. It contributes to the goals of Fundamental 3 by covering basic processes and their practices, assembly, and some first hand experience with tolerances.

Sophomore Year

Introduction to Engineering Design. This course introduces the engineering design process through lecture and experiential learning. Emphasis on the design process provides a structure in which the various phases of design can occur in a logical and efficient order so as to achieve the most successful outcome. An important element of this course is the design-build-test experience which lets the students practice design skills and experience a good portion of the design process. Students are required to work in teams in completing the design project and develop a design prototype that is tested against the requirements at the end of the semester. The students must also prepare a written report with supporting engineering drawings and provide an oral presentation of the design. A recent design project was an automatic device to assist a disabled person to empty the contents of a juice box into a cup.

This course contributes to four of the fundamentals. Because the students are formed into teams and the course includes students from all the engineering disciplines (except chemical engineering) the teams tend are multidisciplinary (Fundamental 1). The students also use the CAD system to develop their drawings (Fundamental 7) which further extends the background knowledge obtained in the freshman course. The students are also taught a design process, which draws heavily from Pahl and Beitz (1984) and from Cross' book (1989) which contributes to Fundamental 8. Finally, because an important part of their design work involves investigating technologies that have not been taught yet, they are given library and literature research assignments to find information needed to complete the design (Fundamental 11).

Junior Year

Elements of Mechanical Design. This course is our version of the commonly found, 'machine elements' course. This course applies the basic concepts of classical solid mechanics to the design of mechanical components including shafts, fasteners, springs, bearings, gears, belts, chains, brakes and couplings. In the past, this course was principally a lecture course which taught the techniques and theories of analysis for these components then described their design. For the past 5 years, we have replaced the last third of the course with an open ended design problem that provides the students with an opportunity to design a mechanical assembly using the theories and methods taught in the course.

This course contributes principally to Fundamental 9 in the component design area. Optimization is not specifically covered here but is covered in an elective course. The students also get a chance to use their solid modeling knowledge in the development and documentation of their design. The solid modeling system allows them to computationally model their design which is one of the topics of Fundamental 4 but FEA is not done. Again, the more advanced computational methods of analysis are covered in elective courses.

Senior Year

Design of Mechanical Systems. This course represents our 'capstone' course and is a continuation of the *Introduction to Engineering Design* course in that it also requires the students to complete the steps of the design process. The major difference in this course is the scope of the problem and the analysis required is usually more sophisticated than the work done at the sophomore level.

The students work in teams of 5-7 students (Fundamental 2), facilitated by the faculty, and develop a paper design by the end of the semester. The students employ a design process (Fundamental 8) of requirements evaluation, concept generation and selection, analysis and final embodiment of the design. The lectures in the course cover the use of advanced design methods, such as QFD and Pugh Concept Selection Method, and advanced design related topics such as design for manufacture (Fundamental 3) and design for safety.

Because of the nature of the projects, the construction of a working prototype is often not possible but students are encouraged to make extensive use of computer and physical models (Fundamental 4) in the development of their ideas. Two oral and written reports are required as part of the course (Fundamental 1).

The semester is broken into two halves, the first being a reverse engineering project where the students must completely reverse engineer an existing product (Fundamental 11) to determine its cost, design requirements, underlying analysis, materials, manufacturing methods and weaknesses. The second half of the semester is spent redesigning the product to a modified set of requirements.

The product realization process is a main focus of the early lectures and students must cost their designs taking into account basic business factors such as labor rates, overhead and profit (Fundamental 1). The students also work in teams but the teams are not truly cross functional in that all the students are mechanical engineering students. However, because of the scope of the projects the students tend to specialize in the different areas that need to be addressed by the team (manufacturing, materials, etc.)

Related Electives

There are several design and manufacturing related electives available to the students as listed in the bottom of Table 3. I will single out two of the courses that have a strong relationship to the fundamentals given in Table 4 and to the later topic of dimensioning and tolerancing.

Detail Design w/CAD. The purpose of this course is to provide the students with the opportunity to experience the detailed design phase of the design process and learn how to effectively employ CAD to complete detailed design tasks. The course covers the phases of modern design practice and provides a further in-depth experience with solid modeling (again using ProEngineer as the CAD tool). The course also provides the students with an operational proficiency to identify critical features of a design, i.e. to relate design functions to features, and set their appropriate specifications, i.e. tolerances. The course is divided into two main sections, an overview of CAD and solid modeling, and the use of CAD to detail a design. Students are given an existing product, such as a common stapler, which they must reverse engineer and analyze the critical functions. The product is then modeled using ProEngineer and a complete set of detailed design drawings of the product is created.

This course has a strong relationship to the topics covered in Fundamentals 7 and 10. The first half of the course provides an introduction and overview to solid modeling and different CAD representations of objects. The second half of the course covers the assembly and tolerance issues under Fundamental 10. We will discuss this aspect of the course in more detail later in the paper.

Advanced Manufacturing Lab I, II (AML I,II).This two course sequence is an interdisciplinary laboratory whose purpose is to expose undergraduates majoring in mechanical, electrical, and industrial engineering to five areas of manufacturing technology. These areas are: robotics, computerized numerical control, metalworking, stamping, heat treating, plastic injection molding, and non-destructive testing. The first semester (AML I) concentrates on hands-on experiences in the lab with the technologies mentioned above. The students learn the procedures and the methods for operating the equipment involved and gain an appreciation for what can be accomplished with them.

An additional component of the first semester is the completion of a written project proposal for the manufacture of small products. The second semester (AML II) concentrates on the actual implementation of the design and manufacturing plan developed in the first semester. Each student team is given a budget and must complete the design and manufacture of the product including appropriate molds, dies and fixtures. The completion of the project is marked by a manufacturing run of 100-200 copies of their design.

Because the students work in teams which are made up of students from several engineering disciplines, the students get some exposure to team participation principles (Fundamental 2). But the principle focus of the course is the manufacturing processes which has a direct relationship to Fundamental 3. Because they must actually manufacture and assemble their design, aspects of Fundamental 10 get addressed as well.

Assessment

It would appear from Table 5 that Fundamentals 5 and 6 are not covered. However, these are covered by other courses not described here. Fundamental 5, Statistics, is covered in a required course, *Modeling and Analysis of Uncertainty* (third year in Table 3).

Fundamental 6, Systems design, and in particular the integration of mechanical, electronic and computers, is being addressed by the elective course, *Mechatronics*. This is a relatively new course in our curriculum which is an adaptation of an older course entitled, *Computer Aided Experimentation*. The new course has both a lab element and a design element to the course and has become quite popular with both undergraduate and master's level students. A more detailed description is given in Craig (1992).

Although Fundamental 1, the Business Context of Engineering Design, is briefly covered as a topic in DMS there is a required course, *Engineering Economics* (fourth year in Table 3), which does cover the economics aspects of business in more detail. This is one areas of the curriculum that is being worked on with the Lally Management school at Rensselaer. They are in the process of developing interactive, multimedia case studies on the relationship between design and other business functions for their curriculum which we are considering using with our students.

In general, we are addressing many of the topics that are outlined in Table 4 but there is certainly room for improvement. One improvement would be to include business students into the design teams to provide a more truly interdisciplinary team experience. Some schools are doing this at the graduate level but it is a bit more difficult to do at the undergraduate level due primarily to the larger number of student and the lack of flexibility in undergraduate programs.

WHERE DO DIMENSIONING AND TOLERANCES FIT IN?

Having addressed the issues of design education fundamentals we move to the question of dimensioning and tolerancing in the curriculum. A closer scrutiny of the design education fundamentals in Table 3 indicates that dimensioning and tolerances are an important part of two of the fundamentals, manufacturing (Fundamental 3) and assembly design and tolerances (Fundamental 10). As Table 5 indicates, these fundamentals get some coverage in the current curriculum. In this section we will provide a bit more detail on the dimensioning and tolerance related topics covered in those courses.

As we did before let's consider what should be the fundamental of dimensioning and tolerancing that should be addressed by the undergraduate program. Not claiming to be an expert on teaching dimensioning and tolerancing principles, I will however postulate that there are three levels of proficiency that need to be acquired by design students: the syntax of dimensions and tolerances, the semantics, and the use of tolerances to create a quality design.

Understanding the syntax of dimensioning and tolerances enables a student read a drawing and extract some first level information from a design. For instance, can a student find the overall dimensions of a part, or where is a certain feature located will depend on some basic syntactic skills. Much the same way a traveler in a foreign country, who has acquired a few words of the language can perhaps find his/her way around a city.

Semantic understanding allows a student to look at a drawing and example part data to determine whether the part satisfies the required dimensions and tolerances specified. There is that additional level of understanding that allows the student to actually interpret the drawing and extract the designers intent.

The final level of understanding is the ability to use dimensioning and tolerancing standards to express design intent that leads to a quality design. At this level of understanding we expect the students to be able to relate dimensions and their tolerances to important design functions in the design. This will require the ability to analyze assemblies of parts and trace the tolerance chain through the assembly.

As we saw in reviewing past curriculums, the students received explicit instruction on the syntax and semantics of dimensions on drawings. Additionally, though the manufacturing courses and some of the other practice oriented courses, there was the opportunity to experience the relationship between design and manufacturing.

Some of the courses in the current curriculum described earlier provide similar instruction and experiences that students of the older curriculum experienced. For each of these I will provide some examples of the work or related topics covered.

Engineering Graphics and CAD

This course, which is the basic drawing and visualization course for all undergraduates, provides the first formal introduction to dimensioning for the students. This portion of the course is one of five units presented in the course, the principle objective being to introduce the students to accepted practices for dimensioning piece parts. The topics covered include identifying size and location dimensions, the use of datums, dimensioning holes, arcs and angled surfaces. Following a short lecture and interactive learning session on dimensioning, the students complete an assignment that requires them to dimension a drawing containing the elements given above. These assignments are done first as hand exercises which are then followed by computer exercises where the students dimension parts modeled in previous exercises. It should be noted that the emphasis here is on dimensioning not on tolerances.An example dimensioning assignment using ProEngineer is shown in Figure 1.

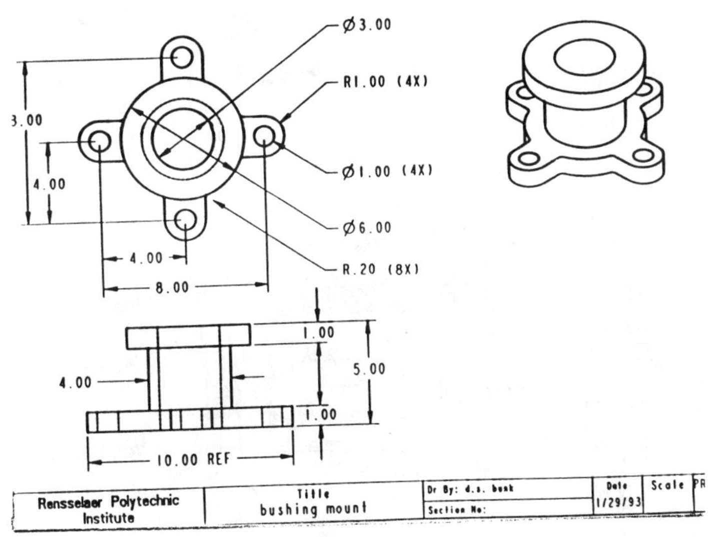

FIG. 1 EXAMPLE DIMENSION ASSIGNMENT FROM ENGINEERING GRAPHICS

Introduction to Engineering Design

As mentioned previously, this is the first course on design given to all undergraduates. Although the focus of the course is learning the design process, the documentation requirements for the course design project provide an opportunity for the students to practice and expand their dimensioning knowledge. They are also provided with the opportunity to develop skills in modeling assemblies since the design projects lend themselves to this.

An introduction to GDT is also provided within the lecture portion of the course. The topics covered include: Form tolerances, basic GDT syntax, datums, the 3-2-1 rule, and MMC/LMC. Although the material is introductory, the students are encouraged to try to employ these ideas in the engineering drawings of their design. One reason the material is kept at an introductory level is that the students in this course come from many different engineering disciplines. Extensive instruction in GDT is not as relevant for some of the other disciplines, such as civil and electrical, as it is for mechanical engineering. Students coming out of this course should be at the basic semantic level, that is they should be able to read dimensions and GDT symbols off a drawing. No instruction is given on tolerance analysis.

Design of Mechanical Systems

As the capstone design course in our curriculum, we have the opportunity to provide additional instruction on topics not sufficiently covered in other areas course. With respect to dimensioning and tolerancing, additional handouts and assignments are provided to review previous material on dimensioning and GDT. The coverage of the material is focused more toward the semantic level. As in the *Introduction to Engineering Design* course, the preparation of engineering drawings and assembly drawings provides an opportunity for the students to further practice and expand their knowledge of dimensioning and tolerancing but no formal tolerance analysis instruction is provided.

Detail Design w/ CAD

One of the main features of this elective course is it's emphasis on instruction in tolerances, and in particular, the relationship between design function and specifying tolerances. Half the course lectures and exercises deal with such topics as tolerance chains, functional surfaces, assembly modeling, functional groups

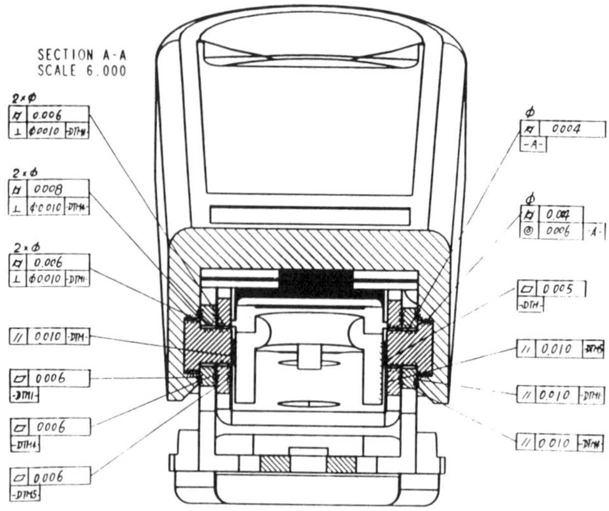

FIG. 2 FUNCTIONAL SURFACES OF A STAPLER

of an assembly, geometric dimensioning, functional dimensioning, form tolerances, datums, and location tolerances. The major exercise of the course, modeling an existing product, provides an opportunity for the students to use the CAD tool to create a set of detailed design drawings. These drawings generally contain much more detailed dimensioning and tolerancing then found in the other required design courses. But more importantly, the students are required to analyze their product and justify their tolerance choices using the methods taught in the course. The students are required to justify their choices during a 15 minute presentation to the course instructor. Figure 2 show sample student defining the functional surfaces at the back pivot of a staples. An example of a 1-D tolerance analysis on a channel of a stapler is given in Figure 2. A detailed part drawing of the base-die of the stapler is given in Figure 4.

The course uses the book by Krulikowski (1991) as a required text in the course and makes extensive use of ProEngineer assembly modeling capabilities. An interesting aspect of this course is that the students are required to create a portfolio of their work which is presented during the last week of classes to which job recruiters are invited.

Advanced Manufacturing Lab I,II

As mentioned above, the AML course sequence provides a hands-on manufacturing experience for undergraduate students. Specific instructio on tolerances relative to the manufacturing processes taught in the course is provided. The students use this information in the design of the product and its tooling giving them first hand experience in the interaction between manufacturing and design which often cross over the issue of tolerances. Most student projects make use of automation in their assembly process and they find out first had the need to understand tolerances and the effect improperly defined tolerances can have on a manufacturing system.

We see from the brief descriptions given above that the instruction the students received in dimensioning and tolerancing progresses through the syntactic and semantic elements in the required courses. In the elective courses, they reach the higher level of understanding of how tolerances can impact the quality of a product. Given that mechanical engineering is a multidisciplinary curriculum by it very nature, I believe that this is the proper approach. All students achieve some basic proficiency, and those more interested in design and manufacturing have the opportunity to acquire more advanced skills.

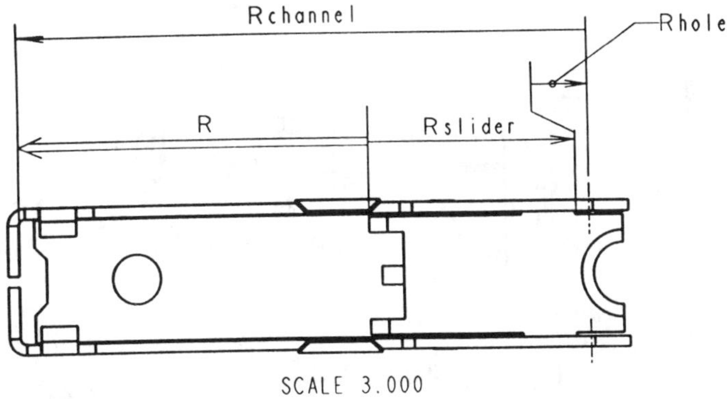

$R = Rchannel - Rslider - Rhole$

$R_{min} = Rchannel_{min} - Rslider_{max} - Rhole_{max} = 1.85 - 0.675 - 0.051 = 1.124 > L\ staples$

Rmin must meet a minimum value (the length of the staples) in order to allow the staples to fit in.

FIG. 3 EXAMPLE 1-D TOLERANCE ANALYSIS FROM DETAILED DESIGN

CONCLUSIONS

We have seen in this paper that the Rensselaer mechanical engineering curriculum has seen a definite shift in emphasis away from design and manufacturing related topics to the engineering sciences. The curriculum of the 1950's was a more hands-on, practice oriented curriculum than the one being offered in the 1980's. We are now seeing a shift back toward a more balanced approach and more design and manufacturing related courses are starting to appear in the curriculum of the 1990's.

Dimensioning and tolerancing was one of those topics that was swept out with the changes from 1950's to 1980's that is also making a comeback in the 1990's. We have shown that it is an important part of the basic mechanical engineering curriculum and that there are opportunities to introduce students to the basic syntactic and semantic issues in the required courses. More proficiency can be gained in elective courses for students with a strong interest in design and manufacture.

The current level of teaching dimensioning and tolerancing, as well as other design related topics, in mechanical engineering curriculums around the country is uneven. A contributing factor is the background and training of the typical faculty member has also changed in the past 40 years. Today's faculty is by and large a product of the 1980's curriculum that did not stress design and manufacturing. Additionally, we have many design faculty who have very little or no industry design experience. Add to this the current emphasis by academic administrations on performing research, and we find ourselves faced with a situation where there are very few qualified faculty throughout the country that can teach topics in these areas.

Clearly some help is needed in personnel, teaching materials, and tools.

Some automated tools do exist that could help here (ACS's VSA comes to mind) but there is very little room within the curriculum for adding additional tool instruction. We have been able to make effective use of ProEngineer because it is becoming tightly integrated into the curriculum. The students learn it early in the curriculum and have an opportunity to continue to use it throughout the 4 years. Adding built-in functions for tolerance analysis to advanced engineering tools such as this would allow for easier integration into the curriculum. If there was a built-in tolerance analysis capability within ProEngineer for instance, it would be easier to require students to perform tolerance analysis in the required design courses rather then putting this off to an elective course. Because students would already have a basic familiarity with the CAD tool it would be an easy matter of teaching the students another set of menu items rather then trying to get them up to speed on the whole tool.

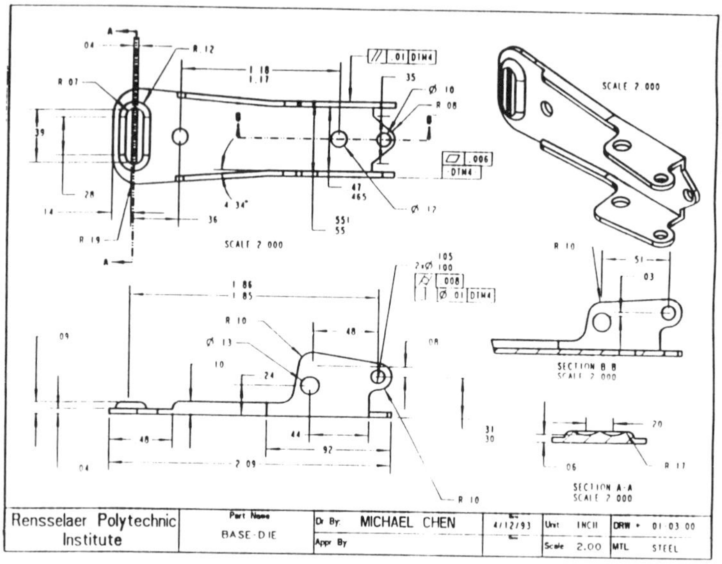

FIG. 4 DETAILED DRAWING OF BASE-DIE OF STAPLER

The best situation however would be to have tools that allowed for tolerance synthesis. The teaching approach to tolerancing would change from analysis to having the students concentrate on the design function that have to be performed. Then once these are understood and specified, turn the problem over to a tolerance synthesis tool for determination of an economical set of tolerances that maintain the design function.

In conclusion, this paper has provided an overview of the design curriculum at Rensselaer and its relationship to a proposed set of design education fundamentals. Included among these fundamentals are dimensioning and tolerancing related topics and we have shown how these are addressed by current required and elective courses. I conclude that under the current constraints of a four year program there is ample opportunity for providing adequate coverage of these topics and more extensive coverage awaits the development of more advanced tools and methods.

REFERENCES

Craig, K., 1992, "Mechatronics at Rensselaer: Integration Through Design," *Computers in Engineering-1992*, ASME, New York, Vol. 2, pp. 247-252.

Cross, N., 1989, *Engineering Design Methods*, J. Wiley & Sons, New York, NY.

Dixon, J. R., 1991a, "Engineering Design Science: The State of Education", *Mechanical Engineering*, Feb. 1991, pp. 64-67.

Dixon, J. R.,1991b, "Engineering Design Science: New Goals for Engineering Education", *Mechanical Engineering*, Mar. 1991, pp. 56-62.

Fischer, C.A., et. al.,1993, "Design Methodology and New Paradigms for Design", *Innovations in Engineering Design Education*, ASME, New York, pp. 81-84.

Krulikowski, A., 1991, *Fundamentals of Geometric Tolerancing and Dimensioning*, Delmar Publishing Inc., Albany, New York.

Pahl, G., and Beitz, W., 1984, *Engineering Design*, London: Springer-Verlag.

CRTD-Vol. 27, International Forum
on Dimensional Tolerancing and Metrology
ASME 1993

CERTIFICATION IN TOLERANCING AND METROLOGY

Edward H. Boyer
Ford Motor Company
Dearborn, Michigan

ABSTRACT

The use of the word "certification" raises the hair on the back of some peoples neck. Its use and misuse can bring numerous horror stories to mind. This discussion will review current needs in industry regarding effective use of tolerancing and how it impacts metrology. While this is not necessarily a "new" topic, some technology advances have seriously undermined previous advances in real world applications.

Worldwide competition has obsoleted some previous methodology in order to maintain competitiveness. Many of these changes were long overdue, but some changes have created "holes" in the development cycle. The smooth continuity from product design to production, and subsequently to the end user, can be very traumatic if it's not done correctly. The resultant program costs, litigation, and "assistance" from senior management could be phenomenal.

As a result of these changes, we have had to take a fresh look at "What is it that must be accomplished?" and "How do we prevent problems from happening?" for everything we do. Hi-Tech "toys" like CAD/CAM, computer simulation, robotics, software analysis, etc. are just money down the drain unless we know how to apply them to make the technology work for us, not against us.

Basic design practices with a strong understanding of applying geometric controls are a MUST to initiate a viable design. Likewise, unless manufacturing can produce the intended design, it's just lines on paper (or a screen). The use of sophisticated metrology equipment in an ideal environment will NOT provide correct data unless it is used correctly. Statistical Process Control doesn't mean a thing unless the right characteristics are identified and measured correctly. Identifying educational needs, expertise requirements and hands on applications experience are some of the methods being employed at Ford. Certification can play an important role in assisting us in this formidable task.

CERTIFICATION DEFINITION

Certification is a scary word to some people. Different people will react in various ways, many of which are negative. Let's take a look at the definition for certification.

According to Webster, Certification is defined as "Having a written statement attesting some fact, esp the status and qualifications of the person holding it". This is not to be confused with Certifiable which is defined as "Fit to be certified insane".

The Oxford dictionary defines Licentiate as "Holder of a certificate of competence in a profession". However, don't confuse it with the word Licentious which means "Sexually Immoral".

ENGINEERING NEEDS

Basically certification means a written statement attesting some fact (status, qualification, competence). The key words are "Some Fact". What is it that needs to be "Certified"? To answer this we need to look at what our needs are. With respect to tolerancing and metrology, some of the objectives Ford is working on for achieving Customer Satisfaction are:

- Product Design - Determine and protect customer wants and needs and encompass them within the product design.

- Function - Identify explicit functional characteristics which must be protected by drawing definition.

- Drawing Specification - Utilize a uniform method of dimensioning and tolerancing to depict design requirements which have a common interpretation throughout Ford operations worldwide.

- Feasibility - Ensure resultant design description is attainable by manufacturing and assembly source at a cost and quality level which is acceptable by the customer.

- Process Control - Identify designated characteristics which must be monitored to enable statistical control of the process at acceptable quality levels. These characteristics must be measurable and driven by explicit product requirements to ensure "correct" data is obtained. Bear with me at this point. I'll come back to defining "correct" a little later.

- Communication - Develop ongoing lines of communication between Product, Manufacturing, Assembly, Outside Suppliers and Customers to refine the product definition and controls as data is obtained. Continuous Improvement feedback both during design development and after production.

ENGINEERING TOOLS

Very briefly, we just covered a tremendous amount of material. I could spend days talking about each one, however, let's just keep our discussion at a macro level, for now. What is it that must be accomplished to achieve my six objectives. This is a question each of you must address individually for your Company. Since I'm from Ford, let's take a look at some of the tools Ford is using.

Product Design - QFD, FMEA, Regulatory Requirements, Team Expertise, Warranty, History, Competition

Function - Team Expertise Engineering

Drawing Specification - ANSI Y14.5M-1982 adopted by Ford Worldwide, GDT Application

Feasibility - Cross Functional Team Approval, Designated Characteristics (Product and Process) Identified

Process Control - Designated Characteristics, SPC, Control Plan Approval, Measurement Strategy and Methodology Approval

Communication - Team Approved Control Plan with revisions based on Data updates

Notice the major elements which affect our success.

- CORRECT understanding and application of GDT.
- CORRECT interpretation and measurement of GDT.

What is a good definition of CORRECT? I consider it correct if the TEAM approved it and callouts adhere to a common standard. Beyond that, it's an education issue.

GDT PROGRAM

In 1983, these elements were expressed to Ford personnel by Mr. Edward S. Roth, Productivity Services. He opened our eyes to the link we needed for communicating between Product and Manufacturing Operations via an excellent GDT standard, ANSI Y14.5M-1982, in a manner we had never seen before. Not just what GDT is, which we had already been using for 25 years, but how to effectively apply it. He also showed us the importance of GDT being interpreted correctly by manufacturing. This is where Ford has placed significant resources to promote success.

Several conditions were needed to initiate a viable program in GDT at Ford.

- GDT Standard - We had to adopt a common standard for Dimensioning and Tolerancing and apply it throughout Ford Operations, worldwide. This we did by endorsing ANSI Y14.5M-1982 totally.

- Corporate Expertise - A task force which eventually became a permanent group was established to develop Ford Training expertise, training programs, and establish a focal point for consulting and assistance to line operations, worldwide.

- Local Expertise - Local experts in GDT were needed to provide day to day guidance within the operations. Some of the local experts have also become instructors in GDT for their area.

- Massive Training Program - A very energetic training program was launched and still continues to upgrade everyone's understanding, application and interpretation of GDT. This included personnel from Product, Design Services, Quality, Suppliers, Manufacturing Processing, and Tooling/Gage Design in joint training sessions. This was a Ford first at that time, 1983.

This leads to the levels of expertise required to meet the various individual job functions at the various operations. Differing capability and expertise was needed to effectively function in their positions. This was true from management levels all the way to floor operators in the plants.

GDT EXPERTISE

I previously spoke of GDT as a language. Consider ANSI Y14.5M-1982, as the "DICTIONARY". You can't expect to hand someone a new Webster or Oxford dictionary and expect them to create a Best Selling Novel. You must be able to master the "LANGUAGE" to create a novel and then not everyone will be able to have a "Best Seller". The same holds true for GDT training and skill levels. The following four levels of expertise were identified to ensure correct implementation of GDT.

- **First** - understand the "VOCABULARY". All the symbols and the definitions must be taught. This is the basic. (Who in your activity requires this knowledge?)

- **Second** - usage of the symbols to form "Grammatically Correct Sentence Structure". Feature control frames must follow stringent rules to ensure correct interpretations can be made. Datum reference frames must be depicted and used correctly. Typically this is referred to as correct Syntax. (Who in your activity requires this knowledge?)

- **Third** - the part definition must depict a "STORY", which is design intent. Mentally, the callouts must develop a cohesive plan for what will happen "PICTORIALLY" when that part is inspected for acceptance. Will the callout give the desired requirement for meeting design intent? This is a very tough question. Metrology and measurement strategy is also included to enable the design to be considered feasible. (Who in your activity requires this knowledge?)

- **Fourth** - manufacturing personnel must be able to "READ" the drawing callouts and clearly understand the requirements conveyed. They must also be able to correctly apply that interpretation to the metrology equipment selected. This is also a very difficult area which requires a deep understanding of GDT definitions. Subtle differences in data measurement methods can totally change the result. For example, do they understand the difference between measuring an axis for Parallelism versus Position on a CMM machine? (Who in your activity requires this knowledge?)

As you can see, it sounds like I'm an English Professor writing a book. That's exactly what must be done with GDT. GDT forms the building blocks for a product to meet customer satisfaction. There are also many engineering areas which are dependent on the GDT expertise levels mentioned. A few key ones that come to mind are: CAD/CAM, Design Simulation, Processing, Assembly, SPC, Metrology. The resultant GDT definition can also be influenced by each of those disciplines. Unfortunately, there are a lot of misconceptions where GDT is thought to be in conflict with some disciplines. I've even encountered instances were a part was completely designed using GDT and someone subsequently directed the engineer to remove all the GDT callouts in lieu of block tolerancing because they were not going to use hard gages. CMMs would be used for all part measurements. How does that make you feel? It's totally wrong. I have to blame it on their degree of understanding of GDT. It's an education problem. The CMM is not the problem, but its applied usage is.

How about SPC, Statistical Process Control? There is also confusion out there that GDT conflicts with SPC. Again it's an education problem. SPC is NOT product acceptance to the drawing. It is controlling the process characteristics statistically to ensure process capability is within specification limits. People must be trained to understand how to measure features RFS and relate the data correctly to the defined Virtual Condition and MMC condition. We are wasting a lot of money and time to collect mountains of statistical data which is junk. In some cases we reject good parts because the MMC tolerance was not considered. Many people cling to the old block tolerances which translated directly to measuring a distance between point A and point B. It's considered easier to analyze. In my opinion, that's like shooting yourself in the foot. Remember the driving force for the product design is to meet customer satisfaction. If I waste my time and money collecting data which has no added value, why am I doing it? This is a very serious problem we all must face.

CMM operators must be able to "READ" GDT callouts and correctly apply them to the measurement device they are using. Let's not forget the programmers of the CMM also. ANSI is developing standards for the mathematical representation and metrology of GDT. That's great, and I'm looking forward to their publication. But, how good is the data if a feature plane or cylinder algorithm is based on a primary datum plane which is derived from three, as cast surfaces? The operator in question recognized that the data varied on the cast pads so he took three measurement points at each pad and averaged them. The datum plane was then derived from the averages of the three pads. By the way, this was a finished, machined part that was being inspected. Are you comfortable with that method? I'm not. We have a lot of work to do in raising everyone's expertise in GDT. I could talk for days on horror stories, just as you could. But, what are we doing about it?

GDT TRAINING PROGRAM

Let me briefly review some of the things Ford is doing. Remember earlier I mentioned the forming of a Corporate GDT group and local GDT experts and trainers at operation level? That in itself was no easy task, but it was essential to implementing an effective GDT program at Ford. Since 1983, over 12,000 employees and suppliers have gone through at least one of our training courses.

The following training courses were developed to handle our immediate needs to implement ANSI Y14.5M-1982:

Management Awareness

Local management must be given sufficient knowledge to understand the principles of GDT, rationale for when to use it, and what can be accomplished through effective implementation. Management must also provide the necessary support for their subordinates to be trained and motivated to use GDT in their area. Cross Functional teams must also be endorsed.

GDT Fundamentals

Convey basic symbology and definitions with sufficient expertise to enable students to select desired control and correctly note the control on the component drawing. There is also considerable coverage on the measurement requirements associated with verification of each control. Some of the students gripe that it's a gaging class, but we feel everyone needs to understand the consequences of a specific callout. At the end of the class, I tease them by stating that they now have sufficient knowledge of GDT to make them dangerous. Only the controls considered necessary should be utilized, and this is a team decision.

Applied GDT

This course is built around the cross functional team of a specific component. All the team members are in the same class. The various GDT controls are reviewed with emphasis on the differences between each when they are used. Then we discuss the component's design intent and function, proposed processing, and measurement/gaging strategy. It's truly amazing at how diverse the interpretations are from an existing drawing. We then revise the drawing dimensioning and tolerancing strategy based on the verbal information previously described to comply with design intent and be compatible with processing and inspection criteria. This is where everyone can plainly see the advantages to team development of the GDT definition. All parties are actively involved and it becomes a team effort.

These product review sessions also continue spontaneously from other students convening their individual teams when they return to their office. This is also where our GDT consultants spend half of their time, facilitating the GDT discussion. Special training classes are also provided when operations request help in specific areas of confusion. In some cases it's a check on their application expertise by reviewing several drawings in the class and "READ" the description defined.

Other Special Courses

Likewise, several metrology courses were developed to train inspectors and operators in the use of various types of inspection equipment. These courses are handled by our metrology experts in the Company.

One area for training which is planned, but not yet implemented, is in using the CMM to evaluate GDT callouts. We have a CMM available on which a course will be designed for operators to get hands on experience in applying the techniques reviewed in previous classes and to develop improved methodologies which everyone can utilize. Ideally we would like to have standard measurement criteria for each GDT control. I hope B89.3.2 will provide us with the help we need.

CERTIFICATION

Where does Certification come in? Well, for one, Ford is in the business of building cars and trucks. It would sure be nice if our resources could be directed to improving design and manufacturing techniques, rather than training our personnel in dimensioning and tolerancing. Some new engineers have not been taught to read a blueprint. CAD operators have not been taught how to use GDT. Knowledge in ANSI Y14.5M is already a functional qualification for many positions. If certification levels of GDT can be identified, those individuals certified would certainly receive preferential hiring, given other qualifications are equal. Some job functions require the basic knowledge conveyed in ANSI Y14.5M, while others will need more advanced application knowledge. We need to be able to hire personnel trained for the job they will be filling.

As a major automotive Company, we can expect to provide training for unique Company procedures and methodologies, however, controlled levels of expertise and trainer competence to a national standard should not be our responsibility. Dimensioning and tolerancing is a functional requirement for all engineers, designers, metrology and quality personnel. Everyone that uses a detail drawing must be able to read and interpret the design depicted. The national standard, ANSI Y14.5M-1982 should be part of all affected college curriculums.

In summary, training in GDT is a cornerstone for many of the successes Ford has been able to achieve and it's just the tip of the iceberg. Ongoing lines of communication between product and manufacturing locations worldwide is rapidly becoming the NORMAL way of doing business. We still have a long way to go, but we made some fantastic advances over the last 9 years. Who knows what we'll accomplish by the year 2000? Thank you.

SESSION 10
THE CURRENT STATE OF AFFAIRS

Moderator: Herbert Voelcker
Cornell University

Reporters: Steffen Parratt
Cornell University
Dean Robinson
Cornell University

INTRODUCTION

This session, the last in the scheduled series, opened with short, invited presentations by a Panel representing four communities involved with tolerancing and metrology, specifically:

the Industrial User community,
the Standards Community,
the Research Community, and
the CAD/CAM Vendor Community.

These talks, and the questions they stimulated, were followed by questions and discussion pertinent to Session 9, and then by general discussion.

The Panelists provided the statements printed below, and the Reporters produced the remaining text from written notes and tape recordings; all of this material has been edited lightly by Herb Voelcker, acting for the *Proceedings* Editor, Vijay Srinivasan. The questions, responses, and remarks attributed to individuals have not been checked by those individuals because of the compressed publication schedule for the *Proceedings*. The Reporters and Editor apologize for any misinterpretations.

PANEL PRESENTATIONS

FOR THE INDUSTRIAL USER COMMUNITY

by
Gregory Hetland
Hutchinson Technology, Inc.

This International Forum is a major success, in my opinion. The effort expended by many individuals on enhancing the current Y14.5 standard, on developing new supporting standards such as Y14.5.1 and B89.3.2, and on staging the conference is commendable. Their dedication and willingness to share their expertise, and also the support provided by their companies, deserve wide recognition.

The current level of proficiency in dimensional tolerancing and metrology in industry at large is unsatisfactory. The percentage of ANSI Y14.5M users who really know how to interpret and apply the standard is quite low. I believe that this lack of understanding is the single greatest cause of inefficiency in the mechanical industries today, and requires aggressive effort to remedy.

In 1992 I had a rare opportunity to be a member of an Industrial Metrology Panel that made presentations to professors from several American universities. The aim of the presentations and discussion was to expose views, philosophies, and needs relative to both measurement technology and education. It's clear that education in GD&T is off to a slow start in the universities, but it *has* started in a few engineering programs. A central problem: where to find time for GD&T in four-year programs already crammed with required courses. A theme that emerged in last year's Panel, and also at this Forum, is the need for the major players – industry, government, and universities – to reevaluate and integrate the tools that are essential for effective and efficient design and manufacturing. GD&T per the Y14.5M standard is certainly one such tool, because it is a contractual language that is not understood very well. All of the burden for GD&T training cannot be put solely on industry. GD&T should be a foundation tool used in many mechanical education areas, such as drafting, design, manufacturing engineering, CAD/CAM, quality technologies, and so forth.

The coverage of inspection equipment at the Forum was limited to contact type coordinate measuring machines. Algorithm issues and sampling strategies have been, and will continue to be, mandatory items of focus to reduce bias and subjective interpretation of the data. Both seem to be getting a good deal of attention at all levels, and were at this Forum as well. (It's worth noting that algorithm issues and sampling strategies are primary issues in all CMM applications, but they are not the only issues. At Hutchison Technology we make computer disk drive components and work routinely in sub-micron regimes. Factors which are often negligable in many mechanical applications are not negligable to us, because every tenth of a micron (0.0001mm) equates to 10% of the tolerance on many features of size on tooling we produce.)

The new Y14.5.1 standard (Mathematical Definition of Dimensioning and Tolerancing Principles) is intended to define mathematical models to support Y14.5M; it will not (as I understand it) provide explicit means to ensure valid and consistent measurement results. The new B89.3.2 standard (Dimensional Measurement Methods) is intended to cover compliance with specifications which comply with Y14.5M. It is important not only that both standards be completed, but that support documents be created (or additional context be added to these standards) to bridge any interpretation gaps between them. This is intended to reduce measurement bias and variability. It's worth noting that some of the participants are software developers, and these may face possible loss of competitive advantage if they bring all of their knowledge to standards development. This will require cooperation between the necessary groups. Fortunately, it does not seem to be a matter of "if it will happen", but "when it will happen".

Sampling strategies for coordinate measuring machines is another area in which companies and CMM manufacturers must work together to define sound conventions. Here again there is a possible loss of technological advantage for specific user companies and CMM manufactuerers that have run their own tests.

CMM testing at universities is a fairly new effort. The test results described at the Forum are not unlike tests I've personally run over the years, and I'm aware of a number of other companies and CMM manufacturers that have run the same. Even though I had no problem with the validity of the results presented here, it is worth noting that test data can be misleading, or certainly limited in scope, if taken with a single CMM in a single environment, using one specific probing system and typically only one or two operators.

I suspect that a large percentage of the information being gathered currently in companies and universities already exists somewhere. Perhaps what we need is a person or group willing to collect such data – provided that companies will agree to share data, and will provide the information needed to assess its relevance and validity. We're all looking for the same information, so let's stop re-creating the wheel!

As I said as the outset, the Forum was a success in all respects, in my opinion. It was a valuable learning experience for me, and a chance to make valuable contacts in related fields. I hope that the next International Forum is not five years away, and that communication on these matters is published more frequently in industry magazines. My long term hope is that some day in the not too distant future the whole world will be working to the same set of standards.

Questions and Discussion

[unknown]: Could you elaborate on your finding that taking certain numbers of CMM data points led to poor results?

Hetland: In our experiments, we evaluated sizeand other parameters based on different numbers of data points. In general, we see a point of diminishing returns, after which increasing the number of samples brings no advantage. However, we found that a plot of size vs. number of data points oscillates slightly as it converges, and certain numbers of samples lead to larger errors than adjacent numbers. (For example, 12 points might be worse than 11 or 13 points.) We don't know why this occurs, but it seems to be very repeatable for a given probe and machine. The effect occurs with size and position measurements, on both cylinders and spheres.

Tom Charlton: In the slides from Bob Hocken's CAM-I work, you saw similar effects. It would be very interesting to compare your data to Hocken's simulations, to see if there is a correlation.

\- - - - - -

FOR THE STANDARDS COMMUNITY

by

Henrik Nielsen
Cummins Engine Company

When first asked by Fred Parsons if I planned to attend the Forum, my answer was 'No' — because the sessions listed in the program did not seem to address the central problems. These problems, as I see them, are the gaps between the virtual world of standards for design intent, such as Y14.5 and Y14.5.1, and the real world of assessing workpieces and standards for gaging, which is the world that B89.3.2 struggles with. When Fred then asked me if I wanted to be a panelist, I jumped at the chance to air my views.

My first comment is that I have enjoyed myself a lot more during the Forum than I anticipated, because we have had some very interesting presentations. But the central problems remain: most presentations were within one field, and did not try to bridge the gaps that are causing us problems.

To define the problem, let me first define the purpose of our combined efforts as I understand it. The central goal is to produce functional workpieces: everything we do is in support of that goal.

The first prerequisite for doing this in a consistent manner is to be able to "see" and "hear" what is happening and to communicate that. Thus metrology functions as the eyes and ears of the organization; GD&T provides a language to communicate (some of) this information. This is first done in product development, where measurements are made to assess the attributes of prototypes. When a prototype performs satisfactorily, the information gained is transferred to drawings that specify the attributes needed in each component to insure that the product functions properly.

These drawings are then used to communicate to production or to subcontractors the requirements for each workpiece. Drawings should be viewed and treated as legal documents. They are contracts between designers and manufacturers. If our drawing definitions are not clear enough for us to communicate what we see and hear, we may as well be blind and deaf ... and I think this is the situation we are presently in.

The essential problem as I, a metrologist, see it is that the call-out definitions in the Y14.5 and Y14.5.1 drawing standards are not measurable. This means that when the result of product development is transferred into drawings, "noise" is introduced, and again when real workpieces are checked for compliance with the drawing, this very same problem introduces uncertainty whose magnitude is impossible to estimate.

Any member of Y14.5 and Y14.5.1 will tell you that these standards are not gaging standards. Clearly this is true, and is as it should be. Unfortunately these non-gaging standards contain definitions and stipulations that are almost impossible for gaging standards to interface with, and thus they have a dramatic impact on gaging issues. Let me summarize the two biggest issues.

The first: Y14.5.1 stipulates that it "*incorporates by reference the smoothing functions defined in ANSI B46.1-1985, Surface Texture*." The intent of the statement is good and I accept it, but unfortunately the filters (smoothing functions) incorporated in the cited standard are only intended to isolate the shortwave component (roughness) while distorting the longwave component (waviness and form) due to phase-shift. The upcoming new version of the B46 standard is even worse, in that it offers several choices for performing smoothing, and they all lead to different results.

The second issue is that Y14.5.1 refers to an infinite set of data points. This is not only unnecessary if an appropriate smoothing function is employed, it also makes it impossible to estimate how far the result of a given actual measurement is from the "true" value. This difficulty in turn makes call-outs more inaccurate than they need to be.

I suggest that these two definitions be replaced with a statement to the effect, "The surface features toleranced in this standard are limited to those of longer wavelengths." This would transfer to the B89.3.2 gaging standards group the responsibility for deciding what wavelengths to include, how the smoothing should be done, and how many points are required to describe adequately a single wave. B89.3.2 could then get on with its job instead of chasing its tail trying to conform to the non-measurable definitions of Y14.5.1.

An alternate approach might be to develop a standard "Y14.5.n: Interpretation of Y14.5 for contractual purposes". This would allow designers to use Y14.5 and Y14.5.1 to exchange information about their intentions, while 'Y14.5.n' would be used by subcontractors and their customers to determine what it takes for an actual workpiece to be acceptable.

Again, while the Forum addressed many interesting issues, I don't think it took us any closer to solving these crucial interface problems, because they were never addressed.

Questions and Discussion

Ted Hopp: You say that because Y14.5.1 is "not measurable", B89.3.2 is precluded from doing its job. I would ask, is the thermodynamic definition of temperature "measurable", in terms of velocities of atomic particles, etc.? That also probably is not measurable, in the same way that 14.5.1 is not measurable. Yet there are practical methods for measuring temperature and methods for estimating uncertainty. Why can't you do the same with B89.3.2?

Neilsen: Our problem, as I see it, is (as the gentleman before me said) the more points you take, the bigger the value of form error you get. So, we have a curve like this (trending upward as the number of points approaches infinity) and the thing that we're asked to measure is when this curve goes to infinity. I don't know of any solid way of estimating, from somewhere here out to infinity, where that curve will go. That's my problem.

Hopp: I would suggest that if your curve doesn't flatten out to a horizontal, then there won't be any way to mathematize the definition in 14.5.1.

_ _ _ _ _ _

FOR THE RESEARCH COMMUNITY

by

Neal Juster
University of Leeds (U.K.)

From the presentations and discussions of the two and a half days of this Forum, it is clear that a number of advances have been made in the years since the NSF-sponsored workshop on tolerancing in Orlando in 1988. The most visible advances are the new version of the dimensioning and tolerancing standard ANSI Y14.5 now entering review, progress in the mathematical definition of Y14.5 in the new draft Y14.5.1standard, the new dimensional measurement standard B89.3.2 (still under development), and the work of the STEP community in developing Part 47 ofSTEP (tolerances), also under development. In addition, this Forum has shown that there is a better and more common understanding of the issues that surround dimensional tolerancing and metrology between and within the research, standards, vendor, and user communities than existed in 1988. It must be stated, however, that there are still many bridges to build and problems to solve.

Outside of the progress made in the production of standards, there have been a number of presentations that have highlighted how specific items identified by the Orlando workshop have progressed. Hocken stated that we now have a better understanding of how CMM algorithms behave with respect to the number of sampling points taken, and Hopp reported on progress in the testing of these algorithms. Turner showed that tolerance analysis packages, rare in 1988, are now commercially available from a number of vendors and that, if applied with care and knowledge, they can be used to reduce product cost and improve product quality. Allen indicated that many leading mechanical CAD/CAM systems now allow tolerance information to be represented and manipulated, a concept hardly in evidence in the 1980s.

The Forum has also shown how the theory of tolerancing has developed since 1988. The results of the Orlando workshop showed that in 1988 there were at least six approaches to tolerance modelling published inthe research literature: offset or non-parametric tolerancing (Requicha), virtual boundaries (Srinivasan and Jayaraman), vectorial tolerancing (Wirtz), M-space (Hoffman; Turner and Wozny), parametric tolerancing and variational geometry (Hillyard and Braid; Lin, Gossard and Light) and constraint propagation (Bernstein and Preiss). At this Forum, Walker introduced the new ANSI standard, Y14.5.1, stating that it was based on a combination of the offset and virtual boundary approaches. Henzold presented the case for using vectorial tolerancing as a basis for CMM algorithms, stating that the ISO standard TR 10 360-1, "Coordinate Metrology – Part 1: Definitions and Applications of Fundamental Geometric Principles" (currently under preparation), is based on the vectorial tolerancing approach. Henzold also stated that although vectorial tolerancing is the most appropriate representation for many applications, the approach used in Y14.5.1 is better for many others.Turner highlighted the fact that most commercial systems are based on parametric tolerancing. Guilford showed that the M-space approach (a parametric representation) is still in active use at Rennsselaer Polytechnic Institute. Only constraint propagation was not mentioned at the Forum, and as far as I am aware, this approach has notappeared in any literature since 1988. From the work presented at the Forum it can be seen that three major approaches to tolerance modelling have been distilled since 1988; virtual boundaries, vectorial tolerancing, and parametric tolerancing.

It is clear from the Forum that although as a community we are beginningto understand how tolerances can be modelled and components inspected, we still have comparatively little understanding of how to set limits on dimensions of size or how best to call-out geometric tolerances such that the products we manufacture will still function correctly even if they are produced at the extreme limits of the tolerance callouts. This has been termed tolerancing for function. Hoeprich introduced studies of how variations from nominal geometry effect the fatigue life of rolling element bearings, and Grant showed how surface texture influenced the failure of a crankshaft bearing. Outside of these two studies and a few others reported in the literature, it appears that most designers can only use the maxim, "This tolerance worked last time, so I will use it again." If we are to gain a better understanding of tolerancing for function, then the research community needs to study and model the interactions between design, manufacture, and inspection. I suspect that when we better understand tolerancing for function, we will find the current standards inadequate.

Taylor gave us a vision of a possible future (the present at Renishaw) where the designers understand the capability of the availablemanufacturing processes and how to inspect the products using in-process probing. This design and manufacturing process does not require conventional drawings nor the use of tolerancing standards. Is this the way we as a community should be moving?

\- - - - - -

FOR THE CAD/CAM VENDOR COMMUNITY

by

James Hurt

Structural Dynamics Research Corporation

Thank you for inviting me to be the final speaker at this 1993 Forum. My talk today has two parts: a summary of what has been covered in the form of quotations from previous speakers, and then a message on one area where we can improve on the current situation.

Part 1: Summary of this Forum

The following summary is in the form of selected quotations from earlier speakers. These are as I recorded them during each speaker's presentation, and I apologize in advance for any misquotations or misintrepretations.

Paul McKim: *"Standards development has been lagging behind technology development."*

This raises the rhetorical question: hasn't the development of everything else been lagging behind technology development?

Al Neumann: *"The current version of Y14.5 is 1982. A lot has happened since then."*

Fred Parsons: *"Know what you don't know, else you really don't know what you do know."*

Fred was quoting someone else whose name I did not record.

Georg Henzold: *"What you cannot measure within a reasonable certainty you cannot produce intentionally."*

Georg was also quoting someone else.

Ari Requicha: *"If we do not know precisely what a tolerance means, then how can we control the manufacturing process, inspect the part, or analyze parts for assemblility?"*

These last three quotes remind me of a quotation from a famous American philosopher:

Yogi Berra: *"If you don't know where you're going, you'll end up somewhere else."*

Herb Voelcker: *"Tolerance specifications can affect how you assemble parts."*

I wish to expand on this by adding that how you assemble parts can affect the design intent of tolerance specifications.

Fred Parsons: *"Measurements for process control may be entirely different from those for part acceptance."*

Archie Anderson: *"The value of CAD training is much better if the trainer has a design or a manufacturing background rather than a computer background."*

Josh Turner: *"The earlier tolerancing is given consideration, the greater its impact on cost and reliability."*

Josh Turner: *"Every company should be using one of the dimension driven tolerance analysis systems."*

I'll don my SDRC cap to remind you that SDRC provides one of these dimension driven tolerance analysis systems.

Rich Walker: *"How do the current dimensioning and tolerancing standards meet design intent?"*

Ari Requicha: *"Should design intent be captured more directly?"*

Rich Walker: *"The language barrier is our most pressing problem."*

Neal Laurance: *"STEP is just now starting to address tolerancing."*

Part 2: My message

The last four quotes are all related to my message, which I'll deliver in a three-step process: I'll tell you what I'm going to tell you, then I'll tell you, and finally, I'll tell you what I told you.

<u>First step</u>: I'm going to tell you that it is time for the CAD and geometric tolerancing communities to get together, to cooperate more fully, and to cross-fertilize ideas and concepts.

<u>Second step</u>: Both the CAD and the geometric tolerancing communities are concerned with the

communication of design concepts. Each has been focusing on particular parts of this communication issue, yet there is no significant difference between these various parts.

Design concepts may be things that do not yet exist, yet must be communicated from individual to individual. This is not an easy process, especially when there is not yet a physical part to look at. It is much easier to describe that Rennishaw part used by Ben Taylor in his talk when you can pass the part around. Communicating a design concept involves communicating a shape, its functionality, and its required degree of perfection before there is a real part to hold in your hand.

Part of this communication issue – the communication of shape – has been the domain of CAD. Another part, the required degree of perfection, has been the domain of geometric tolerancing. The past cooperation between these two communities is best described by paraphrasing Burt Lancaster in the movie, **The Hallelujah Trail**, when he is faced with the problem of keeping several warring factions apart. To paraphrase: the CAD and geometric tolerancing communities have been following 'somewhat parallel paths' while maintaining 'detached contact'.

Both the CAD and geometric tolerancing communities use many of the same terms, especially the term 'design intent'. The communication of design intent will be much improved if both communities cooperate much more closely than they have in the past.

When I first prepared this talk, shortly after Herb Voelcker invited me to give it, I wrote out a long speech. When I started cutting the speech back to an acceptable length, I discovered that it was a long list of areas where one or the other or both of these two communities would benefit from improved cooperation. I am going to give you only two of these areas here.

For the CAD community, geometric tolerancing is much more than annotation on a drawing. It is much more than something you attach to geometry. Geometric tolerancing is trying to tell you why the geometry has that shape and why the geometry is where it is.

For the geometric tolerancing community, CAD has demonstrated that anything less than a solid model is an incomplete representation of the basic geometry. Solid models, like real parts, do not have views. You can only specify a view by using datums to orient that view. How many tolerancing symbols derive meaning from the view they are presented in yet do not have datums?

There are many other areas where both communities would benefit by increased cooperation and coordination. I've given these two merely as examples, and they probably are not the most significant examples.

<u>Third step</u>: What I have been trying to say is that increased cooperation and coordination between the CAD and geometric tolerancing communities will generate benefits for all concerned. And these benefits, which are mainly in the area of improving the communication process, will accrue to all our customers.

Questions and Discussion

Herb Voelcker: What mechanisms do you propose to foster increased communication?

Hurt: Well, I didn't have any in my original 3-hour speech. It has to be the usual mechanisms of getting disparate people together, inviting people like George Allen and me to conferences like this. We need to invite you to our CAD conferences. I don't see any easy way to do it. We have to start working at it.

QUESTIONS AND DISCUSSION ON SESSION 9

Bill Grant (to Per Bennich): You seemed to focus on the missing links in the chains of standards. What about contradictions between links in the same chain, and between chains? To what extent do these exist in ISO chains of standards and how do we deal with them?

Bennich: Your observation is correct. I focused on gaps because they are "visual" and easier to convey. The contradictions are there. One task of the ISO Joint Harmonization Group is to describe these and make plans to remove them. There is a database of known contradictions and gaps in the so-called M3 document (approx. 60-70 pp.). Many countries are contributing, and the next edition is due in August.

Hopp (to Bennich): If the drawing documents design intent, where should engineering intent of downstream processes be documented? (E.g., statistical tolerancing, acceptance limits, and so forth.)

Bennich: Above all, design intent must be clearly defined. This is critical for making changes or improvements to the product – especially if changes are made by new employees who are not familiar with the original design intent. Too often, when design intent is not documented, changes to the product lead to less stable designs and other problems. Downstream processes can be documented apart from design intent, in separate documents.

Grant: Companies often have manufacturing drawings that supplement the design drawing. Such drawings might take into account the existing machine tool, etc.

Voelcker (to Bennich): You said, "The drawing should show or define the function...." Would you provide some specific examples of "function"?

Bennich: Most standards address only fit, but there are other important considerations. For example, bearing surfaces must create a certain pressure, and other surfaces must be painted. Many machine elements have special requirements on surface texture, form, and dimension. These factors control the useful lifetime of the product. If you designed a car by attending only to assembly, the car wouldn't run, or might run for just two hours.

We need tighter and more scientific designs, and we are discovering that the use of specialized tolerances to control factors such as material properties, surface characteristics, etc., can lead to better predictions about product life. Ten or twenty years ago we didn't know these things, and current ISO standards reflect knowledge from that era.

Voelcker: I predict that the word "function" will cause a lot of trouble unless it is better defined. For example, power or energy conversion is a function, which might be implemented with a nuclear power generation station. Another function is "support", in the sense that a bearing supports a shaft. These are very different levels of function. Unless we are more careful in defining or delimiting what we mean by "function", there will be a lot of misunderstanding in this community.

Grant (to Gary Gabriele): Your design and manufacturing curriculum is very welcome. Yet, I noticed that your course sequence (curriculum) has little on metrology relative to assessing actual component dimensions. Do you think this is important for your curriculum, or do you feel this is best accomplished by industrial metrologists providing training to design engineers?

Gabriele: True, the little metrology we have is in two manufacturing courses, in which students make parts and do some inspection. In most universities, you won't find the equipment or expertise needed to cover metrology well. This might be best handled by other means such as on-the-job training. Undergraduate education is meant to teach the fundamentals.

Hopp (to Gabriele): Dixon's taxonomy of design fundamentals seems confusing, with overlaps and no clear rationale for the classification. Is design really systematized enough to be "teachable" as a distinct field of study?

Gabriele: Design is definitely teachable. The fundamental hurdle for the students is learning the process of attacking open-ended problems. Later, the students need to consider the impact of their design decisions, and Dixon's work is central here.

Hopp: These things may be hard to teach outside of a particular context.

Gabriele: The context is the design problem itself. We've had good results teaching via reverse engineering, where the students must compete against an existing product by considering manufacturing cost, production volume, and so forth. Without this context, lectures on design would go right past them.

Paul McKim (to Edward Boyer): Y14 is developing criteria and procedures for a certified Geometric Dimensioning and Tolerancing Professional. What is your view of the ASME approach to certification?

Boyer: We are currently training people in certification because this wasn't available from outside sources. In general, the training we were getting from outside focused on the interpretation of symbols in terms of geometric controls. There was no tie to metrology. Ultimately, we'd like to certify people at an applied level, where they know the symbols, the syntax, and inspection methods.

McKim: Would you make certification a criterion for employment?

Boyer: Yes, for some jobs. We already require credentials such as GD&T knowledge for certain jobs, although more engineers *say* they know GD&T than actually do.

GENERAL DISCUSSION

Ken Sheehan: I borrowed Henrik's slide (on the gap between the "virtual world" of Y14.5.1 and the "reality world" of B89.3.2) to make a point. In 1988 at Orlando, this gap wasn't even recognized. If we do need a new document that bridges the gap by addressing measurement methods, how can we avoid making commercial value judgements with respect to inspection processes and equipment? How can B89.3.2 avoid the same trap?

Voelcker: Methods for bridging the "virtual world", "reality world" gap can be drawn from mathematics. Y14.5.1 is a collection of method-free mathematical definitions. So the virtual world is mathematical functions. Numerical analysis tells you how to evaluate functions that can't be solved in closed form, or in other words, how to solve "real" problems. There is no mystery here, and also no unique solutions. All solutions have an associated cost and accuracy.

Sam Levy: Y14.5.1 gives the design intent, but someone must define what constitutes acceptance of the design intent. Without this, people will argue over what constitutes acceptance of the drawing. We need to bridge this gap.

Ari Requicha: On the measurability of Y14.5.1: if we make no smoothness assumptions, we won't get anywhere. There's always the example of the surface "jumping" at a point, and so on. There's an obvious answer, to me. We say, we'll give you a set of measurement methods that depend on the smoothness of surfaces. You must have a way of characterizing smoothness. We require Lipschitz conditions on functions, and stuff like that.

Rich Walker: We're trying to bring Y14.5 to the level of law, which is to bring it to a mathematical basis. We've made the first step with Y14.5.1. Even if we haven't bridged the gap between tolerancing and metrology, this doesn't invalidate the first step.

Levy: The methods of inspection need to be agreed on. If we don't specify them, there will be 15 ways to inspect the part, and it will always be a problem. We've got to allow what constitutes acceptance of the part to be on the drawing. That's the bridge.

Ray Wheeler: Drawings are qualifying documents. If we add information about how to verify the part, the drawing becomes more restrictive, and this affects suppliers. We leave the inspection and manufacturing criteria up to others. The drawing should remain a qualifying document, and not be complicated further.

Voelcker: I've brought the report from the 1988 Tolerancing Workshop in Orlando*, to help us assess progress since then. That report had three recommended "action items" and six "research challenges": [quoted from report]. My personal scorecard is about 50%. We have made progress on tolerance mathematization, measurement methods, and standards, and some limited progress in education. There has been little or no progress in characterizing process variability, in part because there are no organizations (e.g. standards committees) that are publicly responsible for the area.

Vijay Srinivasan: Andre Clement asked me to tell you that CIRP holds forums on tolerancing every two years. The proceedings of the most recent forum (in April of 1993 in Paris) have been published in book form. If you are interested in obtaining a copy, you can contact me or Professor Clement for more information.

---

* Research Needs and Technological Opportunities in Mechanical Tolerancing, V.A. Tipnis, Ed.; Results of an NSF-Sponsored International Workshop, Orlando, Fl, September 28-October 2, 1988; available as CRTD-15 from ASME, United Engineering Center, New York, NY.

ISSUE-ORIENTED PANELS

INTRODUCTION

In the afternoon of June 19, 1993 four issue-oriented *ad hoc* panels were organized by interested individuals, as indicated below.

Statistical Tolerance Analysis in 3D	by Colin Wearring
Computational Metrology	by Bruno Brazauskas
Relationship Between ANSI & ISO Standards	by Georg Henzold
Where Do We Go From Here?	by Vijay Srinivasan

The panels discussed specific issues, and drafted their conclusions and recommendations, which were communicated to the *Proceedings* Editor, Vijay Srinivasan, by the reporters of the panels. These communications were then edited for uniformity and style, but most of the contents were preserved.

Due to the shortness of time, and difficulties in logistics, the comments attributed to various individuals in the following sections have not been checked by them. The Reporters and Editor apologize for any misinterpretations.

The Steering Committee for the Forum sincerely thanks Colin Wearring, Bruno Brazauskas, Georg Henzold, Vijay Srinivasan, Bill Grant and Richard Jennings for organizing, and reporting on, these panels.

STATISTICAL TOLERANCE ANALYSIS IN 3D

Attendees:

Roberto Altschul	Boeing Corporation
Norm Cutler	Polaroid Corporation
Paul Drake	Texas Instruments
John Hertrich	Applied Computer Solutions
Neal Juster	University of Leeds, England
Steffen Parratt	Cornell University
Ari Requicha	University of Southern California
Hiromasa Suzuki	University of Tokyo, Japan
Colin Wearring	Applied Computer Solutions

The discussion focused on different interpretations of the Y14.5 tolerancing standard as they were applied in tolerance analysis of assemblies. The case of a linear tolerance analysis was used as an analogy for the three dimensional case. In a linear tolerance analysis the worst case (limit stack) and statistical (Root Sum of Squares) technique are established and accepted. The panel was asked if a similar approach could be reasonably established for the three dimensional case.

Worst Case vs Statistical

Initially, the panel discussed some issues and ideas on the difference between worst case and statistical analyses, as they were deemed to be fundamental to the question posed.

Since tolerancing exists to account for variability in manufacturing and assembly, doesn't tolerance analysis imply a statistical behavior of the toleranced parameter? Why is it necessary to specify a "statistical" tolerance? Wouldn't one use worst case tolerance analysis to identify the most important tolerances and use these as candidates for statistical analysis?

Economic factors seem to necessitate the use of statistical analysis. In multipart assemblies (e.g. a linear stack of 100 bars) the use of worst case analysis will generate extremely conservative results for individual components. In general, this approach will not be economically feasible for many situations.

Worst case analysis implies that "any component in my warehouse will fit any assembly". This can only be ensured by measuring (without significant error) every component. Practically, worst case tolerance analysis implies a very low statistical probability that the component will fall outside specification. This can be expressed as a "safety factor" (C_{pk}).

Standardizing Statistical Tolerance

With the conclusion that statistical tolerance analysis was necessary, the discussion focused on which statistical representation would be appropriate for a feature contained inside the three dimensional zone specified by the Y14.5 standard. The following ideas appeared to meet with general agreement by the panel participants.

Process information is the best way to establish the statistical behavior of features for tolerance analysis. Currently there is no well defined way to interpret process information (measurements) into a format which is well suited to the modeling of three dimensional behavior. In the absence of process information, a Gaussian distribution of the feature parameters (size, form, orientation and location) is an appropriate starting point. The combined effect of these parameters on the feature should follow Y14.5 specifications for the tolerance zone.

Because it is desirable to use process information in tolerance analysis, the parameters used for feature variation should be chosen so that the parameters can be measured. If the desired parameters cannot be measured directly, the data may be transformed to the appropriate parametric space. Once the Gaussian model has identified important characteristics, process information should be utilized in successive tolerance models to evaluate the impact of actual process performance on assembly variation.

Tolerance specifications can (should) include the specification of Probability Density Functions (PDFs) in order that they be adequately specified for tolerance analysis and statistical performance. The necessary and sufficient information for this purpose is not accurately defined at this time. If the PDF for a tolerance is unknown, there may be extreme PDFs which could be used to estimate the "world" of possible manufacturing functions.

Summary

The panel seemed to agree that tolerance analysis should be based on a statistical approach, and utilize measured process information to estimate the statistical characteristics. The techniques for this procedure are not well established for the three dimensional case, but, if possible, the parameters used in the tolerance model should be measurable.

Using Gaussian PDFs to describe the parameters of feature variation is an appropriate starting point, but subsequent tolerance models should be constructed using measurements of actual process information. Since all tolerances are expressions of variation, the specification of PDFs for tolerances could (should) be established as a fundamental tolerance characteristic.

COMPUTATIONAL METROLOGY

Attendees:

Uppliappan Babu	University of North Carolina - Charlotte
Bruno Brazauskas	Inspection Technologies
Ted Hopp	NIST
Chunye Ma	University of Michigan
Elsayed Orady	University of Michigan
Jay Raja	University of North Carolina - Charlotte
Charles White	Datum Inspection Services

The intention of the panel was to address the following questions: Beyond the effort of software developers, is algorithm testing necessary? What can be done to respond to Dr.Hopp's "Research Challenges", which were presented to the forum? Time constraints dictated that these questions were not addressed to the fullest.

State of Affairs

The panel first discussed the fact that a great number of analysis packages with geometric fitting routines are available. The number goes well beyond that of different CMM manufacturers. There are numerous after-market measuring software packages which have been developed independently and privately. We must also include the measuring systems developed within companies which address their particular needs.

Although some are responsive, most measuring equipment manufactures are less than cooperative when approached about specific algorithms, hiding behind a proprietary veil. They are also unlikely to share the results of algorithm testing and information on the data sets which were used. There was a feeling within the panel that the major manufacturers and developers probably subjected their algorithms to rigorous testing — however, this may be only wishful thinking on the part of the panel.

The panel was less confident in its speculation about the testing being done by independent developers, reasoning that some of these small companies may lack both the capital and the time necessary when trying to get their product to the market. It is also interesting to note that some software packages have geometric engines which have been passed or sold from company to company who, in turn, have added functions or merely changed the user interface. Often, in these cases, the software distributor either has no knowledge of what testing previously went into the product, or assumes that it was done beforehand.

Guide to the Users

Where does this leave the user? The user of computerized coordinate measuring equipment is placed in the position where he has to determine not only proper measuring methodology, but also whether or not the algorithms process the data in an acceptable manner according to his point gathering techniques. Also, the user needs to know that, within acceptable limits, his results will have correlation with those of other users.

One path available to end users is to inquire about PTB (The German Standards Institute) certification. Software packages marketed in Europe evidently need this certification. Participants on the panel admittedly had little first hand exposure to the

PTB. PTB will, apparently, work with groups during algorithm development or they will function simply as a testing facility prior to release. The PTB may utilize a number of groups of data sets during its approval or testing process. This approach is, however, limited to software released in Europe and only has significance to the U.S. market to the extent that the same software package is released in Europe and in the U.S.A.

Another option for the end user is comparative testing. Basically, this is the input of point data into various programs running applicable fitting routines, and comparing results. Because not all measuring systems allow for the input of electronic data, the pint data must be manually keyed into some packages. In addition to the effort required to undertake this in-house testing, the user needs access to data sets which reflect the proposed methodology and for which there are predictable and universally acceptable results.

NIST is presently working to put the mechanics in place for the testing and certification of algorithms. Central to their plans is ATS (Algorithm Testing System), a software package which can assist in the assessment of geometric fitting routines. They hope that ATS will form the basis of a U.S. standard for software performance evaluation. ATS contains a set of guideline algorithms, a data set generator, and a method for analyzing results from different algorithms. NIST is expecting to have the software finished this summer, and plans on launching their algorithm services later this fall.

The need to know that software package has been through strenuous testing or certification should be an important issue for the end user. However, the user also needs to have access to data sets which mirror his proposed methodology so he may minimize his measurement uncertainty. Finally, it was suggested that it might be beneficial to establish an informal committee to further explore these issues.

RELATIONSHIP BETWEEN ANSI AND ISO STANDARDS

Attendees:

Archie Anderson	Trikon Corp.
Frank Bakos	Eastman Kodak Co.
Per Bennich	PB Metrology
Jim Bryan	Bryan Associates
Tom Charlton	Brown and Sharpe
Bill Grant	Cummins Engine Co.
Georg Henzold	Siemens AG
Alex Krulikowski	GM Power Train Div.
Paul McKim	Caterpillar Inc.
Robert Nicholaisen	Pittsburgh State University
Dean Robinson	Cornell University
Fritz Schachter	Minitek GmbH
Paul Scott	Rank Taylor Hobson Ltd.
Rich Walker	Westinghouse Corp.
Bob Wilhelm	University of North Carolina - Charlotte

Happenings in Europe

The organizer of the panel, Georg Henzold, opened the discussion with the statement that the European Economic Community has decided to tear down the barriers of trade among the EC member nations. As a result, there needs to be more complete standards for this new trade community which will facilitate commerce. A new standards function entitled the European Committee for Standardization (Comité Européen de Normalisation -CEN) has been formed to administer the development of the necessary standards across a very broad range of products, services, manufacturing etc. If CEN approves a standard then all EC member nations must accept this standard as a national standard and declare obsolete any national standards that cause conflict with it. CEN does not want to standardize instead of, or in parallel to, ISO but in conjunction with the ISO standards.

The mechanism used by CEN in accomplishing its objectives is as follows. If it determines a need for a standard in an area where no ISO standard exists, CEN can draft a suggestion and submit it to ISO. If an ISO standard exists, then CEN will ask member nations if they would accept the existing ISO standard as a CEN standard. If they are able to so accept, then the ISO standard is accepted as a CEN standard — for example, the review of ISO 1101 has just been completed by ISO TC10 and is ready for vote. Preliminary disposition in CEN 290 (the technical CEN committee on dimensional metrology) is to accept the standard. The only ISO standards that have not been accepted this way are those that are known to be up for review in ISO. In areas where ISO or CEN standards exist, the member nation standard groups are to stop work on their national standard, and focus on the CEN standard.

General Discussions

This was followed by some spirited discussions on the relationship between ANSI and ISO. Jim Bryan observed that his suggestion to ASME and ANSI to stop work on the B5 machine tool standard, and focus on the ISO standard, was rejected. Paul McKim, who is in the ASME board on international standards, responded that they are trying to impress on the ANSI committees the importance of accepting ISO standards as ASME/ANSI standards. Any copyrighting problem will be solved by a licensing of ISO standards by ASME.

Tom Charlton, while applauding the spirit behind this, felt that the rank and file of ANSI committees will take exception if EC declares American standards null and void. Bill Grant observed that ISO standards tend to focus more on working around political issues (each country wants to avoid too much change in the accepted practice), and not on commercial issues. As a result, the solution that is most often accepted is the solution that is most correct technically. Rich Walker agreed that the U.S. should participate more in ISO, but felt that in ISO whoever speaks the loudest and provides the most work gets his way.

At this point, Georg Henzold observed that (1) we cannot continue to proliferate so many national standards across the world, and (2) generally the best technical solution is the one that is accepted in ISO. When asked how the U.S. can have equal input into CEN, he replied that it can be done through ISO participation.

Specific Issues

The discussions then turned to specific topics. In dimensional metrology (ISO TC3 and ANSI B89), surface texture (ISO TC57 and ANSI B46), and dimensional tolerancing (ISO TC10 and ANSI Y14) there are parallel efforts in ISO and ANSI. Rich Walker suggested that in these three areas the correlation can be reviewed and action taken. Georg Henzold felt that the Joint Harmonization Group (JHG) is the mechanism to do this.

Bill Grant stated his belief that Y14.5.1 will in large part be accepted by the JHG and then by the ISO TCs. He felt a cooperative spirit within ISO when American standards or drafts are brought to TC meetings. Jim Bryan concurred with this observation, and said that this was true for machine tool standards (ANSI B5). Robert Nicholaisen cautioned the panel that Y14.5 and ISO 1101 do not map one to one — ISO is a series of standards that map to Y14.5. Per Bennich said that it was one of the reasons for his development of chains of standards concept which exposed gaps and conflicts in the existing ones. The panel then addressed the following four questions.

1. What happens if ANSI standards are different from those of ISO? The issue is whether this will affect commerce. Tom Charlton felt that in some cases it does not, because businesses make products to meet both the standards. Paul Scott said that in some cases it hurts U.S. companies abroad because they cannot find suppliers who will meet the ANSI standards. Therefore every effort shout be made to make the ISO standards acceptable to both US and EC countries. The US member bodies to the ISO committees must present their needs and be dogged about them in the ISO meetings.

2. How much can ANSI differ from ISO? If a US standard differs from ISO, then it should be noted explicitly in the standard where the difference is. Bill Grant stated that at least for US surface metrology standards, they must not preclude in their language any viable technique that vendors have and that is in general use in industry. ASME has set the precedence that if a vendor threatens law suit for the exclusion of the technique used by his instrument, then ASME will not go into litigation but will instruct the standards committee to rectify the omission. Therefore, ANSI standards include much more than ISO standards, which only standardize the "best practice" techniques that can be made to be universally acceptable. In such cases, Bill Grant suggested that sections of ANSI can be exactly the same as the ISO standard. However, Rich Walker, Tom Charlton and Bill Grant observed that ANSI committees do not always agree to include word-for-word the corresponding ISO section. Sometimes, this is due to proprietary reasons as some technologies are not allowed for open use for standardization. The general view was that ISO standards do not suffer from this problem.

3. How can ISO be made to meet US requirements? The best way is for US participants to provide the work by submitting the technology in draft document form and be present in the meetings to drive it. Tom Charlton warned that there are some fundamental problems between ISO TC 3/WG 10 and ANSI B89.4 on CMM checking and that it probably will not get resolved easily. In spite of this, it was felt that in future U.S. chief technologists should work closely with their ISO colleagues, and ISO should have periodic meetings in the U.S. A key to the acceptance of the U.S. position by ISO is to have the U.S. position well documented and submitted to ISO.

4. How can ANSI alter its policies to match that of ISO? A strong U.S. participation in ISO should help promulgate ISO thoughts into ANSI — for example, they can be installed as parts of ANSI standards as a unit. In addition, the work of Paul McKim and his international committee will help in this endeavor and should continue. It was recommended that towards this end a resolution be made that ANSI will accept ISO documents wherever possible as they stand as unit sections in the U.S. standards.

As a final remark, Georg Henzold stated that the view of CEN as establishing Fortress Europe is totally mistaken. He observed that Europe still intends to do commerce with the rest of the world, and therefore ISO will be guiding EC. CEN standards hopefully will help outside commerce to deal with one standard rather than with 12 different nations in 12 different languages.

WHERE DO WE GO FROM HERE?

Attendees:

Bob Campbell	Harper College
Howard Clark	ASME Research
André Clement	Dassault Systemes
Greg Hetland	Hutchinson Technology Inc.
Jim Hurt	SDRC
Richard Jennings	ICAMP Inc.
Curt Larson	Trikon Corp.
Sam Levy	IGTI
Yuyi Lin	University of Missouri - Columbia
Luc Mathieu	CNAM Paris
Chell Roberts	Arizona State University
Vijay Srinivasan	IBM Research
Herb Voelcker	Cornell University

A Strawman Proposal

Vijay Srinivasan opened the session with the distribution of a strawman proposal titled "Quo Vadis?". In that proposal he observed that:

1. The technical areas covered by the Forum — Dimensional Tolerancing & Metrology — are becoming very important to our industries.
2. His recent experience with standards committees such as Y14.5.1 convincingly showed that changes are necessary in how the technically affected communities (industrial folks, standards committees, researchers, and educators) work together.
3. This Forum, in some ways a follow-on to the 1988 Orlando Workshop, is a USEFUL Forum that we should plan to continue in some fashion.
4. Some of us MAY want to do more, e.g., establish a working technical community affiliated with a society, to provide an organized technical study and research arm to support and complement the work of the standards committees, educators, and so forth.

He then proposed that we

1. Treat ASME Codes and Standards as our prime customer.
2. Coordinate the Forum follow-up with the future ISO, Y14.5, Y14.5.1 and B89.3.2 activities.
3. Stay within some sort of ASME CRTD umbrella for now.

Codes and Standards as Customer

The strawman proposal triggered some discussions. Herb Voelcker said that the "instigation" for this Forum came from the codes and standards community. He observed that most of the codes and standards developers are industrial representatives. The ASME Technical Divisions are dominated by academe who do not view codes and standards as "customers" for their output.

Bob Campbell asked if the discussions of the standard committees can be made available to outsiders. He suggested that some sort of electronic bulletin board would be useful. Vijay Srinivasan responded that currently the standard committee discussions are held in confidence by the committee members. Any dissemination of the information to outsiders needs special permission. The panel felt that some means should be found to inform the technical community at large on the issues and solutions debated by the standards committees without endangering the process by which standards are developed.

Herb Voelcker asked whether the codes and standards groups are receptive to new ideas. He observed that possible discussion groups, "parallel" to the standards committees, could be useful. He saw two ways to approach standards development — one was enumerative, in which for each section of the existing document bring in further refinements and elaborations; the alternative would be to identify an underlying set of principles, which would yield a much shorter document and make the task of codifying much simpler. He felt that users can be "trained" in the current GD&T practices, but they are not "teachable" at the university level.

Organizational Issues

Richard Jennings observed that ASME is already a *de facto* home for the tolerancing and metrology activities since it sponsors the relevant standards and committees, and helped organize the current Forum. He recommended cooperative efforts with societies such as SME to take advantage of SME programs on gauging and inspection. This would broaden the participation base.

Vijay Srinivasan suggested that the ASME Center for Research and Technology Development might provide a good organizational base. Howard Clark agreed that it was possible, and said that CRTD had also worked well in the past with SME in co-sponsoring activities and meetings.

The discussions then focused on organizing technical meetings. Jim Hurt felt that holding periodic technical meetings open to the public will be desirable, particularly if the cost can be kept low by holding them in universities. Greg Hetland also felt that periodic dissemination of technical information to the public is important. He observed that there are too many independent activities that do not communicate with each other. He emphasized the need for a "plan" to integrate all these activities.

Sam Levy observed that the small business community is being overlooked in all the standardization activities. He appealed for cooperation and support for his efforts to develop "simplified version of GD&T" so that it can be used at 2-year college level.

Summary

It was decided that a recommendation would be made to the Forum Steering Committee that an ASME Research Committee on Dimensional Tolerancing and Metrology be formed. This new committee will have the responsibility to organize the next Forum.

Participants
1993 International Forum on Dimensional Tolerancing & Metrology

Mr. George Allen
Systems Engineer
EDS Unigraphics
10824 Hope Street
Cypress, CA 90630
Speaker/Moderator

Mr. Roberto E. Altschul
Boeing Computer Services
P.O. Box 24346
MS 7L-22
Seattle, WA 98124-0346
Speaker/Moderator

Mr. Archie R. Anderson
Consulting Engineer
Trikon Corporation
12116 Townline Road
Grand Blanc, MI 48439
Speaker/Moderator

Mr. Uppliappan Babu
ME & ES Department
UNC Charlotte
Precision Engineering
CARC, Highway 49
Charlotte, NC 28223
Student

Mr. Michael Bahtiarian
ADE Corporation
77 Rowe Street
Newton, MA 02166

Mr. Frank Bakos
Technician,
GD&T Specialist
Eastman Kodak Company
Manufacturing Research
& Engineering
Kodak Park, Building 23
Rochester, NY 14652-4393
Speaker/Moderator

Mr. Jose I. Barrera
AC Rochester (GM)
/ASQC
2401 Brookside Drive
Flint, MI 48503

Mr. Alan Baumgartner
Ford Motor Company
15100 Mercantile Drive
Dearborn, MI 48120

Dr. Per Bennich
PB Metrology Consulting
Birkevej 11
Vaerloese DK-3500
Denmark
Speaker/Moderator

Mr. Theodore D. Benoit
Sr. Design Stds. Analyst
Pratt & Whitney
400 Main Street
M/S 163-13
East Hartford, CT 06108

Mr. Neil S. Bernstein
Cummins Engine Co.
Box 3005
Columbus, IN 47202-3005

Mr. Dean Beutel
Project Manager
Caterpillar, Inc.
P.O. Box 1875
Peoria, IL 61656-1875
Speaker/Moderator

Mr. Malcolm Blair
Steel Founders' Society
of America
455 State Street
Des Piaines, IL 60016

Mr. Robert L. Bourland
General Motors
Powertrain Division
Ecourse & Wiard Road
Ypsilanti, MI 48198-6198

Mr. Edward H. Boyer
Engineering & Materials
and Standard
Ford Motor Company
Suite 700C
30425 Vernon Drive
Birmingham, MI 48025
Speaker/Moderator

Mr. Bruno Brazauskas
Inspection Technologies
Incorporated
111 East Ten Mile Road
Madison Heights, MI 48071

Ms. Dragica Bresovski
901 Elmgrove Road
Rochester, NY 14653-5111

Mr. Curtis W. Brown
Allied Signal Inc.
Kansas City Div
2000 E. 95th Street
P.O. Box 419159, MC2B27
Kansas City, MO 64141
Speaker/Moderator

Mr. James B. Bryan
Bryan Associates
5196 Golden Road, #4
Pleasanton, CA 94566-6443
Speaker/Moderator

Mr. Robert G. Campbell
Harper College
1200 West Algonquin
Road
Palatine, IL 60067

Participants
1993 International Forum on Dimensional Tolerancing & Metrology

Ms. Kirsten Carr
Ford Motor Company
Mail Drop E-1158, SRL
P.O. Box 2053
Dearborn, MI 48121

Mr. Greg Caskey
National Institute of Standards & Technology
B113, Building 220
Gaithersburg, MD 20899

Mr. Thomas Charlton
Manager
Research & Development
Brown & Sharpe Manufacturing Co.
Precision Park
200 Frenchtown Road
North Kingstown, RI 02852
Speaker/Moderator

Mr. Kenneth W. Chase
Mechanical Engineering Department
Brigham Young University
242 Clyde Building
Provo, UT 84602

Mr. Lin-Lin Chen
Iowa State University
2078 Black Engineering
Ames, IA 50011

Mr. Shuo-Yan Chou
Iowa State University
2078 Black Engineering
Ames, IA 50011

Dr. Howard E. Clark
Director of Research
American Society of Mechanical Engineers
1828 L Street, N.W.
Suite 906
Washington, DC 20036
Staff

Mr. Andre' Clement
Scientific Advisor
Dassault Systems
92156 Suresnes
24-28 Avenue du General de Gaulle
BP 310
Surensnes 92156 France

Mr. Richard E. Coombes
Project Engineer
Caterpillar Inc.
Standards Engrg Tech Ctr
Building F
P.O. Box 1875
Peoria, IL 61656-1875

Mr. Terry W. Coplen
McDonnell Douglas Aerospace - East
2404 Stoney Brook Drive
O'Fallon, MO 63366

Mr. Thomas D. Crampton
Mott Community College
1401 E. Court Street
Flint, MI 48503

Mr. Chris C. Cuba
Texas Instruments
6500 Chase Oaks Blvd.
P.O. Box 869305
Plano, TX 75086

Mr. Norm Cutler
Polaroid Corporation
1 Uplano Road
Norwood, MA 02162

Ms. Lori Darcy
Herman Miller Inc.
MS 0441
P.O. Box 302
Zeeland, MI 49464-0302

Mr. Donald E. Day
Monroe Community College
RD3, 4755 Egypt Road
Canandaigua, NY 14424

Mr. Paul Drake
Texas Instruments, Inc.
P.O. Box 650311
MS 3928
Dallas, TX 75265

Mr. Robert J. Dunn
Polaroid Corporation
1 Upland Road
Norwood, MA 02062

Mr. Robert E. Early
GM Technical Center
30300 Mound Road
A/MD-16
Warren, MI 48090

Mr. Faryar Etesami
Department of Mechanical Engineering
Portland State University
P.O. Box 751
Portland, OR 97207

Mr. Shaw C. Feng
National Institute of Standards & Technology
Building 220, Room A127
Gaithersburg, MD 20899

Mr. Charles W. Ferguson
Steelcase Inc.
P.O. Box 1967
Mail Code CD-3W-26
Grand Rapids, MI 49501

Mr. Lowell W. Foster
President & Director
Lowell W. Foster Associates, Inc.
3120 East 45th Street
Minneapolis, MN 55406

Participants
1993 International Forum on Dimensional Tolerancing & Metrology

Prof. Gary Gabriele
Dept. of Mechanical Engrg.
Rensselaer Polytechnic Institute
110 8th Street
Troy, NY 12180-3590
Speaker/Moderator

Mr. John M. Gallagher
Sandia National Laboratories
Organization 2645
P.O. Box 5800
Albuquerque, NM 87185

Dr. Marion B. Grant
Cummins Engine Company
MS 18210
P.O. Box 3005
Columbus, IN 47202-3005
Speaker/Moderator

Mr. William Greenwood
Sandia National Laboratories
P.O. Box 5800
Organization 2645
Albuquerque, NM 87185

Mr. James Guilford
Design & Manufacturing Institute
Rensselaer Polytechnic Institute
CII 7015
Troy, NY 12180-3590
Speaker/Moderator

Mr. Manuel Gutierrez
American Society of Mechanical Engineers
345 East 47th Street
MS 39P
New York, NY 10017-2392
Staff

Mr. Georg Henzold
Siemens AG
Rossbrudhring 25
D-4030 Ratingeu 5
Germany
Speaker/Moderator

Mr. Albert L. Herpich
Xerox Corporation
800 Phillips Road - 147-11B
Webster, NY 14580

Mr. John Hertrich
Applied Computer Solutions
300 Maple Park Boulevard
St. Clair Shores, MI 48081

Mr. Gregory A. Hetland
Hutchinson Technology Incorporated
40 West Highland Park (cc. 648)
Hutchinson, MN 55350

Dr. Robert Hocken
Precision Engineering Lab.
Univ. of North Carolina
Mech. Engrg. & Engrg. Sci.
Smith Bldg - Room 245C
Charlotte, NC 28223
Speaker/Moderator

Mr. Michael R. Hoeprich
The Timken Company
1835 Dueber SW
P.O. Box 6930
Canton, OH 44706-6930
Speaker/Moderator

Mr. Benny Hogston
Supv., Measurement Technology Center
General Motors
Powertrain Division
M/C 712
Ypsilanti, MI 48198

Mr. Richard B. Hook
President
METCON
45 Wood Cove Drive
Coventry, RI 02816
Speaker/Moderator

Dr. Theodore H. Hopp
Leader, Machine Intelligence Group
National Institute of Standards & Technology
Metrology Building, A127
Gaithersburg, MD 20899
Speaker/Moderator

Mr. Chijen Hsu
Parametric Modeling
EDS - Unigraphics Product Development
10824 Hope Street
P.O. Box 6011
Cypress, CA 90630

Mr. Ming C. Huang
Electronic Data Systems Corporation
750 Tower Drive
Troy, MI 48098

Mr. James Hurt
Technical Fellow
SDRC
2000 Eastman Drive
Milford, OH 45150-2789
Speaker/Moderator

Mr. Mark Iannuzzi
Applied Computer Solutions
300 Maple Park Boulevard
St. Clair Shores, MI 48081

Mr. John Jackman
Iowa State University
2078 Black Engineering
Ames, IA 50011

Participants
1993 International Forum on Dimensional Tolerancing & Metrology

Mr. Paul F. Jackson
Ford Motor Company
250 Roseland
Canton, MI 48187-3955

Mr. Richard J. James
Polaroid Corporation
1 Upland Road
Norwood, MA 02062

Mr. Kevin Jehng
EDS
10824 Hope Street
Cypress, CA 90630

Mr. Richard Jennings
ICAMP, Inc.
186 Bolton Center Road
Bolton, CT 06043

Ms. Norma Johnston
Business Manager,
Research
American Society of
Mechanical Engineers
1828 L Street, N.W.
Suite 906
Washington, DC 20036
Staff

Mr. Alan K. Jones
Senior Principal Scientist
The Boeing Company
P.O. Box 24346
M/S 7L-21
Seattle, WA 98124-0346

Mr. Gene Jordan
2669 Bond Street
Rochester Hills, MI 48309

Dr. Neal P. Juster
Department of Mechanical
Engineering
University of Leeds
Leeds LS2 9JT UK

Mr. Raj Kawlra
GM Tech Center
Mfg. Bldg. A/MD-03
30300 Mound Road
Warren, MI 48090

Mr. David Kies
United Technologies
Automotive
5200 Auto Club Drive
Dearborn, MI 48126

Mr. John Kovach
GM Technical Center
30300 Mound Road
Warren, MI 48090-9040

Mr. Harry (Bud) Krayer
Xerox Corporation
800 Phillips Road
MS 102-10A
Webster, NY 14580

Mr. Alex Krulikowski
General Motors
Powertrain Division
Ecorse and Wiard Road
M/C 732
Ypsilanti, MI 48198-6198

Mr. Nikhil Kundu
Perdue University
2424 California Road
Elkhart, IN 46514

Dr. Chris Kuyatt
National Institute of
Standards & Technology
Building 245, Room C210
Gaithersburg, MD 20899
Speaker/Moderator

Mr. C. A. Langton
Ford Motor Company
36200 Plymouth Road
Livonia, MI 48150

Mr. Curt Larson
Trikon Corporation
2295 N. Opdyke
Building A, Suite E
Auburn Hills, MI 48326

Dr. Neal Laurance
Senior Staff Scientist
Ford Motor Company
2247 Research Lab.
P.O. Box 2053
Dearborn, MI 48121-2053
Speaker/Moderator

Mr. Barry L. Lawler
General Motors/EDS #742
800 Tower Drive
Troy, MI 48007

Mr. Jinkoo Lee
University of Michigan
Ann Arbor
2170 Cram Place #15
Ann Arbor, MI 48105
Student

Mr. Samuel Levy
Int'l Geometric
Tolerancing Inst.
2943 Cortina Drive
San Jose, CA 95132

Prof. Yuyi Lin
University of Missouri
Columbia
E2412 EBE
Department of MAE
Columbia, MO 65211

Mr. Mathieu Luc
CNAM - Paris
292 Ave Saint Martin
Paris 75141 CEDEX03
France

Participants
1993 International Forum on Dimensional Tolerancing & Metrology

Mr. Chunye Ma
Graduate Student/
Research Asst.
University of Michigan
118 MSEL
4901 Evergreen Road
Dearborn, MI 48128
Student

Mr. Patrick J. McCuistion
Ohio University
120 Stocker Center
Athens OH 45701-2979

Mr. Jessie B. McKelry
Mott College
1401 E. Court Street
Flint, MI 48503

Mr. Paul McKim
Senior Staff Engineer
Caterpillar Incorporated
Standards Coordination
100 N.E. Adams Street
Peoria, IL 61629-6480
Speaker/Moderator

Mr. Duane Moore
Cooper Tire & Rubber Co.
725 West Eleventh Street
Auburn, IN 46706

Mr. Claude Morel
Ecole Polytechnique,
genie mecanique
C.P. 6079, succursale A
Montreal Quebec
H3C 3A7 Canada

Mr. Ron Morford
Cooper Tire & Rubber Co.
725 West Eleventh Street
Augurn, IN 46706

Mr. Edward P. Morse
Measuring Systems Div.
Brown & Sharpe
Manufacturing
Precision Park
200 Frenchtown Road
North Kingstown, RI 02852
Student

Mr. Ashraf Nassef
McMaster University
Dept of Mechanical Engrg
1280 Main Street West
Hamilton, Ontario L8S 4L7
Canada
Student

Mr. Mark A. Nasson
C. S. Draper Laboratory,
Incorporated
MS #51
555 Technology Square
Cambridge, MA 02139

Mr. Henrik S. Neilsen
Cummins Engine Company
M/C 18210
Box 3005
Columbus, IN 47202-3005
Speaker/Moderator

Mr. Alvin G. Neumann
President/Director
Technical Consultants, Inc.
585 Gunwale Lane
Longboat Key, FL 34228-
3709
Speaker/Moderator

Mr. Robert Nickolaisen
Pittsburg State University
1501 S. Joplin
Shirk Hall
Pittsburg, KS 66762

Dr. Elsayed A. Orady
University of Michigan
4901 Evergreen Road
Dearborn, MI 48128

Dr. W. L. Otto
Concurrent Technologies
Corporation
1450 Scalp Avenue
Johnstown, PA 15904
Speaker/Moderator

Mr. Jahir Pabon
Schlumberger - SLCS
8311 North RR 620 Road
Austin, TX 78726

Mr. Steve Panton
Ford Motor Company
AMTAC
15100 Mercantile Drive
Dearborn, MI 48120

Mr. Kevin R. Paradis
Division Quality Control
Engineer
Ford Motor Company
P.O. Box 1586
17000 Oakwood Boulevard
Dearborn, MI 48121

Mr. Dong-Keun Park
Iowa State University
2078 Black Engineering
Ames, IA 50011
Student

Ms. Jean M. Parks
Xerox Corporation
800 Phillips Road
147-15A
Webster, NY 14580

Participants
1993 International Forum on Dimensional Tolerancing & Metrology

Mr. Steffen W. Parratt
Graduate Student
Cornell University
Cornell Programmable
Automation
195 Engrg & Theory Ctr
Ithaca, NY 14853-3801
Student

Mr. Brian Parry
The Boeing Company
M/S 9U-EA
P.O. Box 3707
Seattle, WA 98124

Mr. Frederick Parsons
Senior Research Engineer
Federal Products Co.
1144 Eddy Street
Providence, RI 02940
Speaker/Moderator

Mr. Anthony R. Peach
Programmer
Autodesk Inc.
2320 Marinship Way
Sausalito, CA 94965

Mr. Allen Pelletier
General Motors NATP
5607 Newking Street
M/C 2201-107
Troy, MI 48098-2666

Mr. Vic Peri
Trikon Corporation
2295 N. Opdyke
Building A, Suite E
Auburn Hills, MI 48326

Mr. Frank Peters
217 1/2 South Patterson
State College, PA 16801
Student

Mr. Richard Phillips
VSM Program Coordinator
General Motors NATP
27800 Dequindre
M/C 3700
Warren, MI 48092-5601

Mr. Steven Phillips
National Institute of
Standards & Technology
B113, Building 220
Gaithersburg, MD 20899

Mr. Ernest Pobocik
Ford Motor Company
18937 Seavitt
Allen Park, MI 48101

Mr. George Quinn
Ryerson Polytechnic
Institute
350 Victoria Street
Toronto, Ontario M5B 2K3
Canada

Dr. Jay Raja
ME & ES Department
UNC Charlotte
Precision Engineering
CARC, Highway 49
Charlotte, NC 28223

Mr. William F. Reiter
Mechanical Engrg Dept
Oregon State University
Rogers Hall 214
Corvallis, OR 97331

Prof. Aristides A.G.
Requicha
Computer Science Dept
Univ. of Southern
California
Los Angeles, CA 90089
Speaker/Moderator

Mr. Chell Roberts
Arizona State University
Department of IE
Tempe, AZ 85287

Mr. Dean Robinson
Cornell University
Cornell Programmable
Automation
195 Engrg & Theory Ctr
Ithaca, NY 14853-3801
Student

Dr. Eric Sandgren
TRW,P.O. Box 8008
Sterling Heights, MI 48311

Mr. Friedrich Schaechter
Minitek Feinmech. Prod.
G.M.B.H.
Draschestr. 31
Vienna A 1232 Austria

Mr. Ron Schall
Engrg Test & Devlpmt Ctr
Ford Motor Company
35500 Phymouth Road
Livonia, MI 48150
Speaker/Moderator

Mr. Nelson Schoeler
CERTI/UFSC
P.O. Box 5053
Florianopolis SC88 041 970
Brazil

Mr. Leslie J. Schonberg
Ford Motor Company
7004 Heather Heath
W. Bloomfield, MI 48322

Mr. Paul J. Scott
Rank Taylor Hobson Ltd.
P.O. Box 36
2 New Star Road
Leicester NEW 750
England

Participants
1993 International Forum on Dimensional Tolerancing & Metrology

Mr. Steve Scott
Cooper Tire & Rubber Co.
725 West Eleventh Street
Auburn, IN 46706

Dr. Vadim Shapiro
Computer Science Dept
General Motors Research Laboratories
30500 Mound Road
Box 9055
Warren, MI 48090-9055

Mr. Kenneth L. Sheehan
Brown and Sharpe
200 Frenchtown Road
N. Kingstown, RI 02852

Dr. Vijay Srinivasan
T.J. Watson Research Ctr.
IBM Research Division
Room 2-150
P.O. Box 218
Yorktown Hts, NY 10589
Speaker/Moderator

Mr. Richard W. Stockdale
Lockhead Missle & Space
San Jose City College
800 Gordon Avenue
San Jose, CA 95127

Ms. Lisa Strachan
Admin Assistant, Research
American Society of Mechanical Engineers
1828 L Street, N.W.
Suite 906
Washington, DC 20036
Staff

Mr. Joe Streng
Saginaw Division GMC
3900 Holland Road
Saginaw, MI 48601

Mr. Douglas Sutphin
Inspection Technologies
111 East Ten Mile Road
Madison Heights, MI 48071

Mr. Hiromasa Suzuki
Associate Professor
The University of Tokyo
Dept. of Graphics & Computer Science
3-8 Komaba, Meguro,
Tokyo 153 Japan

Mr. Joachim G. Taiber
Institute of Machine Tools & Manufacturing IWF
Swiss Federal Institute of Technology
Leonhardstr. 27 Zurich
CH-8092 Switzerland
Student

Mr. Bill Tandler
Multi Metrics, Inc.
865 Lemon Street
Menlo Park, CA 94025

Mr. Benjamin Taylor
Renishaw PLC
The Old Vicarage
Trellech
Gwent NP5 4PA UK
Speaker/Moderator

Mr. Robert Tilove
Computer Science Dept
General Motors Research & Development Center
30500 Mound Road
Box 9055
Warren, MI 48090-9055

Mr. Frederick A. Tolmie
Rockwell International
400 Collins Road NE
Cedar Rapids, IA 52498

Commander Raymond P. Tremblay
U.S. Army ARDEC
Attn: Smcar-Bac-S
Picatinny Arsenal, NJ 07806-5000

Prof. Joshua Turner
Rensaelear Polytechnic Institute
CII 7015
Troy, NY 12180-3590
Speaker/Moderator

Mr. Gary L. Vander Sande
U.S. Army ARDEC
Att: SMCAR-BAC-S, B-6
Picatinny Arsenal, NJ 09806-5000

Mr. Jerry Varani
Ford Motor Company
Room 2228, Building 3
P.O. Box 2053
Dearborn, MI 48236

Mr. Herbert B. Voelcker
Cornell University
MAE, Upson Hall
Ithaca, NY 14853-7501
Speaker/Moderator

Mr. Donald L. Vossler
Manager
EDS Unigraphics
10824 Hope Street
Cypress, CA 90630

Mr. Tho Van Vu
Computer Science Corp.
37660 Durham Road
Sterling Heights, MI 48312

Participants
1993 International Forum on Dimensional Tolerancing & Metrology

Mr. Houssaine Waled
McMaster University
Dept of Mechanical Engrg
1280 Main Street West
Hamilton, Ontario L8S 4L7
Canada
Student

Mr. Richard Walker
Marine Division
Westinghouse Electric
Corporation
401 East Hendy Avenue
Sunnyvale, CA 94022
Speaker/Moderator

Mr. Terence M. Watson
Brown & Sharpe Mfg. Co.
27300 Haggerty Road
Suite F-10
Farmington Hills, MI 48331

Mr. Colin Wearring
Applied Computer
Solutions
300 Maple Park Boulevard
St. Clair Shores, MI 48081

Mr. Ron Webber
Inspection Technologies
111 East Ten Mile
Madison Heights, MI 48071

Mr. Albert Weidmann
Oerlikon-Contraves AG
Birchstrasse
Zurich CH-8050
Switzerland

Mr. Ray Wheeler
Sundstrand Power Systems
4400 Ruffin Road
P.O. Box 85757
San Diego, CA 92186-5757

Mr. Charles H. White
Datum Inspection Services
5202 S. 39th Street
Suite 1
Phoenix, AZ 85040

Mr. Robert Wilhelm
Precision Engrg Program
Univ. of North Carolina
Dept of Mechanical Engrg
& Engrg Science
Highway 49
Charlotte, NC 28223

Mr. Peter Williams
Ryerson Polytechnic
Institute
350 Victoria Street
Toronto, Ontario M5B 2K3
Canada

Mr. Paul R. Wreede
E-Systems
P.O. Box 6056
Greenville, TX 75403-6056

Mr. Hong-Tzong Yan
Cummins Engine Company
Mail Code 50183
Columbus, IN 47202

Mr. Tai Hung Yang
Iowa State University
2078 Black Engineering
Ames, IA 50011
Student

Mr. Philip M. Zaretzki
Electronic Data Systems
Corporation
750 Tower Drive
P.O. Box 7019
Troy, MI 48098

AUTHOR INDEX

CRTD-Vol. 27
Proceedings of the 1993 International Forum on Dimensional Tolerancing and Metrology